Adaptive Optics Progress

Adaptive Optics Progress

Editor

Lefteris Tyler

Adaptive Optics Progress

Edited by **Lefteris Tyler**

ISBN: 978-1-68117-208-8
Library of Congress Control Number: 2016934755

© 2017 by
SCITUS Academics LLC,
www.scitusacademics.com
Box No. 4766, 616 Corporate Way,
Suite 2, Valley Cottage,
NY 10989

Preface

Adaptive optics (AO) is a technology used to improve the performance of optical systems by reducing the effect of wavefront distortions: it aims at correcting the deformations of an incoming wavefront by deforming a mirror in order to compensate for the distortion. It is used in astronomical telescopes and laser communication systems to remove the effects of atmospheric distortion, in microscopy, optical fabrication and in retinal imaging systems to reduce optical aberrations. Adaptive optics works by measuring the distortions in a wavefront and compensating for them with a device that corrects those errors such as a deformable mirror or a liquid crystal array. Although active optics can ensure that a telescope's main mirror always retains a perfect shape, the turbulence of the Earth's atmosphere distorts images obtained at even the best sites in the world for astronomy. Astronomers have turned to a method called adaptive optics. Sophisticated, deformable mirrors controlled by computers can correct in real-time for the distortion caused by the turbulence of the Earth's atmosphere, making the images obtained almost as sharp as those taken in space. Adaptive optics allows the corrected optical system to observe finer details of much fainter astronomical objects than is otherwise possible from the ground. Active and adaptive optics technology has emerged from the laboratory and is being applied to improve the performance of optical imaging and laser systems. In the last few years, development of both systems and components has accelerated. Many new concepts and devices have appeared, among which are high-performance deformable mirrors, new types of wave front sensors, and more sophisticated wave front processing algorithms. Equally important, a better understanding of the system design aspects of adaptive optics has been reached, particularly of the need for optimizing each system according to its application.Current developments in adaptive optics for ground-based astronomy include the use of IR wavelengths, partial wavefront compensation using natural guide stars, and the use of laser guide stars to allow all-sky coverage with full compensation at visible wavelengths. The book entitled Adaptive Optics Progress offers the applications of adaptive optics throughout the fields of communicationas well as its original usages in astronomy and beam propagation.

Table of Contents

Chapter 1   Adaptive Optics for Visual Simulation................................ 1

Chapter 2   Adaptive Optics and Optical Vortices................................ 31

Chapter 3   Adaptive Optics Technology for High-Resolution Retinal Imaging... 83

Chapter 4   Devices and Techniques for Sensorless Adaptive Optics 125

Chapter 5   A Solar Adaptive Optics System.. 151

Chapter 6   Impact of Liquid Crystals in Active and Adaptive Optics. 171

Chapter 7   Embedded Adaptive Optics for Ubiquitous Lab-on-a-Chip Readout on Intact Cell Phones... 189

Chapter 8   The Influence of Filter Slit on the Imaging Performance of the Solar Grating Spectrometer Based on Adaptive Optics.. 207

Chapter 9   Optical Power Allocation for Adaptive Transmissions in Wavelength-Division Multiplexing Free Space Optical Networks... 221

Chapter 10   The Human Eye and Adaptive Optics.............................. 245

Chapter 11   Radiation Hydrodynamics Using Characteristics on Adaptive Decomposed Domains for Massively Parallel Star Formation Simulations... 285

Index... 329

Chapter 1

Adaptive Optics for Visual Simulation

Enrique Josua Fernández

Laboratorio de Óptica, Instituto Universitario de investigación en Óptica y Nanofísica (IUiOyN), Universidad de Murcia, Campus de Espinardo (Edificio 34), 30100 Murcia, Spain

ABSTRACT

A revision of the current state-of-the-art adaptive optics technology for visual sciences is provided. The human eye, as an optical system able to generate images onto the retina, exhibits optical aberrations. Those are continuously changing with time, and they are different for every subject. Adaptive optics is the technology permitting the manipulation of the aberrations, and eventually their correction. Across the different applications of adaptive optics, the current paper focuses on visual simulation. These systems are capable of manipulating the ocular aberrations and simultaneous visual testing though the modified aberrations on real eyes. Some applications of the visual simulators presented in this work are the study of the neural adaptation to the aberrations, the influence of aberrations on accommodation, and the recent development of binocular adaptive optics visual simulators allowing the study of stereopsis.

1. THE EYE AND ITS ABERRATIONS

Vision is a fascinating puzzle involving different aspects. In a first stage, light emitted or reflected by external objects reaches the eye. The image of the object is projected onto the observer's retina by the optics of the eye. The information contained in the image is to be processed and transmitted from the retina to the brain by neural cells. The brain is finally in charge of the psychological interpretation of the scene for a useful perception of our surrounding reality. Each part of the complete process of vision is complex enough for requiring a separate analysis. The first part of vision pertains solely the optical aspects of formation of the images, and it is usually termed the optical or first stage. In this work, several aspects related to the optical stage will be shown, in particular, the connection of vision and the optical quality of the eye, and how adaptive optics has dramatically changed our methods and approaches for the study and understanding of vision.

It seems appropriate starting with a description from an optical point of view of the eye. The eye is a complex system showing a tremendous histological richness. However, when studying the formation of images on the retina from the real world only some parts of the eye are relevant [1]. In the following, some fundamental parts of the eye are revisited. Some of them are shown in Figure 1.

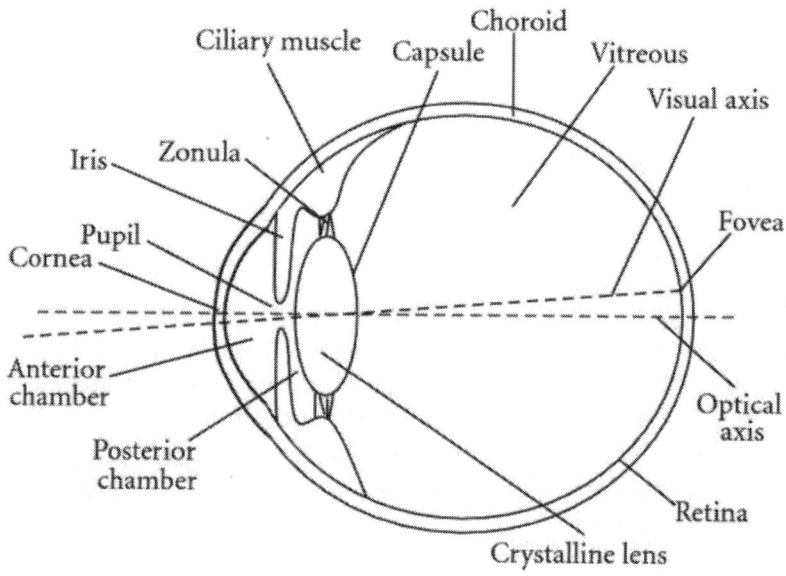

Figure 1

The first element that light encounters coming from the object or scene is the cornea. The latter provides approximately two-thirds of the total optical power of the eye. Considering solely the optical properties, the cornea might be characterized by the radii of curvature of its anterior and posterior surface, thickness, and refractive index, which exhibits an average value of 1.38 in the normal eye. After the cornea, the light finds the anterior chamber. This is a space filled with the aqueous humor, essentially compounded by water, showing a thickness of 3.05 mm and a refractive index of 1.34. The next relevant element for the light in its travel to the retina is the iris. This is a circular muscle which defines the eye pupil, where the light beam is limited through. The iris diameter can change because of a number of factors. Across others, the most relevant changes occur in response to illumination variations in the scene and changes in the distance to the fixation point. After the iris, the posterior chamber is found. This is defined as the space filled with aqueous humor and limited by the crystalline lens, which acts as its posterior limit. The crystalline lens is beyond doubt one of the most fascinating and complex optical elements of the eye. It provides to the eye, together with the cornea, the rest of the necessary optical power for correctly focusing images on the retina. It is compounded by a large number of concentric layers, of different thickness and cellular age. This particular structure grants the lens with a complex distribution of refractive index along its optical axis, sometimes compared with a GRIN, or gradient index, lens. Its external shape resembles a biconvex lens, slightly more curved at the posterior outer surface, showing a total average thickness of 4 mm. For the study of the eye from an optical point of view, a first useful approximation for modeling the lens is taking a single effective refractive index of 1.42. The crystalline lens is enclosed by a thin membrane endowed with elastic properties, mostly compounded by type IV collagen, known as the capsule. The stiffness of the capsule is larger in the young eye, determining the external shape of the crystalline lens. The capsule is suspended equatorially by means of a delicate network of elastic fibers known as the zonula. The most external parts of the fibers are inserted into the ciliary muscle by the ciliary processes. The ciliary muscle is in direct contact with the sclerotica, which is the most outer part of the eye. As it will be shown later in this work with more details, all these elements described previously are interconnected and they are of relevance for the understanding of the accommodation, a fundamental capability of the eye.

The vitreous humor is located right after the crystalline lens. It shows a gel-like texture with a refractive index similar to that of the aqueous humor. Its role is mostly keeping the rest of elements fixed at the right position in the eye and protecting the retina. The retina is formed by a multitude of cells of different kinds, exhibiting a highly organized structure of different layers grouping similar cells. It is responsible for the phototransduction of the optical image into chemical and physical signals which are sent and impulses which are sent to the brain by means of specialized neural cells. In the brain these signals are decoded by a delicate mesh of neurons for the final subjective psychological perception of vision. The zone of the retina where the images are projected for accurate vision is called the fovea. The latter corresponds with the retinal area most populated of photoreceptors. The photoreceptors are the cells responsible for the detection of light.

In the study of the optical aspects and formation of images in the eye, the use of some axes is convenient. The optical axis of the eye is defined as the imaginary line containing the centers of curvature of the different surfaces of the elements compounding the eye. In the human eyes the optical axis does not intersect the fovea [2, 3]. That is the main reason for introducing the visual axis or line of sight. This axis is defined as the imaginary line connecting the fixation point, where the eye is looking at, and the fovea. The angle between the two axes is commonly named as alpha angle, although some alternative definitions can be found in the literature.

In the previous description, all elements of the eye are modeled by ideal surfaces of given radii of curvature and known refractive indexes. This kind of modeling is extremely useful for the understanding of several aspects of image formation. This description of the eye can provide a simple explanation of the generation of blurred retinal images in the presence of refractive errors, approximate image size, and so forth. From this model one could infer that optical images generated on the retina are somehow comparable to those produced by a regular optical instrument, in terms of quality. The reality is nevertheless more complicated [4–6]. In the real eye, none of the surface can be generally described by a simple radius of curvature, not even by a conic surface [7]. The surfaces exhibit deviations from ideal surfaces inherent to any biological tissue. The different surfaces separating media of distinct refractive indexes are not aligned neither their hypothetical centers of curvature lay on a common axis. That makes the different surfaces appear as misaligned to each other in the eye. In addition, the optics of the eye does not remain steady, but on the contrary it presents continuous fluctuations in time [8–11]. Those come mainly from the changes in the crystalline lens induced by the physiological tension of the ciliary muscle. The tear film, the intraocular pressure, the changes, and movement associated with the ocular humors are other factors preventing a steady situation in the eye. All these circumstances result in that the retinal images finally formed on the observers' retina were far from perfect images. Retinal images are affected and degraded by ocular aberrations [4, 5, 12–18]. Those are the deviation of the wavefront of the incoming light from a paraxial situation, where all light rays would converge in a single point on the retina. Therefore, the interest in measuring and characterizing the effect of aberrations and their dynamics on vision is fundamental [12]. In the last years, many efforts have been devoted to understand to what extent these aberrations can affect the perception in vision. In addition, correcting those aberrations would permit to achieve unprecedented resolution in the retinal images recorded in the living eye, with promising applications in the diagnosis and treatment of several retinal conditions. In this scenario, adaptive optics (AO) is playing a major role as it will be shown in the following.

2. ADAPTIVE OPTICS

A first simple definition for adaptive optics could be the optical technique allowing the measurement and subsequent correction of optical aberrations. In spite of the simplicity of the previous definition, the two fundamental concepts sustaining AO are clearly present. A first key point is the measurement of aberrations, that must be performed is a robust manner [8–10], and faster than the typical temporal variation which in intended to be later corrected. The other evident pillar of AO is the correction of aberrations. The AO system must allow for the controlled manipulation of the wavefront, once it has been estimated. The AO aberration correction concept inherently includes the measurement of the effect of the correction over the wavefront so that closed loop control of aberration can be eventually performed. There is an alternative concept generally known as active optics which performs the correction after the measurement of the aberrations, with no possibility of closed loop operation. Consequently, in active optics there is no means of measuring the effect of the correction over the wavefront. Figure 2 shows schematically a diagram of active and adaptive optics.

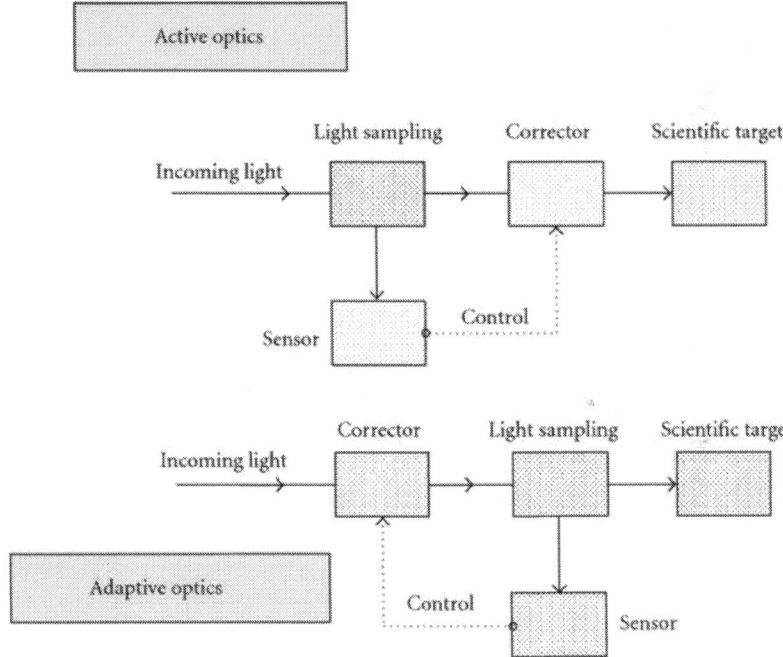

Figure 2

The first proposal of the use of AO technology for civil purposes was suggested in the context of astronomical optics in the year 1953 [19], around thirty years earlier of the experimental demonstration of the technique. Irrespectively of the optical quality and the aperture of the telescopes, light passing through the atmosphere is seriously degraded by effect of the changing temperature, gradients of pressure, winds, and in general any disturbance able to locally alter the refractive index of the medium. Therefore, the images recorded at the ground level are systematical affected and degraded by all these factors. AO was presented as the technology providing the solution to the atmospheric turbulence problem over the astronomical images recorded in the ground-based telescopes. Nowadays, every professional telescope in the world is endowed with AO for the atmospheric aberration correction. The field of astronomical optics was the only one benefiting from AO during a couple of decades, approximately from the eighties to nineties. Probably the elevated cost of the AO systems prevented other fields from exploring the possibilities of AO. It could even be said with some humor that it was especially true in the many other fields unfortunately not enjoying of astronomical funds. In the context of vision sciences, it was not until 1989 [20] when some of the concepts from AO were somehow applied in the eye. In that work, the use of a deformable mirror for the static compensation of the ocular aberrations was reported. The final target of the correction was the improvement of the resolution achieved on the retinal images obtained with a scanning laser ophthalmoscope. The correction of aberrations was restricted to the existing astigmatism.

The progress and evolution of AO in the eye was firstly sustained on the development of reliable methods for measuring the ocular aberrations. Among the many different techniques for subjectively estimating the ocular aberrations, the Hartmann-Shack (H-S) wavefront sensor has been the most widely employed method in the context of vision. That technique was imported from astronomical optics as well. The first proposal of such technique for the study of the eye was reported in 1991 [21]. In that work, the H-S wavefront sensor was proposed for estimating the topography of the living cornea. The experimental demonstration of such technique in the eye, measuring the total aberrations, was definitively published in 1994 [22]. Other works have contributed for notably improving the efficiency of the algorithms [23], improving the technique to the level that it has achieved today. IR range is the preferred illumination for measuring the aberrations in the eye [24–27]. The use of the H-S wavefront sensor allowed for measuring high-order aberrations, consequently opening the door to their subsequent correction [24, 28]. It was in 1994 when the first demonstration of high-order static aberration correction was published [29]. Some years later, in 2001, the first closed-loop aberration corrections in real time performed in the living eye were finally reported [11, 30]. In the work of Liang, static ocular aberration correction was accomplished by means of a deformable mirror, for the first time beyond defocus and astigmatism. Two applications of AO in the eye were surveyed: vision through a virtually perfect optics and retinal imaging of high resolution. These uses of the AO have been widely explored by the scientific community in the last years. Regarding the field of retinal imaging, a number of important results arouse from the application of AO to the different imaging techniques. Almost every technique has been merged with AO in the last decade, providing new insights into the existing knowledge of the retinal structure in vivo. Flood illumination fundus cameras were the first benefiting from the utilization of AO, permitting the in vivo recording of the photoreceptors mosaic, even allowing the classification of the three types of them [31]. AO combined with modern laser scanning ophthalmoscopes is providing a unique tool for studying the intimate morphology of the living retina [32], also unveiling interesting aspects of vision. Last technique benefiting from AO was optical coherence tomography (OCT) [26, 33–35]. OCT combines high axial resolution and fast scanning rates. AO has allowed to study the true three-dimensional architecture of the living retina with important applications in the understanding of many retinal conditions [25, 27, 36–40].

3. ADAPTIVE OPTICS AND VISUAL SIMULATION

The use of AO for providing a virtually perfect vision was the other evident application of the technique [29]. Correcting the ocular aberrations should allow vision for reaching its physiological and perceptual limits. The experiments showed an increase in the visual performance in absence of aberrations. Nevertheless, correcting high-order aberrations in the normal young eye produces a limited impact in vision. Some works have studied the distribution of aberrations beyond defocus and astigmatism [14, 18]. It is in the pathologic eye, for instance, those affected by corneal conditions as keratoconus, where AO correction of high order aberrations can really produce a significant benefit in vision. Now it is accepted that high-order aberrations are of a modest importance in the total picture when normal young eyes are considered. In addition, an AO system uses of a number of elements as the corrector, the wavefront sensor, and so forth which prevent with the current state-of-the-art a system compact enough to be coupled directly to the patient's eye.

AO can operate not only for correcting aberrations but in general for manipulating the wavefront. Aberrations can be partially corrected, selecting which particular terms are modified, and which are left as they appear in the eye for instance. This possibility opens a rich number of experiments for the better understanding of the relationship among optical quality and vision. It was in 2002 when this paradigm was experimentally demonstrated in the eye for the study of vision [41, 42]. In the work of Fernández et al., [41], the AO system was coupled to an additional optical relay for the display of stimuli. Visual testing was enabled simultaneously for the operation of the AO system. The apparatus was presented as a visual simulator. Different wavefronts could be generated and vision through the modified optics tested in parallel.

Figure 3 presents a picture of the first visual simulator system. The complete setup was mounted over an optical table of 1 m2. A membrane deformable mirror with 37 independent actuators acted as the correcting device (OKO-Flexible Optical, The Netherlands). Due to the limited amplitude of deformation of the mirror, large amounts of aberrations could not be programmed [43]. In practice, for normal eyes most of the weight of the aberrations is distributed across defocus and astigmatism. Defocus was controlled by a motorized Thorner optometer of high precision, and astigmatism was eventually compensated by introducing trial lens if required. Once static correction of these large aberrations was further investigated, the deformable mirror was able for the real-time compensation of their temporal fluctuations. An external monitor was coupled to the system for the presentation of visual stimuli.

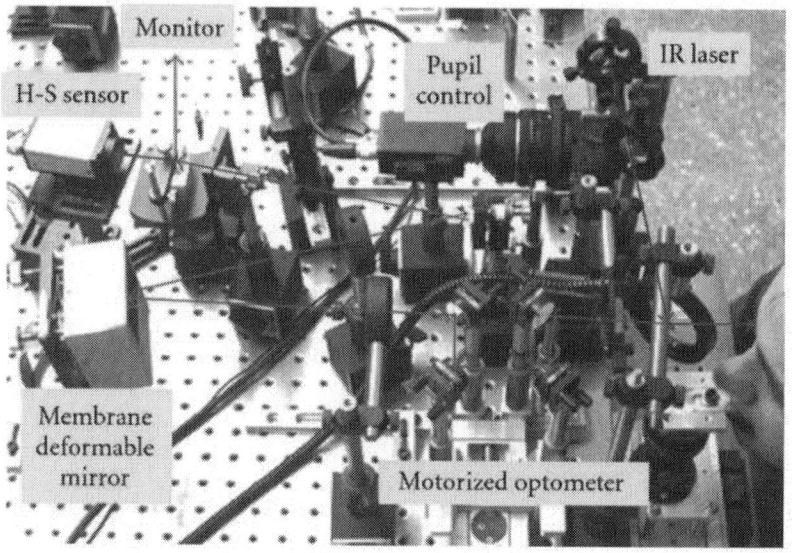

Figure 3

In the first experiment of visual simulation with adaptive optics, a subject with a remarkable coma aberration, in addition of defocus, was asked to complete visual acuity tests while some aberrations were manipulated. Three different situations were explored for understanding the role of aberrations in vision. The coma aberration presented a given axis, causing a marked directionality in the subject's point spread function. In the eye, the existence of such aberration caused an asymmetric blur in the retinal images which should degrade visual performance accordingly. Some amounts of controlled coma aberration were added to the natural aberrations of the subject. From the objective retrieval of the ocular aberration, other optical parameters such as point spread function (PSF) or the modulation transfer function (MTF) could be inferred. Visual acuity was obtained under three cases in the study: vision with natural aberrations; coma aberration added in the same axis than the existing in the subject's eye; the case of 90 deg rotation of the coma aberration, keeping approximately the same modulus than in natural case. The analysis of the MTF revealed that from a purely optical perspective, the quality of the natural and the rotated case were similar; so acuity test obtained from the average of different orientations should not exhibit substantial differences. In the situation of addition of coma aberration, the area of the MTF presented a significant decrease at every frequency, so the poorest acuity was expected in this case. The results were somehow surprising, since adding the same amount of coma aberration to the eye at distinct directions, once corrected of low-order aberrations, caused significant differences in the visual perception of acuity tests. Adding coma aberration in the same direction that the one already present in the eye degraded vision less than the same value at any other direction. This result suggested that vision is coupled to the optical quality of the eye in an unexpected manner. The larger tolerance of vision when retinal images are degraded in a particular direction might be due to a kind of adaptation of the visual system to its own aberrations. Figure 4 shows graphically these findings. The average visual acuity with the standard deviation from the measurements is plotted with the corresponding generated aberrations. Their associated PSFs and the measured RMS of the wavefront are also displayed in Figure 4.

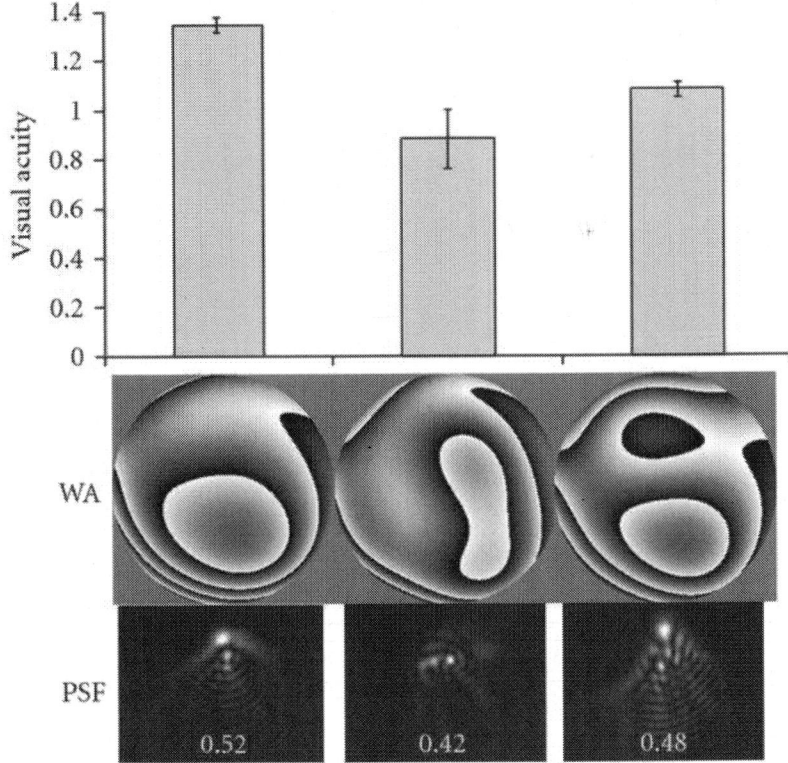

Figure 4

This hypothesis of adaptation to the aberrations was investigated in a different experiment, where the vision of a larger number of subjects was tested with manipulated wavefronts [44]. In such work, the perception of blur was evaluated by using a random pattern showing irregular black spots on a bright background. Monochromatic illumination at 543 nm was used in order to avoid the possible effects of chromatic aberrations. The ocular aberrations of each subject were recorded. Using the retrieved wavefront, an algorithm generated the appropriate shape for the deformable mirror so that the combination of ocular and mirror wavefronts produced a rotated version of the original ocular aberration. The procedure was performed in closed loop, at real time. This operation guaranteed that the subject was continuously viewing the test through the modified aberration pattern. It must be stressed out that the experiment forced the optical quality of the retinal images projected on the observer to be exactly the same, but for orientation. A two-alternatives forced choice test was programmed. Images through the natural aberrations of the eye were displayed together with the same through the modified version of the aberrations. Subjects were asked for changing the level of perceived blur in the manipulated version of the images until equalization of blur with natural viewing was achieved. For accomplishing the task, the subjects multiplied the rotated version of the wavefront by a factor which globally reduced the level of aberrations, simultaneously in all terms. The results were a systematic perception of larger blur when viewing through a rotated version of the own subjects' aberrations. Since the optical quality of the retinal images was similar in all cases, the different perception of blur had to be produced by neural factors. The proposed explanation to this novel effect was a new kind of adaptation to the own individual aberration pattern. Essentially, the brain could be used to extract information from images degraded in a constant manner, imposed by the subject's ocular aberrations. Any change away from this natural shape degrades perception, even if the optical quality levels are similar. The experiment was interesting since it opened the door to a novel perspective for understanding vision. Many other experiments using AO since then have explored a number of open questions related to adaptation of vision to the ocular aberrations [45–50].

The effects of neural adaptation might become more evident in those eyes suffering of larger monochromatic aberrations. A typical example of highly aberrated optics can be found in those patients with keratoconus. This is a relatively well-studied condition producing a severe deformation of the anterior and posterior surfaces of the cornea. As it has been mentioned before, the effects on the retinal images are usually very strong, degrading the vision of the patients up to dramatic limits. Some other works have used adaptive optics in the context of keratoconic eyes, trying to better understand the capability of the eye for adapting to these large aberrations [51–53].

An intriguing question partially arousing from these experiments of adaptation is the possible effect of the aberrations on accommodation. Aberrations change when the eye accommodates, beyond defocus and astigmatism. Therefore, correcting or altering the aberrations with AO for far vision could eventually degrade somehow accommodation or near vision, since the aberrations pattern in the eye might be significantly distinct from the natural shape. Many questions are to be answered. These particular points will be treated in the following.

4. ADAPTIVE OPTICS AND ACCOMMODATION

Accommodation is possibly one of the most studied functions of the visual system. This feature allows the eye to increase its optical power so that near objects can be correctly focused onto the observer's retina. The augment in power is exclusively caused by the crystalline lens. The intimate mechanism of accommodation involves several components. A simplified picture of the accommodation is presented in the following. The blurred retinal image produced by a near object originated, through a number of neural channels, the contraction of the ciliary muscle. The contraction of the muscle reduces the space around the crystalline lens in the equatorial plane. The first consequence is the reduction of the tension in the zonula. The capsule containing the crystalline lens and the lens itself are then able to compress towards their relaxed state. The anterior and posterior surfaces of the crystalline lens experience an increase in their curvature. The total thickness of the lens also enlarges. All these geometrical changes in the crystalline lens produce the global effect of increasing its power. Several aspects concerning the mechanism of accommodation are still unveiled. An intriguing question is how the visual system is able to change the power of the eye in the correct direction. In a situation where the eye is focusing a target situated at middle distance, a change of the stimulus or a variation on its position will immediately produce a blurred image onto the retina [54]. Defocus alone does not provide any cue about the direction of change, since the degradation of the image is symmetrical around the initial image plane [55]. The visual system is, however, known to produce the accommodation in the right direction in most of the situations [56]. In this context, the existing chromatic aberration of the eye has been proposed and demonstrated as a source of asymmetry in the retinal images providing, at least in part, information about the factual direction for accommodation [57–60]. Other optical cues have been proposed, for instance, the microfluctuations in accommodation [61]. More recently the ocular aberrations have been also proposed as a possible source of asymmetry in the defocused retinal images which eventually could provide an additional cue for determining the direction of accommodation [62]. Adaptive optics was applied for studying accommodation, and the possible role of monochromatic aberrations in the visual response to defocused stimuli [34]. The experiment consisted in the correction of monochromatic aberrations during accommodation induced by an abrupt change of 1.5 D in the vergence of the stimulus. The variation in the perceived position of the stimulus was generated by using a motorized Badal optometer. Two subjects with normal accommodation capability and normal values for their ocular aberrations participated in the experiment. The procedure was performed under monocular vision and with monochromatic illumination. These two factors are related with known optical cues for accommodation. The experiment isolated all other optical cues but high-order monochromatic aberrations. Defocus was uncorrected for allowing the subjects to freely change its value. Several parameters related to the accommodation response were obtained with and without adaptive optics aberration correction. The final level of achieved accommodation was measured under the two conditions. No significant variation of this particular parameter was detected, meaning that the precision of the accommodation was independent of the existing ocular aberrations. This was a somehow surprising result, since it showed that the quality of the retinal images, when they are distorted by normal levels of monochromatic aberrations, is not a key factor in the eye for finding the best focused retinal image. Another parameter of the accommodation studied in the experiment was the latency time of the visual response. The latency can be understood as the time gap between the blurring of the stimulus and the instant when the eye begins the accommodation. This latency was also found to remain independent of the correction of monochromatic aberrations. The other temporal factor involved in the dynamic of the accommodation and retrieved in the experiment was the response time. This parameter was defined as the time that the accommodation is changing from the initial value to the final level achieved. For this variable, a significant increment in its absolute value when correcting the monochromatic aberrations was measured. It indicated that the compensation of the aberrations degraded this particular parameter, causing an increment in the required time to achieve the final level of accommodation for focusing the stimulus. That was an unexpected result, since it showed for the first time an important function of the visual system as the accommodation suffering a deterioration directly produced by the correction of the monochromatic aberrations. A possible conclusion of that work is that natural aberrations play a role in some visual functions, as the accommodation, and their correction might not be reasonable in all cases. Others works have studied the possible relationship between accommodation and monochromatic aberrations. Some experiments have reported the lack of effect of correcting aberrations in some subjects [63–65], while other have shown a clear benefit in the dynamic of the accommodation associated to the aberrations compensation [66].

Possibly for some subjects monochromatic aberrations are a useful cue in accommodation, and consequently they employ the asymmetry in the retinal images for enhancing the dynamics. It is reasonable then that those subject experience a degradation in their accommodation when aberrations are compensated. On the other hand, other subjects might not use such information, and consequently the correction of retinal images through the aberration compensation enhances some of their accommodation parameters because of a better retinal image. Depending on the level of aberrations, it might be also reasonable that for some subjects the correction of monochromatic aberrations was irrelevant for the accommodation function. In any case, the question remains unsolved and possible future experiments will provide additional information about the connection between accommodation and monochromatic aberration.

5. BINOCULAR AO

Normal vision is not monocular vision. At the beginnings of the development of AO in the eye, all the experiments were devised under monocular vision. Binocular vision provides a number of advantages for our perception that cannot be neglected [67], for instance, the capability for perceiving depth in the natural scenes. Because of the necessity for objectively measuring binocular refraction, some optometers capable of simultaneous defocus estimation of the two pupils were developed in the recent past [68–71]. They were the antecedent systems to the first binocular wavefront sensors. In 2008, Kobayashi et al. reported a system endowed with two Hartmann-Shack wavefront sensors specifically designed for binocular estimation of the aberrations from the two eyes [72]. The system operated in open view. With this apparatus, the subjects could undergo visual testing while the measurement of their ocular aberrations was taking place. This first approach of replicating the sensor brought about the duplication of the cost of the system, together with an increase in the complexity of the control of the setup. In the same year, another alternative to this first solution for the binocular objective estimation of the ocular aberrations was presented in the work of Hampson et al. [73]. In this other approach, a design allowing the estimation of the aberrations from both eyes employing a single sensor was reported. Using a Hartmann-Shack-based apparatus, the light emerging from each pupil was redirected into the system, keeping along the optical relays the two beams spatially resolved. The two eyes exit pupils were projected on the surface of the wavefront sensor, so that the camera could obtain in a single frame the spots from both pupils. Appropriate postprocessing allowed the retrieval of the ocular aberrations from both eyes. The advantages of such apparatus were evident in terms of cost and complexity.

It was not until 2009 that the first binocular adaptive optics system was reported and successfully applied on real eyes [74]. The most remarkable feature of this system, specifically designed for operating under binocular vision, was the employment of a single correcting device in combination with a single wavefront sensor. The latter operated a similar approach of that reported by Hampson [73], separating the two pupils on the surface of the detector. The correcting device [74, 75] followed an analogous principle for managing the two pupils. Those were simultaneously projected on its surface, still spatially resolved, and independently driven from the computer. The fundamental issue arising from this configuration was procuring enough resolution on the corrector for manipulating aberrations, including high-order aberrations, from both pupils with sufficient accuracy. In order to solve this obstacle, a liquid crystal-based correcting device was incorporated in the system. This kind of correctors was a basic pillar for the understanding of the current state-of-the-art of binocular adaptive optics visual simulation. Figure 5 shows a diagram of a binocular AO visual simulator. Their most important and attractive characteristic was possibly the enormous resolution that they exhibited, particularly as compared with deformable mirrors. In the first binocular visual simulator, a commercial spatial light modulator (LCOS-SLM X10468-04, Hamamatsu Photonics, Japan) with SVGA resolution was incorporated. The total number of pixels was , selecting approximately 58000 pixels for each pupil. The device allowed keeping a reasonable number of pixels dedicated for the control of the aberrations from each eye. The experimental system incorporated an optical relay for displaying binocular stimuli. The capability of the apparatus was demonstrated by testing the impact of different combinations of spherical aberration on vision. For such purpose, the contrast sensitivity of a subject at 7.8 c/deg in green light was obtained for different conditions. The contrast sensitivity was obtained by using a two-alternative forced choice test, showing a panel with the target grating and different contrast and other with homogenous background. The subject answered which one was displaying the grating. The time for displaying the stimuli was 500 ms. The grating subtended 1 deg, assuring isoplanatic conditions at the retina. The contrast of the grating was set randomly. A psychometric curve was obtained for inferring the value of the contrast sensitivity. That was estimated as the detection threshold at 75% of confidence. The experiment was repeated under monocular and binocular conditions to discern the influence of binocular summation. The value of spherical aberration was ±0.2 μm for a pupil of diameter 4 mm. The experiment accomplished in this work was presented as a demonstration of the potential of the technique. A single subject underwent contrast sensitivity testing. Therefore, extrapolation to general conclusions should be taken with care. Still, some interesting results were obtained. Those are grouped in Figure 6.

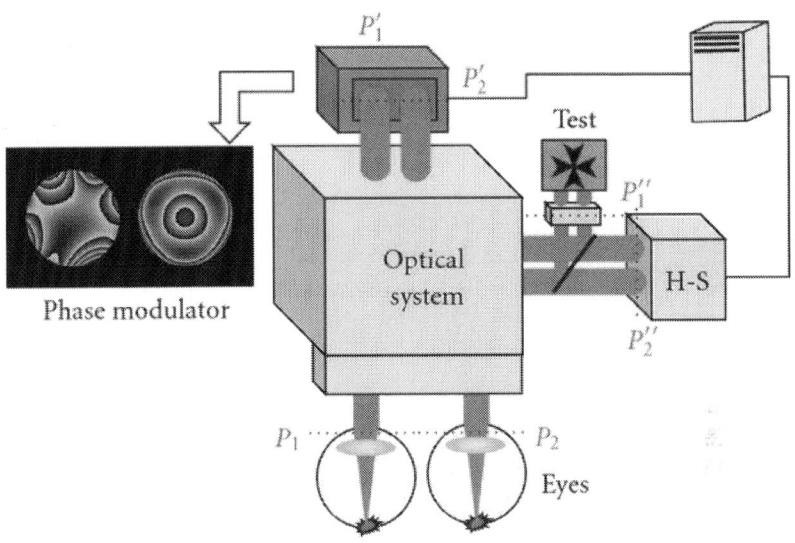

Figure 5

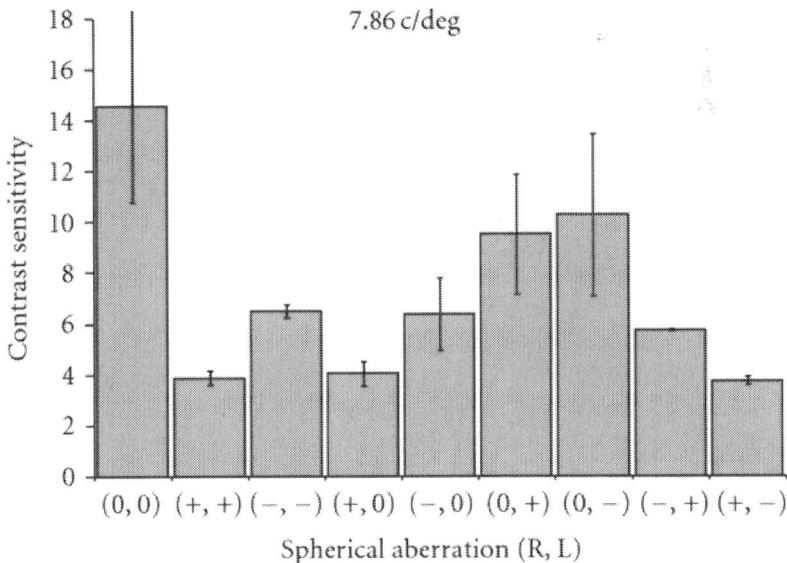

Figure 6

The plot shows the contrast sensitivity obtained for different combinations of spherical aberration. The signs + and − indicate the value of the spherical aberration added on the eyes. The position of the value in the brackets represents which eye was affected for each value, being the first and second values for the left and the right eye, respectively. The subject reported his dominant eye to be the left one. Systematically, keeping the dominant eye with the natural aberrations caused a better visual acuity. That was particularly evident when the results were compared with those corresponding to the reversal cases, where spherical aberration was added to the nondominant eye. Another interesting result was that the degradation of visual acuity was more dramatic when positive spherical aberration was added to the dominant eye, in absolute values. This particular eye exhibited a larger tolerance to negative spherical aberration.

Among the different advantages that binocular vision provides over our visual perception, stereopsis is one of the most evident features. Stereopsis is intimately connected with the perception of depth [76–79]. The sensation of depth can be also generated from a number of monocular cues. The previous knowledge of the size of an object can produce the estimation of its relative distance regarding other parts of the scene, therefore creating the psychological experience of depth. Other classical monocular cues are the motion parallax, the perspective, texture gradient, accommodation, defocus blur, and so forth. Stereopsis is often defined as the capability for perceiving the depth in a scene originated exclusively for the distinct position of the images on each retina, or more commonly referred to as retinal disparity. This kind of retinal parallax, produced by the relatively shifted perspective of the object each eye has been widely studied in the context of visual perception.

Stereopsis involves a delicate neural processing in addition to the purely optical or geometrical stage. Actually, stereopsis has traditionally been studied in the context of psychology of the perception. Some recent attempts for introducing the possible impact of the ocular optical quality over stereopsis have been reported [80, 81]. In this direction, the possible influence on stereopsis due to the changes in the ocular aberrations has been considered [82]. Such changes can be found as a consequence of refractive surgery. Adaptive optics provides the ideal tool for characterizing the actual impact of aberrations on stereopsis. Not only low-order aberrations, as defocus and astigmatism can be tested, but in general any required aberration. In the work of Fernandez et al. [83], the stereopsis was evaluated degrading the wavefront with different aberrations by using a binocular adaptive optics visual simulator. The system incorporated two different internal displays for projecting distinct retinal images over each retina. This permitted the generation of retinal disparity, indistinguishable from an optical perspective to that created with a real scene in front of the subject. The stereopsis was tested through the measurement of the stereoacuity. Stereoacuity provides a numerical estimate of the capability of the subject for detecting changes in depth associated exclusively to retinal disparity. Actually, stereoacuity is the minimum retinal disparity, given as the subtended angle, causing the perception of depth. Using separate displays for each eye guaranteed that no other monocular cue was involved in the experiment. Random dot stereograms were used for obtaining stereoacuity. Random dot stereograms are a useful tool for understanding pure stereopsis, particularly global stereopsis. They were introduced by Julesz in the 70s [84] and since then these have been widely employed for binocular vision research. In the aforementioned work, pure defocus was first added under different conditions. Selecting a pupil diameter of 4 mm, stereoacuity was retrieved for natural vision with low-order refractive correction (including defocus and astigmatism), addition of 1 D of defocus in both eyes and 1 D of defocus in one of the eyes. The results showed a progressive loss of stereoacuity in the different conditions previously presented, in the same order that they have been mentioned. In particular, for one of the subjects the values of stereoacuity were 4, 6, and 8 sec, respectively. The worst case regarding the value of stereoacuity was reported when asymmetric induction of defocus, rather than in the bilateral case. This result confirmed previous studies about the relative impact of blur, caused by defocus or in general low-order aberrations, on stereopsis. There is a large number of works which have considered other factors as the loss of contrast or luminance of the stimuli and their effect on stereopsis. Most of them could be considered from a wide perspective as different outcomes of defocus [85–92]. It is relatively well established that monocular degradation of the retinal images, through defocus, contrast, or luminance, decreases stereopsis more than the corresponding bilateral degradation. The work of Fernández et al. [83] introduced for the first time in the literature the evaluation of the impact on stereopsis of a high-order aberration generated by adaptive optics. The trefoil aberration was selected and applied both binocular and monocular while simultaneous measurement of the stereoacuity was performed. The obtained trend followed that typically occurring for defocus. The addition of trefoil in a single eye produced a larger degradation on stereopsis than the bilateral case. The obtained values of stereoacuity were 4, 13, and 18 sec for natural vision (with no additional aberrations), unilateral and bilateral addition of pure trefoil, respectively. An evolution of this experimental setup was reported by Schwarz [93]. The new apparatus incorporated an additional liquid crystal operating in transmission for the manipulation of the pupils. This other liquid crystal acted as an amplitude modulation device, manipulating pupils' size and their relative position during the measurements, still keeping the capability for phase modulation with the other correcting device.

6. CONCLUSIONS

In this paper, different results and applications of adaptive optics have been covered. In particular, special attention has been paid to visual simulation.

There are some other related topics which have not been presented. In this context, possibly one of the most promising fields is the employment of adaptive optics for the design of new ophthalmic elements [94, 95]. This is one of the applications that have benefitted more from adaptive optics visual simulation in the last years. Due to its extension and complexity, this topic has not been tackled in the current paper, but it is definitely an important application which has already changed some paradigms in optical design. The technique permits for the first time the virtual implementation of the designed optics on real eyes before manufacturing. Adaptive optics can reproduce the wavefront, and a visual simulator apparatus can allow patients to see through the optics, eventually testing the expected visual benefit. Moreover, for customized solutions incorporating the compensation of individual ocular aberrations by visual simulators provides not only the platform for measuring these aberrations but the station capable of testing the effects, again prior to the practical piecing of the optics. That new procedure significantly reduces cost and provides a faster method for optimizing the final element to be coupled to the eye. As an example, it has been successfully demonstrated for intraocular lenses, contact lenses and for studying the effect of spherical aberrations for extending the depth of field [96–101].

From a more basic science perspective, another interesting topic where adaptive optics visual simulators and related systems are helping to answer many fundamental questions is the true impact of aberrations on vision. Once ocular aberrations have been correctly measured, the question arising is their influence on vision. Many works have reported progress in this field from the early beginning of the adaptive optics in vision [29, 49, 102–120].

There are other exciting applications of adaptive optics not covered in the current work. High-resolution retinal imaging is unveiling the intimate structure and function of the retina. Some experiments are connecting the retinal structure and the visual function, answering some basic questions about how we see the world [48, 121–130].

Adaptive optics in visual sciences is in constant evolution. In this paper, several success and applications of the technology have been presented, but probably many others are to come in the near future. Other interesting reviews on adaptive optics have been published, where the interested reader will find a distinct focus and very useful information about the technique with additional valuable references [131, 132]. Adaptive optics has dramatically changed our knowledge about the eye and vision. A natural step would be the incorporation of such modality in the clinical practice, so that patients and clinicians can benefit from that. The state-of-the-art adaptive optics is now mature enough to be incorporated into a rich variety of instruments. The future will probably show a progressive merging of adaptive optics into the most widely used ophthalmoscopes for increasing their resolution up to cellular level. The visual simulators are already in the market for advanced refraction, and they have the potential for becoming the reference phoropters.

REFERENCES

1.   Y. Le Grand and S. G. El Hage, Physiological Optics, Springer, Berlin, Germany, 1980.

2.   J. C. Barry, K. Branmann, and M. C. M. Dunne, "Catoptric properties of eyes with misaligned surfaces studied by exact ray tracing," Investigative Ophthalmology & Visual Science, vol. 38, no. 8, pp. 1476–1484, 1997.

3.   J. Tabernero, A. Benito, V. Nourrit, and P. Artal, "Instrument for measuring the misalignments of ocular surfaces," Optics Express, vol. 14, no. 22, pp. 10945–10956, 2006. ·

4.   P. Artal and A. Guirao, "Contribution of cornea and lens to the aberrations of the human eye," Optics Letters, vol. 23, pp. 1713–1715, 1998.

5.   P. Artal, A. Guirao, E. Berrio, and D. R. Williams, "Compensation of corneal aberrations by the internal optics in the human eye," Journal of Vision, vol. 1, no. 1, pp. 1–8, 2001.

6.   J. E. Kelly, T. Mihashi, and H. C. Howland, "Compensation of corneal horizontal/vertical astigmatism, lateral coma, and spherical aberration by internal optics of the eye," Journal of Vision, vol. 4, no. 4, pp. 262–271, 2004.

7.   J. Tabernero, A. Benito, E. Alcón, and P. Artal, "Mechanism of compensation of aberrations in the human eye," Journal of the Optical Society of America A, vol. 24, no. 10, pp. 3274–3283, 2007.

8.   L. Diaz-Santana, C. Torti, I. Munro, P. Gasson, and C. Dainty, "Benefit of higher closed-loop bandwidths in ocular adaptive optics," Optics Express, vol. 11, no. 20, pp. 2597–2605, 2003.

9.   K. M. Hampson, C. Paterson, C. Dainty, and E. A. H. Mallen, "Adaptive optics system for investigation of the effect of the aberration dynamics of the human eye on steady-state accommodation control," Journal of the Optical Society of America A, vol. 23, no. 5, pp. 1082–1088, 2006.

10.   E. J. Fernández and P. Artal, "Dynamic eye model for adaptive optics testing," Applied Optics, vol. 46, no. 28, pp. 6971–6977, 2007.

11. H. Hofer, P. Artal, B. Singer, J. L. Aragón, and D. R. Williams, "Dynamics of the eye's wave aberration," Journal of the Optical Society of America. A, vol. 18, no. 3, pp. 497–506, 2001.

12. P. Artal, A. Benito, and J. Tabernero, "The human eye is an example of robust optical design," Journal of Vision, vol. 6, no. 1, pp. 1–7, 2006.

13. P. Artal and J. Tabernero, "The eye's aplanatic answer," Nature Photonics, vol. 2, no. 10, pp. 586–589, 2008.

14. J. Porter, A. Guirao, I. G. Cox, and D. R. Williams, "Monochromatic aberrations of the human eye in a large population," Journal of the Optical Society of America. A, vol. 18, no. 8, pp. 1793–1803, 2001.

15. M. P. Cagigal, V. F. Canales, J. F. Castejón-Mochón, P. M. Prieto, N. López-Gil, and P. Artal, "Statistical description of the wave front aberration in the human eye," Optics Letters, vol. 27, no. 1, pp. 37–39, 2002.

16. L. N. Thibos, X. Hong, A. Bradley, and X. Cheng, "Statistical variation of aberration structure and image quality in a normal population of healthy eyes," Journal of the Optical Society of America A, vol. 19, no. 12, pp. 2329–2348, 2002.

17. J. S. McLellan, P. M. Prieto, S. Marcos, and S. A. Burns, "Effects of interactions among wave aberrations on optical image quality," Vision Research, vol. 46, no. 18, pp. 3009–3016, 2006.

18. J. F. Castejón-Mochón, N. López-Gil, A. Benito, and P. Artal, "Ocular wave-front statistics in a normal young population," Vision Research, vol. 42, pp. 1611–1617, 2002.

19. H. W. Babcock, "The possibility of compensating astronomical seeing," Publications of the Astronomical Society of the Pacific, vol. 65, no. 386, p. 229, 1953.

20. W. Dreher, J. F. Bille, and R. N. Weinreb, "Active optical depth resolution improvement of the laser tomographic scanner," Applied Optics, vol. 28, no. 4, pp. 804–808, 1989.

21. S. Goelz, J. J. Persoff, G. D. Bittner, J. Liang, C. T. Hsueh, and J. F. Bille, "New wavefront sensor for metrology of spherical surfaces," in Active and Adaptive Optical Systems, vol. 1542 of Proceedings of SPIE, pp. 502–511, 1991.

22. J. Liang, B. Grimm, S. Goelz, and J. F. Bille, "Objective measurement of wave aberrations of the human eye with the use of a Hartmann-Shack wave-front sensor," Journal of the Optical Society of America A, vol. 11, no. 7, pp. 1949–1957, 1994.

23. P. M. Prieto, F. Vargas-Martín, S. Goelz, and P. Artal, "Analysis of the performance of the Hartmann-Shack sensor in the human eye," Journal of the Optical Society of America A, vol. 17, no. 8, pp. 1388–1398, 2000.

24. E. J. Fernández, A. Unterhuber, B. Považay, B. Hermann, P. Artal, and W. Drexler, "Chromatic aberration correction of the human eye for retinal imaging in the near infrared," Optics Express, vol. 14, pp. 6231–6225, 2006.

25. E. J. Fernández, B. Hermann, B. Považay et al., "Ultrahigh resolution optical coherence tomography and pancorrection for cellular imaging of the living human retina," Optics Express, vol. 16, no. 15, pp. 11083–11094, 2008.

26. E. J. Fernández and W. Drexler, "Influence of ocular chromatic aberration and pupil size on transverse resolution in ophthalmic adaptive optics optical coherence tomography," Optics Express, vol. 13, no. 20, pp. 8184–8197, 2005.

27. E. J. Fernández and P. Artal, "Ocular aberrations up to the infrared range: from 632.8 to 1070 nm," Optics Express, vol. 16, no. 26, pp. 21199–21208, 2008.

28. C. Paterson, I. Munro, and J. C. Dainty, "A low cost adaptive optics system using a membrane mirror," Optics Express, vol. 6, no. 9, pp. 175–185, 2000.

29. J. Liang, D. R. Williams, and D. T. Miller, "Supernormal vision and high-resolution retinal imaging through adaptive optics," Journal of the Optical Society of America A, vol. 14, no. 11, pp. 2884–2892, 1997.

30. E. J. Fernández, I. Iglesias, and P. Artal, "Closed-loop adaptive optics in the human eye," Optics Letters, vol. 26, no. 10, pp. 746–748, 2001.

31. Roorda and D. R. Williams, "The arrangement of the three cone classes in the living human eye," Nature, vol. 397, no. 6719, pp. 520–522, 1999.

32. Roorda, F. Romero-Borja, W. J. Donnelly, H. Queener, T. J. Hebert, and M. C. W. Campbell, "Adaptive optics scanning laser ophthalmoscopy," Optics Express, vol. 10, no. 9, pp. 405–412, 2002.

33. Hermann, E. J. Fernández, A. Unterhuber et al., "Adaptive-optics ultrahigh-resolution optical coherence tomography," Optics Letters, vol. 29, no. 18, pp. 2142–2144, 2004.

34. E. J. Fernández, B. Považay, B. Hermann et al., "Three-dimensional adaptive optics ultrahigh-resolution optical coherence tomography

using a liquid crystal spatial light modulator," Vision Research, vol. 45, no. 28, pp. 3432–3444, 2005.

35. E. J. Fernández, A. Unterhuber, P. M. Prieto, B. Hermann, W. Drexler, and P. Artal, "Ocular aberrations as a function of wavelength in the near infrared measured with a femtosecond laser," Optics Express, vol. 13, no. 2, pp. 400–409, 2005.

36. Y. Zhang, B. Cense, J. Rha et al., "High-speed volumetric imaging of cone photoreceptors with adaptive optics spectral-domain optical coherence tomography," Optics Express, vol. 14, no. 10, pp. 4380–4394, 2006.

37. R. J. Zawadzki, S. S. Choi, S. M. Jones, S. S. Oliver, and J. S. Werner, "Adaptive optics-optical coherence tomography: optimizing visualization of microscopic retinal structures in three dimensions," Journal of the Optical Society of America A, vol. 24, no. 5, pp. 1373–1383, 2007.

38. Merino, C. Dainty, A. Bradu, and A. G. Podoleanu, "Adaptive optics enhanced simultaneous en-face optical coherence tomography and scanning laser ophthalmoscopy," Optics Express, vol. 14, no. 8, pp. 3345–3353, 2006.

39. E. Bigelow, N. V. Iftimia, R. D. Ferguson, T. E. Ustun, B. Bloom, and D. X. Hammer, "Compact multimodal adaptive-optics spectral-domain optical coherence tomography instrument for retinal imaging," Journal of the Optical Society of America A, vol. 24, no. 5, pp. 1327–1336, 2007.

40. M. Pircher, R. J. Zawadzki, J. W. Evans, J. S. Werner, and C. K. Hitzenberger, "Simultaneous imaging of human cone mosaic with adaptive optics enhanced scanning laser ophthalmoscopy and high-speed transversal scanning optical coherence tomography," Optics Letters, vol. 33, no. 1, pp. 22–24, 2008.

41. J. Fernández, S. Manzanera, P. Piers, and P. Artal, "Adaptive optics visual simulator," Journal of Refractive Surgery, vol. 18, no. 5, pp. S634–S638, 2002.

42. P. Artal, E. J. Fernández, and S. Manzanera, "Are optical aberrations during accommodation a significant problem for refractive surgery?" Journal of Refractive Surgery, vol. 18, no. 5, pp. S563–S566, 2002.

43. J. Fernández and P. Artal, "Membrane deformable mirror for adaptive optics: performance limits in visual optics," Optics Express, vol. 11, no. 9, pp. 1056–1069, 2003.

44. P. Artal, L. Chen, E. J. Fernandez, B. Singer, S. Manzanera, and D. Williams, "Neural compensation for the eye's optical aberrations," Journal of Vision, vol. 4, no. 4, pp. 281–287, 2004.

45. L. Chen, P. Artal, D. Gutierrez, and D. R. Williams, "Neural compensation for the best aberration correction," Journal of Vision, vol. 7, no. 10, pp. 1–9, 2007.

46. E. A. Rossi, P. Weiser, J. Tarrant, and A. Roorda, "Visual performance in emmetropia and low myopia after correction of high-order aberrations," Journal of Vision, vol. 7, no. 8, article 14, pp. 1–14, 2007.

47. J. Murray, S. L. Elliott, A. Pallikaris, J. S. Werner, S. Choi, and H. J. Tahir, "The oblique effect has an optical component: orientation-specific contrast thresholds after correction of high-order aberrations," Journal of Vision, vol. 10, no. 11, pp. 1–12, 2010.

48. E. A. Rossi and A. Roorda, "Is visual resolution after adaptive optics correction susceptible to perceptual learning?"Journal of Vision, vol. 10, no. 12, pp. 1–14, 2010.

49. L. Sawides, E. Gambra, D. Pascual, C. Dorronsoro, and S. Marcos, "Visual performance with real-life tasks under adaptive-optics ocular aberration correction," Journal of Vision, vol. 10, no. 5, p. 19, 2010.

50. L. Sawides, S. Marcos, S. Ravikumar, L. Thibos, A. Bradley, and M. Webster, "Adaptation to astigmatic blur," Journal of Vision, vol. 10, no. 12, pp. 1–15, 2010.

51. R. Sabesan, T. M. Jeong, L. Carvalho, I. G. Cox, D. R. Williams, and G. Yoon, "Vision improvement by correcting higher-order aberrations with customized soft contact lenses in keratoconic eyes," Optics Letters, vol. 32, no. 8, pp. 1000–1003, 2007.

52. R. Sabesan and G. Yoon, "Visual performance after correcting higher order aberrations in Keratoconic eyes," Journal of Vision, vol. 9, no. 5, article 6, pp. 1–10, 2009.

53. R. Sabesan and G. Yoon, "Neural compensation for long-term asymmetric optical blur to improve visual performance in keratoconic eyes," Investigative Ophthalmology & Visual Science, vol. 51, no. 7, pp. 3835–3839, 2010.

54. L. M. Smithline, "Accommodative response to blur," Journal of the Optical Society of America, vol. 64, no. 11, pp. 1512–1516, 1974.

55. F. Ciuffreda, "Accommodation and its anomalies," in Vision and Visual Dysfunction, J. R. Cronly-Dillon, Ed., Macmillan, New York, NY, USA, 1991.

56. Stark and Y. Takahashi, "Absence of an odd-error signal mechanism in human accommodation," IEEE Transactions on Biomedical Engineering, vol. 12, no. 3-4, pp. 138–146, 1965.

57. E. F. Fincham, "The accommodation reflex and its stimulus," British Journal of Ophthalmology, vol. 35, pp. 5–80, 1951.

58. P. B. Kruger, S. Mathews, K. R. Aggarwala, and N. Sánchez, "Chromatic aberration and ocular focus: fincham revisited," Vision Research, vol. 33, no. 10, pp. 1397–1411, 1993.

59. R. Aggarwala, E. S. Kruger, S. Mathews, and P. B. Kruger, "Spectral bandwidth and ocular accommodation," Journal of the Optical Society of America A, vol. 12, no. 3, pp. 450–455, 1995.

60. W. Campbell and G. Westheimer, "Factors influencing accommodation responses of the human eye," Journal of the Optical Society of America, vol. 49, pp. 568–571, 1959.

61. W. N. Charman and G. Heron, "Fluctuations in accommodation: a review," Ophthalmic and Physiological Optics, vol. 8, pp. 153–164, 1988.

62. B. J. Wilson, K. E. Decker, and A. Roorda, "Monochromatic aberrations provide an odd-error cue to focus direction," Journal of the Optical Society of America A, vol. 19, pp. 833–839, 2002.

63. Chen, P. B. Kruger, H. Hofer, B. Singer, and D. R. Williams, "Accommodation with higher-order monochromatic aberrations corrected with adaptive optics," Journal of the Optical Society of America A, vol. 23, no. 1, pp. 1–8, 2006.

64. S. S. Chin, K. M. Hampson, and E. A. H. Mallen, "Role of ocular aberrations in dynamic accommodation control," Clinical and Experimental Optometry, vol. 92, no. 3, pp. 227–237, 2009.

65. S. S. Chin, K. M. Hampson, and E. A. H. Mallen, "Effect of correction of ocular aberration dynamics on the accommodation response to a sinusoidally moving stimulus," Optics Letters, vol. 34, no. 21, pp. 3274–3276, 2009.

66. E. Gambra, L. Sawides, C. Dorronsoro, and S. Marcos, "Accommodative lag and fluctuations when optical aberrations are manipulated," Journal of Vision, vol. 9, no. 6, article 4, pp. 1–15, 2009.

67. F. W. Campbell and D. G. Green, "Monocular versus binocular visual acuity," Nature, vol. 208, no. 5006, pp. 191–192, 1965.

68. R. Clark and H. D. Crane, "Dynamic interaction in binocular vision," in Eye Movement and the Higher Psychological Functions, J. W. Senders,

D. F. Fisher, and R. A. Monty, Eds., pp. 77–88, Erlbaum, New York, NY, USA, 1978.

69. Heron and B. Winn, "Binocular accommodation reaction and response times for normal observers," Ophthalmic and Physiological Optics, vol. 9, no. 2, pp. 176–183, 1989.

70. G. Heron, B. Winn, J. R. Pugh, and A. S. Eadie, "Twin channel infrared optometer for recording binocular accomodation," Optometry and Vision Science, vol. 66, no. 2, pp. 123–129, 1989.

71. F. Okuyama, T. Tokoro, and M. Fujieda, "Binocular infrared optometer for measuring accommodation in both eyes simultaneously in natural-viewing conditions," Applied Optics, vol. 32, no. 22, pp. 4147–4154, 1993.

72. M. Kobayashi, N. Nakazawa, T. Yamaguchi, T. Otaki, Y. Hirohara, and T. Mihashi, "Binocular open-view Shack-Hartmann wavefront sensor with consecutive measurements of near triad and spherical aberration," Applied Optics, vol. 47, no. 25, pp. 4619–4626, 2008.

73. S. S. Chin, K. M. Hampson, and E. A. H. Mallen, "Binocular correlation of ocular aberration dynamics," Optics Express, vol. 16, no. 19, pp. 14731–14745, 2008.

74. E. J. Fernández, P. M. Prieto, and P. Artal, "Wave-aberration control with a liquid crystal on silicon (LCOS) spatial phase," Optics Express, vol. 17, pp. 11013–11025, 2009.

75. M. Prieto, E. J. Fernández, S. Manzanera, and P. Artal, "Adaptive optics with a programmable phase modulator: applications in the human eye," Optics Express, vol. 12, no. 17, pp. 4059–4071, 2004.

76. Howard and B. J. Rogers, Binocular Vision and Stereopsis, vol. 29 of Oxford Psychology Series, Oxford Universit Press, 1995.

77. W. Reading, Binocular Vision: Foundations and Applications, Butterworth-Heinemann, Boston, Mass, USA, 1983.

78. R. Fielder and M. J. Moseley, "Does stereopsis matter in humans?" Eye, vol. 10, no. 2, pp. 233–238, 1996.

79. O'Connor, E. E. Birch, S. Anderson, and H. Draper, "The functional significance of stereopsis," Investigative Ophthalmology & Visual Science, vol. 51, no. 4, pp. 2019–2023.

80. J. Castro, J. R. Jiménez, E. Hita, and C. Ortiz, "Influence of interocular differences in the Strehl ratio on binocular summation," Ophthalmic and Physiological Optics, vol. 29, no. 3, pp. 370–374, 2009.

81.   R. Jiménez, J. J. Castro, R. Jiménez, and E. Hita, "Interocular differences in higher-order aberrations on binocular visual performance," Optometry and Vision Science, vol. 85, no. 3, pp. 174–179, 2008.

82.   R. Jiménez, J. J. Castro, E. Hita, and R. G. Anera, "Upper disparity limit after LASIK," Journal of the Optical Society of America. A, vol. 25, no. 6, pp. 1227–1231, 2008.

83.   E. J. Fernández, P. M. Prieto, and P. Artal, "Adaptive optics binocular visual simulator to study stereopsis in the presence of aberrations," Journal of the Optical Society of America A, vol. 27, no. 11, pp. A48–A55, 2010.

84.   B. Julesz, Foundations of Ciclopean Perception, The University of Chicago Press, 1971.

85.   J. V. Lovasik and M. Szymkiw, "Effects of aniseikonia, anisometropia, accommodation, retinal illuminance, and pupil size on stereopsis," Investigative Ophthalmology & Visual Science, vol. 26, no. 5, pp. 741–750, 1985.

86.   P. P. Schmidt, "Sensitivity of random dot stereoacuity and Snellen acuity to optical blur," Optometry and Vision Science, vol. 71, no. 7, pp. 466–471, 1994.

87.   C. Schor and T. Heckmann, "Interocular differences in contrast and spatial frequency: effects on stereopsis and fusion," Vision Research, vol. 29, no. 7, pp. 837–847, 1989.

88.   K. Cormack, S. B. Stevenson, and D. D. Landers, "Interactions of spatial frequency and unequal monocular contrasts in stereopsis," Perception, vol. 26, no. 9, pp. 1121–1136, 1997.

89.   D. L. Halpern and R. R. Blake, "How contrast affects stereoacuity," Perception, vol. 17, no. 4, pp. 483–495, 1988.

90.   G. Legge and Y. Gu, "Stereopsis and contrast," Vision Research, vol. 29, no. 8, pp. 989–1004, 1989.

91.   C. Wood, "Stereopsis with spatially degraded images," Investigative Ophthalmology & Visual Science, vol. 3, no. 3, pp. 337–340, 1983.

92.   Geib and C. Baumann, "Effect of luminance and contrast on stereoscopic acuity," Graefe's Archive for Clinical and Experimental Ophthalmology, vol. 228, no. 4, pp. 310–315, 1990.

93.   C. Schwarz, P. M. Prieto, E. J. Fernández, and P. Artal, "Binocular adaptive optics vision analyzer with full control over the complex pupil functions," Optics Letters, vol. 36, no. 24, pp. 4779–47781, 2011.

94. C. Cánovas, P. M. Prieto, S. Manzanera, A. Mira, and P. Artal, "Hybrid adaptive-optics visual simulator," Optics Letters, vol. 35, no. 2, pp. 196–198, 2010.

95. Manzanera, P. M. Prieto, D. B. Ayala, J. M. Lindacher, and P. Artal, "Liquid crystal adaptive optics visual simulator: application to testing and design of ophthalmic optical elements," Optics Express, vol. 15, no. 24, pp. 16177–16188, 2007.

96. P. A. Piers, E. J. Fenandez, S. Manzanera, S. Norrby, and P. Artal, "Adaptive optics simulation of intraocular lenses with modified spherical aberration," Investigative Ophthalmology & Visual Science, vol. 45, no. 12, pp. 4601–4610, 2004.

97. P. A. Piers, S. Manzanera, P. M. Prieto, N. Gorceix, and P. Artal, "Use of adaptive optics to determine the optimal ocular spherical aberration," Journal of Cataract and Refractive Surgery, vol. 33, no. 10, pp. 1721–1726, 2007.

98. H. Guo, D. A. Atchison, and B. J. Birt, "Changes in through-focus spatial visual performance with adaptive optics correction of monochromatic aberrations," Vision Research, vol. 48, no. 17, pp. 1804–1811, 2008.

99. K. M. Rocha, L. Vabre, N. Chateau, and R. R. Krueger, "Expanding depth of focus by modifying higher-order aberrations induced by an adaptive optics visual simulator," Journal of Cataract and Refractive Surgery, vol. 35, no. 11, pp. 1885–1892, 2009.

100. J. S. Werner, S. L. Elliott, S. S. Choi, and N. Doble, "Spherical aberration yielding optimum visual performance: evaluation of intraocular lenses using adaptive optics simulation," Journal of Cataract and Refractive Surgery, vol. 35, no. 7, pp. 1229–1233, 2009.

101. R. Legras, Y. Benard, and H. Rouger, "Through-focus visual performance measurements and predictions with multifocal contact lenses," Vision Research, vol. 50, no. 12, pp. 1185–1193, 2010.

102. Lundström, S. Manzanera, P. M. Prieto et al., "Effect of optical correction and remaining aberrations on peripheral resolution acuity in the human eye," Optics Express, vol. 15, no. 20, pp. 12654–12661, 2007.

103. D. Williams, G. Y. Yoon, J. Porter, A. Guirao, H. Hofer, and I. Cox, "Visual benefit of correcting higher order aberrations of the eye," Journal of Refractive Surgery, vol. 16, no. 5, pp. S554–S559, 2000.

104. G. Y. Yoon and D. R. Williams, "Visual performance after correcting the monochromatic and chromatic aberrations of the eye," Journal of the Optical Society of America A, vol. 19, no. 2, pp. 266–275, 2002.

105. K. M. Rocha, L. Vabre, F. Harms, N. Chateau, and R. R. Krueger, "Effects of Zernike wavefront aberrations on visual acuity measured using electromagnetic adaptive optics technology," Journal of Refractive Surgery, vol. 23, no. 9, pp. 953–959, 2007.

106. E. Dalimier and C. Dainty, "Use of a customized vision model to analyze the effects of higher-order ocular aberrations and neural filtering on contrast threshold performance," Journal of the Optical Society of America A, vol. 25, no. 8, pp. 2078–2087, 2008.

107. E. Dalimier, C. Dainty, and J. L. Barbur, "Effects of higher-order aberrations on contrast acuity as a function of light level," Journal of Modern Optics, vol. 55, no. 4-5, pp. 791–803, 2008.

108. S. Marcos, L. Sawides, E. Gambra, and C. Dorronsoro, "Influence of adaptive-optics ocular aberration correction on visual acuity at different luminances and contrast polarities," Journal of Vision, vol. 8, no. 13, article 1, 2008.

109. D. A. Atchison, H. Guo, W. N. Charman, and S. W. Fisher, "Blur limits for defocus, astigmatism and trefoil," Vision Research, vol. 49, no. 19, pp. 2393–2403, 2009.

110. D. A. Atchison, H. Guo, and S. W. Fisher, "Limits of spherical blur determined with an adaptive optics mirror," Ophthalmic and Physiological Optics, vol. 29, no. 3, pp. 300–311, 2009.

111. S. L. Elliott, S. S. Choi, N. Doble, J. L. Hardy, J. W. Evans, and J. S. Werner, "Role of high-order aberrations in senescent changes in spatial vision," Journal of Vision, vol. 9, no. 2, pp. 1–16, 2009.

112. J. Li, Y. Xiong, N. Wang et al., "Effects of spherical aberration on visual acuity at different contrasts," Journal of Cataract & Refractive Surgery, vol. 35, no. 8, pp. 1389–1395, 2009.

113. S. Li, Y. Xiong, J. Li et al., "Effects of monochromatic aberration on visual acuity using adaptive optics," Optometry and Vision Science, vol. 86, no. 7, pp. 868–874, 2009.

114. G. M. Pérez, S. Manzanera, and P. Artal, "Impact of scattering and spherical aberration in contrast sensitivity," Journal of Vision, vol. 9, no. 3, pp. 1–10, 2009.

115. H. Rouger, Y. Benard, and R. Legras, "Effect of monochromatic induced aberrations on visual performance measured by adaptive optics technology," Journal of Refractive Surgery, vol. 26, no. 8, pp. 578–587, 2010.

116. P. Artal, S. Manzanera, P. Piers, and H. Weeber, "Visual effect of the combined correction of spherical and longitudinal chromatic aberrations," Optics Express, vol. 18, no. 2, pp. 1637–1648, 2010.

117. H. Guo and D. A. Atchison, "Subjective blur limits for cylinder," Optometry and Vision Science, vol. 87, no. 8, pp. E549–E559, 2010.

118. P. Gupta, H. Guo, D. A. Atchison, and A. J. Zele, "Effect of optical aberrations on the color appearance of small defocused lights," Journal of the Optical Society of America A, vol. 27, no. 5, pp. 960–967, 2010.

119. K. M. Rocha, L. Vabre, N. Chateau, and R. R. Krueger, "Enhanced visual acuity and image perception following correction of highly aberrated eyes using an adaptive optics visual simulator," Journal of Refractive Surgery, vol. 26, no. 1, pp. 52–56, 2010.

120. P. de Gracia, C. Dorronsoro, E. Gambra, G. Marin, M. Hernández, and S. Marcos, "Combining coma with astigmatism can improve retinal image over astigmatism alone," Vision Research, vol. 50, no. 19, pp. 2008–2014, 2010.

121. H. Hofer, B. Singer, and D. R. Williams, "Different sensations from cones with the same pigment," Journal of Vision, vol. 5, no. 5, pp. 444–454, 2005.

122. M. Putnam, H. Hofer, N. Doble, L. Chen, J. Carroll, and D. R. Williams, "The locus of fixation and the foveal cone mosaic," Journal of Vision, vol. 5, no. 7, pp. 632–639, 2005.

123. W. Makous, J. Carroll, J. I. Wolfing, J. Lin, N. Christie, and D. R. Williams, "Retinal microscotomas revealed with adaptive-optics microflashes," Investigative Ophthalmology & Visual Science, vol. 47, no. 9, pp. 4160–4167, 2006.

124. Raghunandan, J. Frasier, S. Poonja, A. Roorda, and S. B. Stevenson, "Psychophysical measurements of referenced and unreferenced motion processing using high-resolution retinal imaging," Journal of Vision, vol. 8, no. 14, article 14, 2008.

125. L. C. Sincich, Y. Zhang, P. Tiruveedhula, J. C. Horton, and A. Roorda, "Resolving single cone inputs to visual receptive fields," Nature Neuroscience, vol. 12, no. 8, pp. 967–969, 2009.

126. E. Dalimier and C. Dainty, "Role of ocular aberrations in photopic spatial summation in the fovea," Optics Letters, vol. 35, no. 4, pp. 589–591, 2010.

127. K. Y. Li, P. Tiruveedhula, and A. Roorda, "Intersubject variability of foveal cone photoreceptor density in relation to eye length,"

Investigative Ophthalmology & Visual Science, vol. 51, no. 12, pp. 6858–6867, 2010.

128. S. B. Stevenson, A. Roorda, and G. Kumar, "Eye tracking with the adaptive optics scanning laser ophthalmoscope," in Proceedings of the Symposium on Eye-Tracking Research and Applications (ETRA '10), pp. 195–198, ACM, Austin, Tex, USA, March 2010.

129. S. Poonja, S. Patel, L. Henry, and A. Roorda, "Dynamic visual stimulus presentation in an adaptive optics scanning laser ophthalmoscope," Journal of Refractive Surgery, vol. 21, no. 5, pp. S575–S580, 2005.

130. Q. Yang, D. W. Arathorn, P. Tiruveedhula, C. R. Vogel, and A. Roorda, "Design of an integrated hardware interface for AOSLO image capture and cone-targeted stimulus delivery," Optics Express, vol. 18, no. 17, pp. 17841–17858, 2010.

131. Roorda, "Adaptive optics for studying visual function: a comprehensive review," Journal of Vision, vol. 11, no. 7, 2011.

132. K. M. Hampson, "Adaptive optics and vision," Journal of Modern Optics, vol. 55, no. 21, pp. 3425–3467, 2008.

Chapter 2

Adaptive Optics and Optical Vortices

S. G. Garanin[1], F. A. Starikov[1] and Yu. I. Malakhov[2]

[1] *Russian Federal Nuclear Center –VNIIEF, Institute of Laser Physics Research, Russia*
[2] *International Science and Technology Center, Russia*

1. INTRODUCTION

The achievement of minimal angular divergence of a laser beam is one of the most important problems in laser physics since many laser applications demand extreme concentration of radiation. Under the beam formation in the laser oscillator or amplifier with optically inhomogeneous gain medium and optical elements, the divergence usually exceeds the diffraction limit, and the phase surface of the laser beam differs from the plane surface. However, even if one succeeds in realizing the close-to-plane radiation wavefront at the laser output, the laser radiation experiences increasing phase disturbances under the propagation of the beam in an environment with optical inhomogeneities (atmosphere). These disturbances appear with the wavefront receiving smooth, regular distortions, the transverse intensity distribution becomes inhomogeneous, and the beam broadens out.

The correction of the laser radiation phase, which is a smooth continuous spatial function, can be performed using a conventional adaptive optical system including a wavefront sensor and a wavefront corrector. The wavefront sensor performs the measurement (in other words, reconstruction) of the radiation phase surface; then, on the basis of these data, the wavefront corrector (for example, a reflecting mirror with deformable surface) transforms the phase front in the proper way. If all components of the adaptive optical system are involved in the common circuit with the feedback, then the adaptive system is known as a closed-loop system. The adaptive correction of the wavefront with smooth distortions has a somewhat long history and considerable advances [1, 2, 3, 4, 5, 6].

When a laser beam passes a sufficiently long distance in a turbulent atmosphere, the so-called regime of strong scintillations (intensity fluctuations) is realized. Under such conditions the optical field becomes speckled, lines appear in the space along the beam axis where the intensity vanishes and the surrounding zones of the wavefront attain a helicoidal (screw) shape. If the intensity in an acnode of the transverse plane is zero, then the phase in this point is not defined. In view of its screw form, the phase surface in the vicinity of such point has a break, the height of which is divisible by the wavelength. Since the phase is defined accurate to the addend that is aliquot to 2π, it is formally continuous but under a complete circling on the phase surface around the singular point one cannot reach the starting place. The integration of the phase gradient over some closed contour encircling such singular point results in a circulation not equal to zero, in contrast to the null circulation at the usual smoothed-inhomogeneous regular phase distribution. The indicated properties represent evidence of strong distortions of the wavefront – screw dislocations or optical vortices. The vortical character of the beam is detected with ease in the experiment after the analysis of the picture of its interference with the obliquely incident plane wave: the interference fringes arise or vanish in the centers of screw dislocations forming peculiar "forks".

Scintillations in the atmosphere especially decrease the efficiency of light energy transportation and distort the information carried by a laser beam in issues of astronomy and optical communications. Scintillation effects present special difficulty for adaptive optics, and their correction is one of key trends in the development of state-of-the-art adaptive optical systems.

However it should be noted that the possibility to control the optical vortices (including the means of adaptive optics) presents interest not only for atmospheric optics but for a new optical field, namely, singular optics [7, 8, 9]. The fact is that optical vortices have very promising applications in optical data processing, micro-manipulation, coronagraphy, etc. where any type of management of the singular phase could be required.

This chapter is dedicated to wavefront reconstruction and adaptive phase correction of a vortex laser beam, which is generated in the form of the Laguerre-Gaussian LG_0^1 laser mode. The content of the chapter is as follows. In Section 1 we specify the origin and main properties of optical vortices as well as some their practical applications. Section 2 is dedicated to a short description of the origination of optical vortices in a turbulent atmosphere and correspondent problems of the adaptive optics. In Section 3 some means are given concerning the generation of optical vortices under laboratory conditions, aimed at the formation of a "reference" optical vortex with the maximally predetermined phase surface, and the experimental results of such formation are illustrated. Section 4 is concerned with vortex beam phase surface registration, based on measurements of phase local tilts using a Hartmann-Shack wavefront sensor and a novel reconstruction technique. In

Section 5 experimental results of correction of a vortex beam are demonstrated in the conventional closed-loop adaptive optical system including a Hartmann-Shack wavefront sensor and a bimorph deformable adaptive mirror. Conclusions summarize the abovementioned research results.

2. ORIGIN, MAIN PROPERTIES AND PRACTICAL APPLICATIONS OF OPTICAL VORTICES

The singularity of the radiation field phase S is identified by the term "optical vortex", which can appear in the complex function $\exp\{iS\}$ representing a monochromatic light wave [10]. The amplitude of scalar wave field A (and, correspondingly, its intensity $I=|A|^2$) in the point of the vortex location approaches zero. The phase of radiation S changes its value by $2\pi m$ with the encircling the singularity point clockwise or counter-clockwise, where m is the positive or negative integer number known as the vortex "topological charge". In the centre of the vortex (intensity zero point) the phase remains indefinite. Such an optical singularity is the result of the interference of partial components of the wave field with a phase shift, which is initial or acquired during propagation in an inhomogeneous medium. In 3D space the points with zero intensity form zero lines. On these lines the potentiality of phase field is violated; and the regions of "defective" (singular) phase can be considered as vortex strings like the regions with concentrated vorticity, which are considered in the hydrodynamics of ideal liquid. We are interested in a case where the lines of zero intensity have a predominantly longitudinal direction, i.e. form a longitudinal optical vortex. The equiphase surface in the vicinity of such a line has the appearance of a screw-like (helicoidal) structure, threaded on this line. In the interference pattern of the vortex wave under consideration with any regular wave, the vortices are revealed through the appearance of so-called "forks" (i.e., branching of interference fringes), coinciding with zero points of the intensity.

Investigations of waves with screw wavefront and methods of their generation were reported as early as by Bryngdahl [11]. The theory of waves carrying phase singularities was developed in detail by Nye and Berry [12, 10], prompting a series of publications dealing with the problem (see [13, 14, 7, 8] and the lists of references therein). The term "optical vortex" was introduced in [15]. Along with the term "optical vortex", the phenomenon is also referred to as "wavefront screw dislocation". The latter appeared because of similarities between distorted wavefront and the crystal lattice with defects. The following terms are also used: "topological defects", "phase singularities", "phase cuts", and "branch cuts".

Thus the indication of the existence of an optical vortex in an optical field is the presence of an isolated point $\{r, z\}$ in a plane, perpendicular to the light propagation axis, in which intensity I (r, z) is approaching zero, phase S (r, z) is indefinite, and integration of the phase gradient

$\nabla_\perp S$ field over some closed contour Γ encircling this point results in a circulation not equal to zero:

$$\oint_\Gamma \nabla_\perp S(\rho, z)d\rho = 2\pi m,$$

(1)

where $d\rho$ is the element of the contour Γ.

The propagation of slowly-varying complex amplitude of the scalar wave field A in the free space is described by the well-known quasi-optical equation of the parabolic type (see, for example, [16]):

$$\frac{\partial A}{\partial z} - \frac{i}{2k}\frac{\partial^2 A}{\partial r^2} = 0,$$

(2)

where $k=2\pi/\lambda$ is wave number, λ is the radiation wavelength, z is the longitudinal coordinate corresponding with the beam propagation axis, $r=e_x x + e_y y$ is the transverse radius-vector. In the process of radiation propagation in the medium vortices appear, travel in space, and disappear (are annihilated).

Laguerre-Gaussian laser beams LG_n^m are related to the familiar class of vortex beams and are used most often in experiments with optical vortices [7-9]. They are the eigen-modes of the homogeneous quasi-optical parabolic equation (2) [17], so they do not change their form under the free space propagation and lens transformations. The correspondent solution of equation (2) in cylindrical coordinates (r, φ, z) has the following form:

$$A(r, \varphi; z) = A_0 \frac{w_0}{w}\left(\frac{r}{w}\right)^m \Phi_m(\varphi) L_n^m\left(2\frac{r^2}{w^2}\right)\exp\left(-\frac{r^2}{w^2} + i\frac{zr^2}{z_0 w^2} - i(2n + m + 1)\text{arctg}\frac{z}{z_0}\right),$$

(3)

The typical transverse size of the beam w in (3) is determined by the relation $w^2 = w_0^2[1+z^2/z_0^2]$ where, in its turn, w_0 is the transverse beam size in the waist and $z_0 = kw_0^2/2$ is the typical waist length. The radial part of distribution (3) includes the generalized Laguerre polynomial

$$L_n^m(x) = \frac{e^x}{x^m n!} \frac{d^n}{dx^n}\left(x^{n+m} e^{-x}\right).$$

(4)

In addition to an item responsible for wavefront curvature and transversely-uniform Gouy phase, the angular factor $\Phi_m(\varphi)$ contributes to the phase part of distribution (3). The angular part of formula (3) is represented in the form of a linear combination of harmonic functions

$$\Phi_m(\varphi) = c_1 \cdot \Omega_m(\varphi) + c_2 \cdot \Omega_{-m}(\varphi),$$

(5)

where $\Omega_m(\varphi)=\exp(im\varphi)$, $\varphi=\operatorname{arctg}(y/x)$ is the azimuth angle in the transverse plane. The c_1 and c_2 constants determine the beam character and the presence of singularity in it.

Let's consider two cases of the angular function $\Phi_m(\varphi)$ distribution from (5) at $m>0$. In the first case, when $c_1=1$ and $c_2=0$, we have $\Phi_m(\varphi)=\exp(im\varphi)$. The form of helicoidal phase surface assigned by the $m\varphi$ function (we do not take into account the wavefront curvature in (3) determining the beam broadening or narrowing only) in the vortex Laguerre-Gaussian laser beam is presented in Figure 1. This phase does not depend on r at the given φ and rises linearly with φ increasing. In the phase surface the spatial break of $m\lambda$ (or $2\pi m$ radians) depth is present. Under the complete circular trip around the optical axis on the phase surface it is impossible to get to the starting point. As it has been commented above, such a shape of the phase factor is what causes the singular, vortex behavior of the beam. The positive or negative sign of m determines right or left curling of the phase helix.

On the optical axis, in the vortex center, the intensity is zero, resulting from the behavior of the radial dependence of (3) and generalized Laguerre polynomial (4). The beam intensity distribution in the transverse plane, as it is seen from (3), is axially-symmetrical (modulus of A depends on r only) and visually represents the system of concentric rings. In the simplest case at $n=0$, $m=1$ (LG_0^1 mode) the intensity distribution has a doughnut-like form that is shown in Figure 2. Figure 2 also presents a picture of the interference of the given vortex laser beam with the obliquely incident monochromatic plane wave. In the picture, fringe branching is observed in the beam center with the "fork" formation (fringe birth) typical for screw dislocation [7-9]. At the arbitrary m number the quantity of fringes that are born corresponds with this number, the double, triple, and other "forks" are formed. The presence of "forks" in the interference patterns of such kind is the standard evidence of the vortex nature of the beam in the experiment.

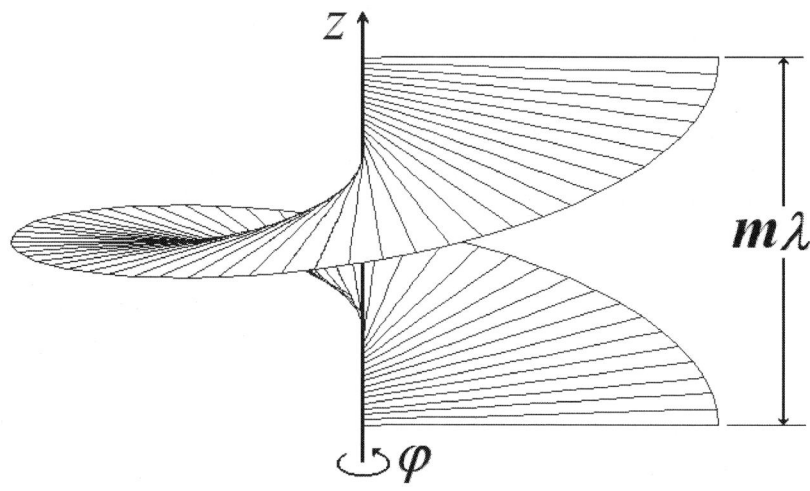

Figure 1. Phase surface shape of a Laguerre-Gaussian beam carrying an optical vortex.

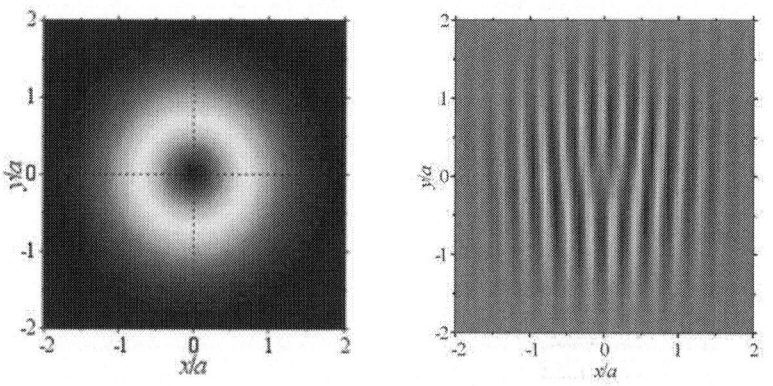

Figure 2. The intensity distribution in the Laguerre-Gaussian laser beam LG_n^m and the picture of its interference with an obliquely incident plane wave at $n=0$, $m=1$, $\Phi_m=\exp(i\phi)$.

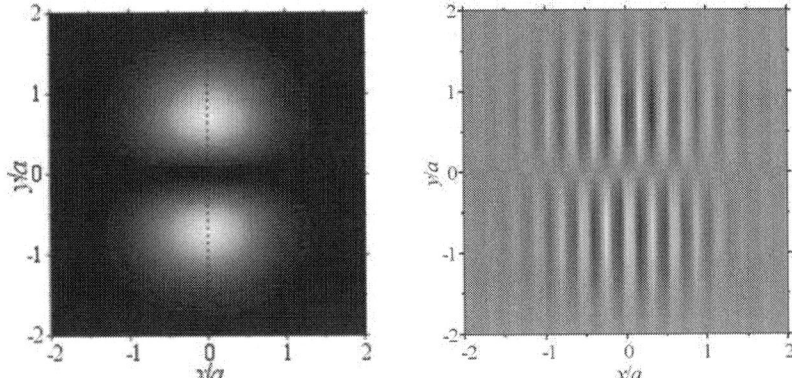

Figure 3. The intensity distribution in the Laguerre-Gaussian laser beam LG_n^m and the picture of its interference with an obliquely incident plane wave at n=0, m=1, $\Phi_m=\sin(i\phi)$.

Let's consider one more case when $c_1=-c_2=1/2i$ in (5), then $\Phi_m(\varphi)=\sin(m\varphi)$. Under such conditions, according to (3) the intensity distribution has no axial symmetry depending on φ. In Figure 3 the beam intensity distribution in the transverse plane at $\Phi_m=\sin\varphi$ for $n=0$, $m=1$ is shown. The phase portrait of the beam is also given in Figure 3. The half-period shift of interference fringes in the lower half plane as compared to the upper one demonstrates the phase asymmetry with respect to the x axis. The fringe numbers in the lower and upper half planes are the same, i.e. the fringe birth does not take place. It is seen that the edge rather than screw phase dislocation occurs here since the phase distribution is step-like with a break of π (instead of 2π!) radians. The given example of the Laguerre-Gaussian beam demonstrates that at $\Phi_m=\sin\varphi$ the beam is not the vortical in nature. There is no singular point in the beam transverse section but there is a particular line (y=0) where the intensity is zero. At m>1 in the beam there are m such lines passing through the optical axis and dividing the transverse beam section by 2m equal sectors. The radiation phase in each sector is uniform and differs in π in the neighboring sectors.

Light beams with optical vortices currently attract considerable attention. This attention is encouraged by the extraordinary properties of such beams and by the important manifestations of these properties in many applications of science and technology.

It is known for a long time that light with circular polarization possesses an orbital moment. For the single photon its quantity equals $\pm\hbar$, where $\hbar$ is the Planck constant. However only relatively recently it was shown [18, 19] that light can have an orbital moment irrespective of its polarization state if its azimuth phase dependence is of the form $S=m\varphi$ where φ is azimuth coordinate in the transverse cross section of the beam and m is the positive or negative integer. The authors [18] supposed that the moment of each photon is defined by the formula $L=m\hbar$. As this phase dependence is the characteristic feature of the helical wavefront (the form of which is presented in Figure 1) so the beam carrying the optical vortex and possessing such phase front has to own the non-zero orbital moment $m\hbar$ per photon. The quantity of m is defined by the topological charge of optical vortex.

The concept of orbital moment is not new. It is well known that multipole quantum jumps can results in the emission of radiation with orbital moment. However, such processes are infrequent and correspond to some forbidden atomic and molecular transitions. However, generating the beam carrying the optical vortex, one can readily obtain the light radiation beam with quantum orbital moment. Such beams can be used in investigations of all kinds of polarized light. For example, the photon analogy of spin-orbital interaction of electrons can be studied and in general it is possible to organize the search for new optical interactions. As the m-factor can acquire arbitrary values, any part of the beam (even one photon) can carry an unlimited amount of information coded in the topological charge. Thus the density of information in a channel where coding is realized with the use of orbital moment could be as high as compared with a channel with coding of the spin states of a photon. Because only two circular polarization states of the photon are possible, one photon can transmit only one bit of information. Presently, optical vortices have generated a great deal of interest in optical data processing technologies, namely, the coding/decoding in optical communication links in free space [20, 21], optical data processing [22], optical interconnects [23], and quantum optics information processing [24].

The next practical application of optical vortices is optical micromanipulations and construction of so called optical traps, i.e. areas where the small (a few micrometers) particles can be locked in [25, 26]. Progress in the development of such traps allows the capture of particles of low and large refraction indexes [27]. Presently, this direction of research finds further continuation [28, 29, 30].

It is also possible to use optical vortices to register objects with small luminosity located near a bright companion. Shadowing the bright object by a singular phase screen results in the formation of a window, in which the dim object is seen. The optical vortex filtration of such a kind was proposed in [31]. Using this method the companion located at 0.19 arcsec near the object was theoretically differentiated with intensity of radiation 2×10^5 times greater [32]. The possibility to use this method to detect planets orbiting bright stars was also illustrated by astronomers [33, 34]. Vortex coronagraphy is now undergoing further development [35, 36]. There are a number of examples of non-astronomical applications [37, 38].

It was proposed to use optical vortices to improve optical measurements and increase the fidelity of optical testing [39, 40], for investigations in high-resolution fluorescence microscopy [41], optical lithography [42, 43], quantum entanglement [44, 45, 46], Bose–Einstein condensates [47].

Optical vortices show interesting properties in nonlinear optics [48]. For example, in [49, 50, 51] it was predicted that the phase conjugation at SBS of vortex beams is impossible due to the failure of selection of the conjugated mode. For a rather wide class of the vortex laser beams a novel and interesting phenomenon takes place which can be called the phase transformation at SBS. In essence there is only one Stokes mode, the amplification coefficient of which is maximal and higher than that of the conjugated mode. In other words, the non-conjugated mode is selected of in the Stokes beam. The principal Gaussian mode, which is orthogonal to the laser vortex mode, is an example of such an exceptional Stokes mode. The cause of this phenomenon is in the specific radial and azimuth distribution of the vortex laser beam. It is interesting that the hypersound vortices are formed in the SBS medium in accordance with the law of topologic charge conservation. The predicted effects have been completely confirmed experimentally [52, 53, 54].

3. OPTICAL VORTICES IN TURBULENT ATMOSPHERE AND THE PROBLEM OF ADAPTIVE CORRECTION

In early investigations [12] it was shown that the presence of optical vortices is a distinctive property of the so called speckled fields, which form when the laser beam propagates in the scattering media. Experimental evidence of the existence of screw dislocations in the laser beam, passed through a random phase plate, were obtained in [55, 56, 57] where topological limitations were also noted of adaptive control of the laser beams propagating in inhomogeneous media.

Turbulent atmosphere can be represented as the consequence of random phase screens. Under propagation in the turbulent atmosphere the regular optical field acquires rising aberrations. These aberrations manifest themselves in the broadening and random wandering of laser beams; the intensity distribution becomes non-regular and the wavefront deviates from initially set surface. These deformations of the wavefront can be corrected using adaptive optics. To this end, effective sensors and correctors of wavefront were designed [1-6]. The problem becomes more complicated when the laser beam passes a relatively long distance in a weak turbulent medium or if the turbulence becomes too strong. In this case optical vortices develop in the beam; the shape of the wavefront changes qualitatively and singularities appear.

The influence of the scintillation effects are determined (see, for example, [2, 4]) by the closeness to unity of the Rytov variance

$$\sigma_\chi^2 \approx 0.56 k^{7/6} \int\limits_0^L C_n^2(z) z^{5/6} (z/L)^{5/6} dz,$$

(6)

where $C_n^2(z)$ describes the dependence of structure constant of the refractive index fluctuations over the propagation path and L is the path length. The regime of strong scintillations is not realized when $\sigma_\chi^2 \ll 1$.

Figure 4 demonstrates the results of numerical simulation of propagation of a Gaussian laser beam (λ=1 μ) in turbulent atmosphere in a model case of invariable structure constant C_n^2 =10^{-14} cm$^{-2/3}$ for the distance of 1 km when the regime of strong scintillations is realized according to (6). The steady-state equation for the slowly-varying complex field amplitude A differs from the equation (2) by the presence of the inhomogeneous term:

$$\frac{\partial A}{\partial z} - \frac{i}{2k}\frac{\partial^2 A}{\partial r^2} + \frac{ik}{2}(\tilde{\varepsilon}-1)A = 0,$$

(7)

where $\tilde{\varepsilon}$ is the fluctuating dielectric permittivity of the turbulent atmosphere. In numerical simulations we use the finite-difference algorithm of numerical solving of parabolic equation (7) described in [58]. It is characterized by an accuracy, which considerably exceeds the accuracy of the widespread spectral methods [59]. The amplitude error of an elementary harmonic solution of the homogeneous equation is equal to zero, whereas the phase error is significantly reduced and proportional to the transverse integration step to the power of six. To take into account the inhomogeneous term of the equation, the splitting by physical processes is employed. The effect of randomly inhomogeneous distribution of dielectric permittivity is allowed for using a model of random phase screens, which is commonly used in calculations of radiation propagation in optically inhomogeneous stochastic media [60]. The spatial spectrum of dielectric permittivity fluctuations is described taking into account the Tatarsky and von Karman modifications of the Kolmogorov model [61].

The fragment of speckled distribution of optical field intensity after the propagation is shown in Figure 4. Dark spots are seen where the intensity vanishes. As it has been noted before, the presence of optical vortices in the beam is easily detected, based on the picture of its interference with an obliquely incident plane wave. The correspondent picture is shown in Figure 4 as well. In the centers of screw dislocations the fringe branching is observed, i.e. the birth or disappearance of the fringes takes place with formation of typical "forks" in the interferogram (compare with Figure 2). There are also zones of edge dislocations (compare with Figure 3). The number, allocation and helicity of the vortices in the beam are random in nature but the vortices are born as well as annihilated in pairs. If the initial beam is regular (vortex-free), then the total topological charge of the vortices in the beam will be equal to zero in each transverse section of the beam along the propagation path in accordance with the conservation law of topological charge (or orbital angular moment) [7-9].

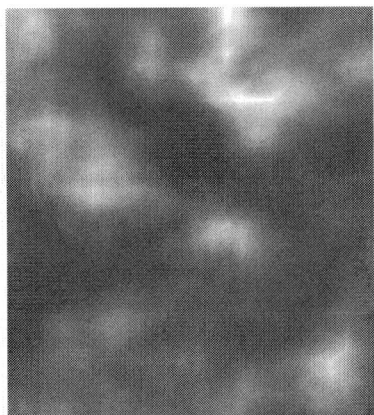 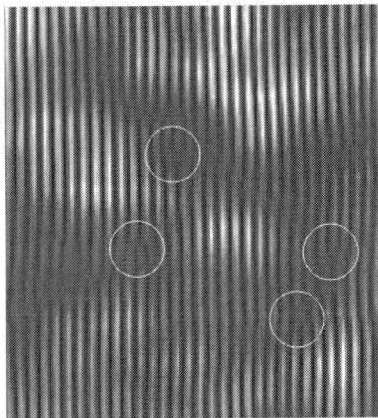

Figure 4. Optical vortices in the laser beam after atmospheric propagation: the speckled intensity distribution and the picture of interference of the beam with the obliquely incident plane wave including "forks" denoted by light circles.

One of the first papers dealing with the appearance of optical vortices in laser beams propagating in randomly inhomogeneous medium was published by Fried and Vaughn in 1992 [62]. They pointed out that the presence of dislocations makes registration of the wavefront more difficult and they considered methods for solving the problem. In 1995 the authors of Ref. [63] encountered this problem in experimental investigations of laser beam propagation in the atmosphere. It was shown that the existence of light vortices is an obstacle for atmospheric adaptive optical systems. After that it was theoretically shown that screw dislocations give rise to errors in the procedure of wavefront registration by the Shack-Hartmann sensor [64, 65]. Due to zero amplitude of the signal in singular points, the information carried by the beam becomes less reliable and the compensation for turbulent aberrations is less effective [66]. Along with [63], the experimental investigation [67] can be taken here as an example where the results of adaptive correction are presented for distortions of beams propagating in the atmosphere.

Since one of the key elements of an adaptive optical system is the wavefront sensor of laser radiation, there is a pressing need to create sensors that are capable of ensuring the required spatial resolution and maximal accuracy of the measurements. In this connection there is necessity need to develop algorithms for measurement of wavefront with screw dislocations, which are sufficiently precise, efficient and economical given the computing resources, and resistant to measurement noises. The traditional methods of wave front measurements [1-6] in the event of the above-mentioned conditions are in fact of no help. The wavefront sensors have been not able to restore the phase under the conditions of strong scintillations [68]. The experimental determination of the location of phase discontinuities itself already generates serious difficulties [69]. In spite of the fact that the construction features of algorithms of wavefront recovery in the presence of screw dislocations were set forth in a number of theoretical papers [68, 69, 70, 71, 72, 73, 74, 75], there were not many published experimental works in this direction. Thus, phase distribution has been investigated in different diffraction orders for a laser beam passed through a specially synthesized hologram, designed for generating higher-order Laguerre-Gaussian modes [76]. An interferometer with high spatial resolution was used to measure transverse phase distribution and localization of phase singularities. The interferometric wavefront sensor was applied also in a high-speed adaptive optical system to compensate phase distortions under conditions of strong scintillations of the coherent radiation in the turbulent atmosphere [77] as well as when modelling the turbulent path under laboratory conditions [78]. In [77, 78] the local phase was measured, without reconstructing the global wavefront that is much less sensitive to the presence of phase residues. The interferometric methods of phase determination are rather complicated and require that several interferograms are obtained at various phase shifts between a plane reference wave and a signal wave. It is noteworthy, however, that in the adaptive optical systems [1-6] the Hartmann-Shack wavefront sensor [79, 80] has a wider application compared with the interferometric sensors including the lateral shearing interferometers [81, 82], the curvature sensor [83, 84, 85], and the pyramidal sensor [86, 87]. The cause of this is just in a simpler and more reliable arrangement and construction of the Hartmann-Shack sensor. However, there have been practically no publications of the results of experimental investigations connected with applications of this sensor for measurements of singular phase distributions.

The problem of a wavefront corrector (adaptive mirror) suitable for controlling a singular phase surface is also topical. In the adaptive optical systems [77, 78] the wavefront correctors were based on the micro-electromechanical system (MEMS) spatial light modulators with the large number of actuators. The results of [77, 78] shown that continuous MEMS mirrors with high dynamic response bandwidth, combined with the interferometric wavefront sensor, can ensure a noticeable correction of scintillation. However, the MEMS mirrors are characterized by low laser damage resistance that can considerably limit applications. The bimorph or pusher-type piezoceramics-based flexible mirrors with the modal response functions of control elements have a much higher laser damage threshold [3-5]. Recently [88] a complicated cascaded imaging adaptive optical system with a number of bimorph piezoceramic mirrors was used to mitigate turbulence effect basing, in particular, on conventional Hartmann-Shack wavefront sensor data. Conventional adaptive compensation was obtained in [88] which proved to be very poor at deep turbulence. The scintillation and vortices may be one of the causes of this.

In the investigations, the results of which are described in this chapter, the development of an algorithm of the Hartmann-Shack reconstruction of vortex wavefront of the laser beam plays a substantial role. The creation of efficient algorithms for the wavefront sensor of vortex beams implies the experiments under modeling conditions when the optical vortices are artificially generated by special laboratory means. Moreover, as long as the matter concerns the creation of a new algorithm of wavefront reconstruction, it is possible to estimate its accuracy only under operation with the beam, the singular phase structure of which is known in detail beforehand. The formation of optical beams with the given configuration of phase singularities and their transformations is one of main trends in the novel advanced optical branch – singular optics [7-9].

Thus, the first stage of the research sees the generation of a vortex laser beam with the given topological charge. In our case the role of this beam is played by the single optical vortex, namely, the Laguerre-Gaussian mode. Further, at the second stage, with the help of the Hartmann-Shack wavefront sensor, the task of registration of the vortex beam phase surface is solved using the new algorithm of singular wavefront reconstruction. Finally, at the third stage, the correction of the singular wavefront is undertaken in a closed-loop adaptive optical system, including the Hartmann-Shack wavefront sensor and the wavefront corrector in the form of a piezoelectric-based bimorph mirror.

4. GENERATION OF OPTICAL VORTEX

As it has been indicated above, to examine the accuracy of the wavefront reconstruction algorithm and its efficiency in the experiment itself a "reference" vortex beam has to be formed with a predetermined phase surface. This is important as, otherwise, it would be impossible to make sure that the algorithm recovers the true phase surface under conditions when robust alternative methods of its reconstruction are missing or unavailable. The Laguerre-Gaussian vortex modes LG_n^m can play the role of such "reference" optical vortices.

To create a beam with phase singularities artificially from an initial plane or Gaussian wave, a number of experimental techniques have been elaborated. There are many papers concerning the various aspects of generation of beams with phase singularities (see, for example, [89, 90, 91, 92, 93, 94, 95, 96, 97, 98, 99, 100, 101, 102, 103, 104]). Among other possibilities, we can also refer to several methods for phase singularity creation in the optical beams based on nonlinear effects [105, 106, 107, 108]. The generation of optical vortices is also possible in the waveguides [109, 110, 111]. The adaptive mirrors themselves can be used for the formation of optical vortices [112, 113]. In this chapter, though, we dwell only on a number of ways to generate the vortex beams, which allow one to form close-to-"reference" vortices with well-determined singular phase structure that is necessary for the accuracy analysis of the new algorithm of Hartmann-Shack wavefront reconstruction.

One method for generation of the screw dislocations is by forming the vortex beam immediately inside a laser cavity. The authors of [114] were the first to report that the generation of wavefront vortices is possible using a cw laser source. It was shown in [115] that insertion of a non-axisymmetric transparency into the cavity results in generation of a vortex beam. It was reported in [116] that a pure spiral mode can be obtained by introducing a spiral phase element (SPE) into the laser cavity, which selects the chosen mode. The geometry of the cavity intended for generation of such laser beams from [116] is shown in Figure 5. Here a rear mirror is replaced by a reflecting spiral phase element, which adds the phase change $+2im\varphi$ after reflecting. As a result of reflection, the phase of a spiral mode $-im\varphi$ changes to $+im\varphi$. The cylindrical lens inside the cavity is focused on the output coupler. This lens inverts the helicity of the mode back to the field described by $\exp\{-im\varphi\}$ and ensures the generation of the required spiral beam. The beam at the output passes another cylindrical lens and its distribution becomes the same as inside the cavity. A pinhole in the cavity ensures the generation of a spiral mode of minimal order, i.e., TEM_{01} mode. It should be stressed that the spiral phase element determines the parameters of the spiral beam within the cavity so that by its variation the parameters of the output beam can be controlled.

This method was tested with a linearly polarized CO_2-laser. The reflecting spiral phase element was made of silicon by multilevel etching. It had 32 levels with the entire height of break λ that corresponds to $m=1$. Precision of etching was about 3% and deviation of the surface from the prescribed form was less than 20 nm. The reflecting coefficient of the element was greater than 98%; the diameter and length of the laser tube were 11 mm and 65 cm, respectively. The lens inside the cavity with focal length 12.5 cm was focused in the output concave mirror with a radius of 3 m. An identical lens was placed outside the cavity to collimate the beam. In Figure 6 we demonstrate the stable spiral beam obtained in the experiment [116]. The vortical nature of the beam is proved not by demonstration of the "fork" in the interferogram but by the doughnut-like intensity distribution in the near and the far zone.

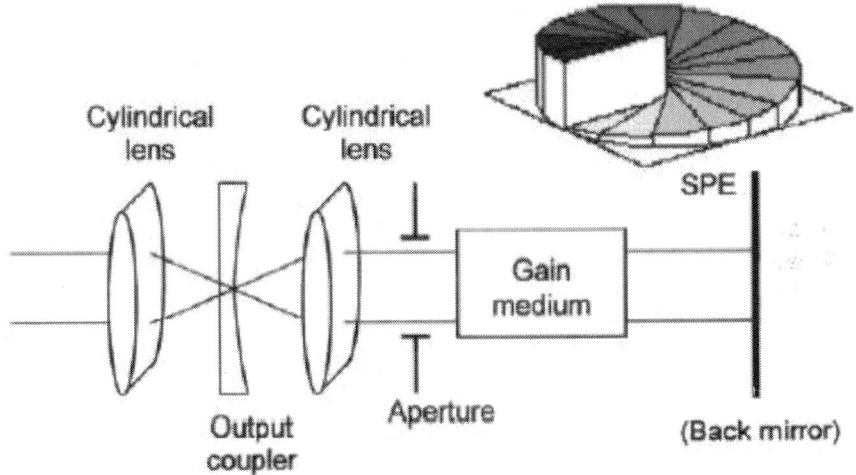

Figure 5. Configuration of a laser cavity intended for generation of spiral beams [116].

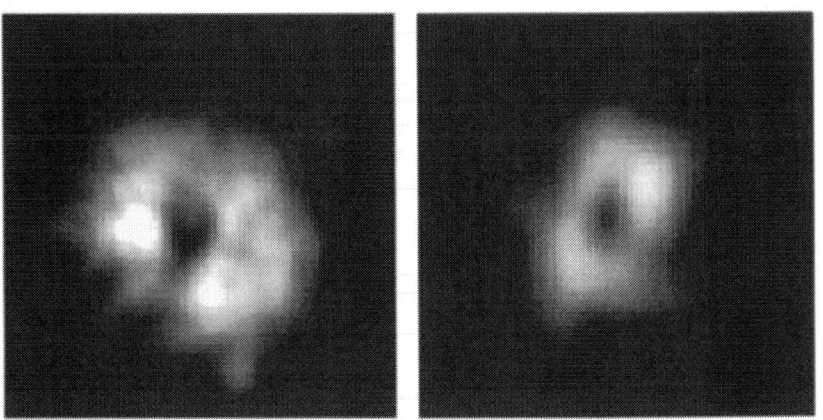

Figure 6. Intensity distribution of vortex beam generated in the laser cavity in near (left) and far (right) zone [116].

A ring cavity with the Dove's prism can also be used to generate vortex beams. It was shown [117] that modes of such a resonator are singular beams.

The next way to generate the optical vortex uses a phase (or a mode) converter. Usually it transforms a Hermit-Gaussian mode, generated in the laser, into a corresponding Laguerre-Gaussian mode. This method was first proposed in [118]. In the experiment the authors used a cylindrical lens, the axis of which was placed at an angle of 45° with respect to the HG_{01} mode for conversion into a LG_0^1 mode. The incident mode appears to the lens as a superposition of HG_{01} mode parallel to the lens axis and a mode perpendicular to the axis. The mode perpendicular to the lens axis passes through a focus, advancing the relative Gouy phase between the two modes of $\pi/2$ as required to form the doughnut mode from uncharged HG_{01} and HG_{10} modes.

In Ref. [91] an expression was derived for an integral transformation of Hermit-Gaussian modes into Laguerre-Gaussian modes in the astigmatic optical system, and it was shown theoretically that passing the beam through the cylindrical lens can perform the conversion. The theory of a $\pi/2$ mode converter and a π mode converter, produced by two cylindrical lenses, was described in more details in [92]. Padgett et al. describe in a tutorial paper [119] how a range of Laguerre-Gaussian modes can be produced using two cylindrical lenses starting from the corresponding Hermit-Gaussian modes, and present the clear examples, showing the intensity and phase distributions obtained. The initial higher order Hermit-Gaussian modes can be produced in the laser with intracavity cross-wires. The authors of [96] used a similar technique with a Nd:YAG laser operating at the 100 mW level.

Even in the absence of the required initial HG_{10} mode in the laser emission, it is easy to produce artificially a similar configuration by introducing a glass plate in a half of the TEM_{00} beam and achieving the necessary π phase shift, and then to apply mode conversion. An example of such doughnut beam creation was reported in [120]. The efficiency of conversion was about 50%. It is possible to use the cylindrical lens mode converter [121] but with production of the initial higher order Hermit-Gaussian mode, by exciting it in an actively stabilized ring cavity, matching in the Gaussian beam from their titanium sapphire laser. The efficiency was up to 40% for the LG_0^1 mode, and higher order doughnuts could also be obtained easily. The method proposed in [122] is based on the formation of a pseudo HG_{NM} mode, propagating the Gaussian beam of a number of edges of thin glass plates and forming the edge dislocations with the following its astigmatic conversion into a Laguerre-Gaussian mode.

It was reported in [123] that in the event of ideal conversion, the efficiency of Hermit-Gaussian mode transformation into Laguerre-Gaussian mode is about 99.9%. The spherical aberration does not reduce the efficiency factor. Typically cylindrical lenses are not perfect and their defects give rise to several Laguerre-Gaussian modes. The superposition of components can be unstable and this means a dependence of intensity on the longitudinal coordinate. If special means are not employed the precision of lens fabrication is about 5%, in this case the efficiency of beam transformation into Laguerre-Gaussian mode is 95%. Imperfections of 10% result in drop of efficiency down to 80%.

In [124, 54] the formation of the Laguerre-Gaussian LG_0^1 or LG_1^1 modes was performed at the output of a pulsed laser-generator of Hermit-Gaussian HG_{01} or HG_{21} modes with the help of a tunable astigmatic $\pi/2$-converter based on the so-called optical quadrupole [125]. It consists of two similar mechano-optical modules, each of which incorporates the positive and negative cylindrical lenses with the same focal length and a positive spherical lens. The mechanical configuration of each module can synchronously turn the incorporated cylindrical lenses in the opposite directions with respect to the optical axis, which ensures its rearrangement. In the initial position the optical forces of cylindrical lenses completely compensate each other and their axes coincide with the main axes of the intensity distribution of the laser. The distance between the modules is fixed so that the spherical lenses in different modules are located at a focal distance from each other and form the optical Fourier transformer.

To study the phase structure of radiation, in [54] use was made of a special interferometer scheme, where the reference beam was produced from a part of the original Laguerre–Gaussian LG_0^1 or LG_1^1 mode (see Figure 7). As a result, each of the modes interfered with a similar one, but with a topological charge of the opposite sign (the opposite helicity), i.e., with LG_0^{-1} or LG_1^{-1} mode. The interference fringe density depended on the thickness of the plane-parallel plate 1 and could be additionally varied by inclining the mirrors 2 (see figure 7).

The peculiarity of the interference of two Laguerre–Gaussian modes, having the opposite helicity of the phase, manifests itself in the branching of a fringe in the middle of the beam and formation of a characteristic "fork" with an additional fringe appearing in the centre, as compared with the case of a vortex mode interfering with a plane reference wave (see Figure 2). Such branching of fringes indicates the vortex nature of the investigated beam, while the absence of branching is a manifestation of the regular character of the beam phase surface.

Figure 8 displays the experimental distributions of intensity of the laser mode LG_0^1 in far field and its picture of interference with an obliquely incident wave in the form of LG_0^{-1} mode.

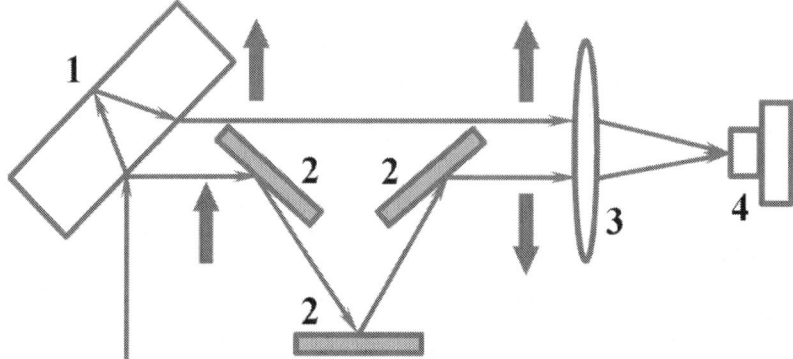

Figure 7. The optical scheme for registration of the phase portrait of a laser beam [54]: 1 – dividing parallel-sided plate, 2 – mirrors, 3 – lens, 4 – CCD camera.

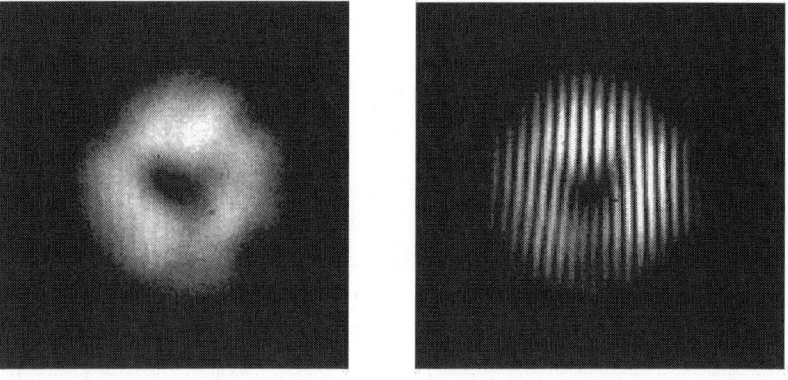

Figure 8. The experimental distribution of intensity and phase portrait of the laser mode LG_0^1 obtained at usage of a phase converter [54].

The invention of a branched hologram [89, 93] uncovered a relatively easy way to produce beams with optical vortices from an ordinary wave by using its diffraction on the amplitude diffraction grating. The idea of singular beam formation is based on the holographic principle: a readout beam restores the wave, which has participated in the hologram recording. Instead of writing a hologram with two actual optical waves, it is sufficient to calculate the interference pattern numerically and, for example, print the picture in black-and-white or grey scale. The amplitude grating after transverse scaling can, when illuminated by a regular wave, reproduce singular beams in diffraction orders.

Using the description of the singular wave amplitude (2), one can easily calculate the pattern of interference of such wave with a coherent plane wave tilted by the angle γ with respect to the z axis. The calculated interference pattern depends on the angle γ between the interfering waves and corresponds to two well-known holographic schemes: on-axis [126] when $\gamma=0$ and off-axis [127] holograms. The spiral hologram (or spiral zone plate) realized under the on-axis scheme suffers from all the disadvantages inherent to the on-axis holograms, namely, the lack of spatial separation of the reconstructed beams from the directly transmitted readout beam. Therefore the on-axis spiral holograms have not found wide application unlike the off-axis computer-generated holograms [128].

Under interference between the plane wave and the optical vortex with unity topologic charge the transmittance of amplitude diffraction grating varies according to

$$T = \left[\frac{1}{2}\left[1-\cos\left(\left(\frac{2\pi x}{\Lambda}\right)-\text{arctg}(\frac{y}{x})\right) \right] \right]^2,$$

,

(8)

where $\Lambda=\lambda/\gamma$ is the grating period. When the basic Gaussian mode passes through the grating and is focused by the lens, the $LG_0{}^1$ and $LG_0{}^{-1}$ modes occur in the far field in the 1st and −1st orders of diffraction, respectively. The period of the grating should be equal to 100-200 microns to separate the orders of diffraction properly in the actual experiment.

The two simplest ways to fabricate the amplitude diffraction gratings in the form of computer-synthesized holograms are as follows. The first involves the printing of an image onto a transparency utilized in laserjet printers. The second approach consists in photographing an inverted image, printed on a sheet of white paper, onto photo-film. Fragments of images of the gratings with the profile (8) obtained upon usage of the laser transparency with the resolution of 1200 ppi as well as the photo-film are shown in Figure 9 [129, 130, 131]. The usage of the photo-film is more preferable since it gives higher quality of the vortex to be formed and greater power conversion coefficient into the required diffraction order.

The experimental set-up scheme for formation of the optical vortex with the help of computer-synthesized amplitude grating is shown in Figure 10. The experimental set-up consists of a system for forming the collimated laser beam ($\lambda=0.633$ μ), a Mach-Zehnder interferometer and a registration system of the far field beam intensity and interference pattern. The system for forming the collimated beam includes the He-Ne laser 1 and a collimator 2 consisting of two lenses forming the Gaussian beam with a plane wave front. The Mach-Zehnder interferometer consists of two plane plates 3 and two mirrors 4. The computer-synthesized amplitude diffraction grating 5 is inserted into one of the arms of the interferometer. The lens 6 focuses the radiation onto the screen of CCD-camera 7. When blocking or admitting the reference beam from the second arm of the interferometer, the CCD camera registers the far field intensity of the vortex beam or its interference pattern with the reference beam, respectively. Varying the angle between both these beams allows one to vary the interference fringe density.

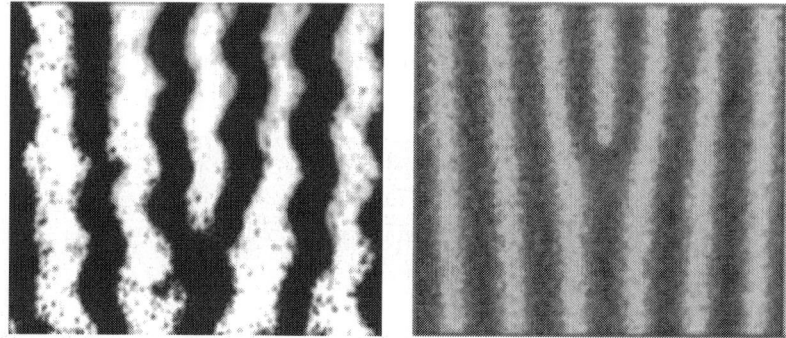

Figure 9. Magnified fragment of the amplitude grating in the experiment with laser transparency (left) and photo-film (right).

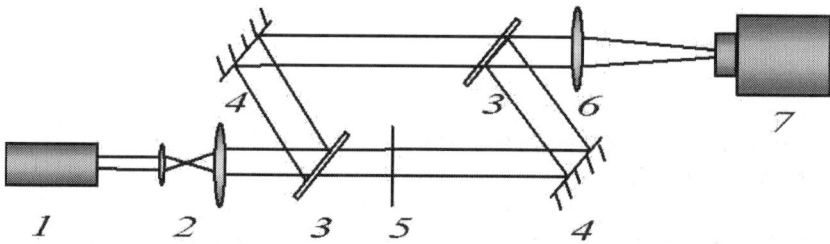

Figure 10. The set-up scheme for formation of the optical vortex:1 – He-Ne laser; 2 – collimator; 3 – optical plane plate, 4 – reflecting plane mirror; 5 – amplitude grating forming the optical vortex; 6 – lens; 7 – CCD camera.

After passing the beam through the optical scheme, the central peak (0-th diffraction order) is formed in the far-field zone. It concerns the non-scattered component of the beam that has passed through the grating. Less intensive two doughnut-shaped lateral peaks are formed symmetrically from the central peak. The lateral peaks represent the optical vortices and have a topological charge equal to the value and opposite to the sign. In the 1st order of diffraction there is only 16.7 % of energy penetrating the grating in an ideal scenario. In the experiment this part of energy is equal to about 10% owing to the imperfect structure of the grating and its incomplete transmittance.

For registration of the optical vortex it is necessary to cut off the unnecessary diffraction orders. The pictures of doughnut-like intensity distribution of the optical vortex (the lateral peak) in far field and its interference pattern are shown in Figure 11. The rigorous proof of that the obtained lateral peaks bear the optical vortices is the availability of typical "fork" in the interference pattern. The formed vortex in Figure 11, as the vortices in Figures 6 and 8, is rather different from the ideal LG_0^1 mode that is seen from their comparison with Figure 2.

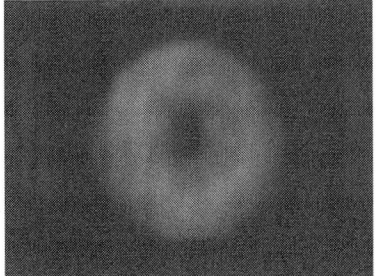 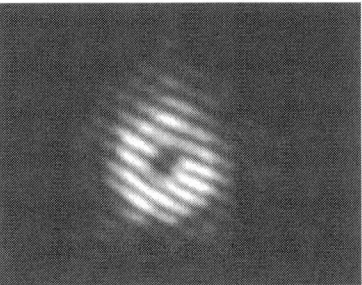

Figure 11. The intensity of the lateral diffraction order in the far field and interference pattern with the plane wave in the case of computer-synthesized amplitude diffraction grating.

Phase transparencies can be used to generate optical vortices. Application of the phase modulator results in phase changes and, after that, in amplitude changes with deep intensity modulation and the advent of zeros. In [132] the optical schematic was described, in which the wave carrying the optical vortex is recorded on thick film (Bregg's hologram) that is used to reproduce the vortex beam. The diffraction efficiency in this schematic is about 99%. A relatively thin transparency with thickness varied gradually in one of the half planes is used in the other method [133]. The efficiency of this method is greater than 90%. A similar method was proposed in [134], but a dielectric wedge was used as the phase modulator. In general, a chain of several vortices is formed as the product of this process. The shape deformation of each vortex depends on the wedge angle and on the diameter of the beam waist on the wedge surface. Varying the waist radius, one can obtain the required number of vortices (even a single vortex).

One more method of the optical vortex generation was proposed in [95]. In this method a phase transparency is used, which immediately adds the artificial vortex component into the phase profile. One such phase modulator is a transparent plate, one surface of which has a helical profile, repeating the singular phase distribution. To obtain Laguerre-Gaussian mode the depth of break onto the surface should be equal to $m\lambda/(n_1-n_2)$, where n_1 is the plate index of refraction, and n_2 is the medium index of refraction. If these conditions are met, the optical vortex appears in the far field. The main difficulty of this method lies in the problems of fabricating such a transparency. A special mask is used in the manufacture of such plates, which is made negative relative to the spiral phase plate to be formed [135]. The mask is made of brass and checked by the control interferometric system of high precision. After fabrication the mask is filled with a polymer substance and covered by glass. The spiral phase plate is formed on the glass as a result of polymerization. The transfer coefficient of such a plate is 0.98. Publications have appeared recently concerning the generation of phase transparencies using liquid crystals [136, 137].

We note that the manufacture of phase modulators is a special branch of optics called kinoform optics. At the heart of this branch lays the possibility to realize the phase control of radiation by a step-like change of the thickness or the refraction index of some structures [138]. Light weight, small size, and low cost are the most attractive features of kinoform phase elements, when compared with lenses, prisms, mirrors, and other optical devices. The kinoforms can be described as optical elements performing phase modulation with a depth not greater than the wavelength of light. This aim is realized by jumps of the optical path length not less than the even number of half wavelengths. These jumps form the lines dividing the kinoform into several zones. In boundaries of each zone the optical path length can be constant (there are two levels of binary phase elements), they can change discretely (n-level phase elements), or they can change more continuously (in an ideal phase element n approaches infinity). With an increase of n the phase efficiency of the element increases as well as its ability to control light properly. The application of kinoforms facilitates a reduction in the number of optical elements in the system by combining optical properties of several elements into one kinoform. Thus, these optical methods offer broad potential for anyone who wants to obtain beams with desired properties and to generate beams with optical vortices.

The fabrication of spiral (or helicoidal) phase plates techniques has progressed in recent years [139, 140, 141]. We will describe the generation of a doughnut Laguerre-Gaussian LG_0^1 mode with the help of a spiral phase plate [129, 130, 142, 143, 144] manufactured with etching of the fused quartz substrate using kinoform technology. Quartz displays a high damage threshold at $\lambda=0.3-1.3$ μ, high uniformity of chemical composition and refractive index n that minimizes the laser beam distortions on passing.

The fabrication of a kinoform spiral phase plate of fused quartz is performed as follows [142]. A quartz plate, 3 cm in diameter and 3 mm in thickness, is taken as the substrate. Both surfaces of the substrate are mechanically polished with a nanodiamond suspension up to the flatness better than $\lambda/30$. Special precautions are made to avoid the formation of surface damage layer that may destabilize subsequent etching. A multi-level stepped microrelief, imitating the continuous helicoidal profile, is fabricated using precise sequential etchings of the surface through a photoresist mask in a mixture based on hydrofluoric acid. At every stage, a level pattern is formed in the photoresist layer using a method of deep UV photolithography. The temperature during etching is stabilized with an accuracy of $\pm0.1^{\circ}$C. The 16- and 32-level spiral phase plates at $m=1$ and 2 have been fabricated. As the calculations show, such stepped plates and an ideal plate with an exactly helicoidal surface give practically the same optical vortices in the far field. As contrasted to the examples of spiral phase plates [140], the plates [142] have a very high laser damage threshold and a working diameter of 2 cm that is larger by an order of magnitude. The high laser damage resistance of such plates allows their use in experiments with powerful laser beams [52-54]. The general view of the spiral phase plate is shown in Figure 12.

The 3D image of the central part of a 32-level spiral phase plate designed for $\lambda=0.633$ μ, $m=1$ is shown in Figure 13. As it is seen from Figure 13, upon motion around the plate axis along a circle (perpendicular to the propeller bosses), the change of the etched profile altitude is linear with a rather high accuracy. The total break height of the microrelief on the plate surface of 1317.5 nm agrees with the calculated value $\lambda/(n-1)=1339.6$ nm, where n is the substrate refraction index, with an accuracy about of one and a half percent. As the measurements show, the roughness of the etched and non-etched surfaces (including the deepest one) is approximately the same. The roughness rms of each step surface equals 1-1.5 nm, amounting to 2-3% of the height of one step of 43.2 nm.

It should be noted that a laser beam in the form of a principal Gaussian mode with a plane wavefront that passes through a spiral phase plate maximally resembles the LG_0^1 mode in the focal plane of a lens (in far field). The part of this mode in the beam exceeds more than 90%; the residual energy is confined in general in the higher Laguerre-Gaussian modes LG_n^1. It should be noted that the proper intensity modulation of the beam incident to the plate can additionally enhance the portion of the LG_0^1 mode.

Figure 12. The photo-image of the 32-level spiral phase plate.

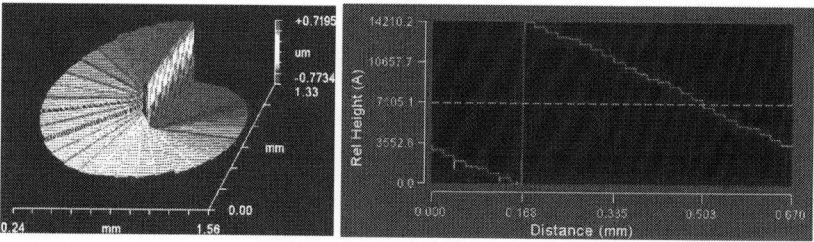

Figure 13. The image of surface in the near-axis region of the 32-level phase plate designed for λ=0.633 μ and its profile shape under motion along the circular line.

To generate a vortex beam with the help of a spiral phase plate, the experimental setup shown in Figure 10 is used. The spiral phase plate is installed into the scheme instead of the amplitude diffraction grating. In this case a vortex is formed in the 0th diffraction order in far field. Figure 14 demonstrates the experimental distributions of laser intensity in the far field and the pattern of interference of this beam with a obliquely reference plane wave. It is seen that the beam intensity distribution has a true doughnut-like shape. The wavefront singularity appears, as before, by fringe branching in the beam center with the forming of a "fork" typical for screw dislocation with unity topological charge.

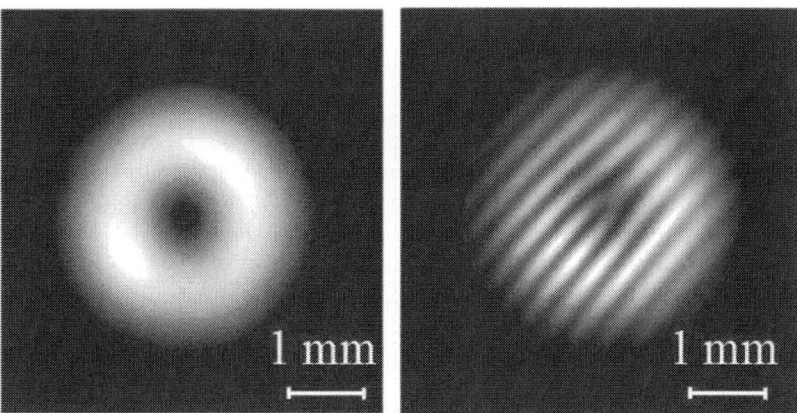

Figure 14. Experimental distribution of intensity of a vortex beam in far field and its interference pattern with obliquely incident reference plane wave in an experiment with a kinoform spiral phase plate.

The experimental data are in good agreement with the results of numerical simulation of the optical system, taking into account the stepped structure of spiral phase plate. The results barely differ from the distribution shown in Figure 2. It should be noted that the vortex quality (similarity to LG_0^1 mode) is very good, caused by the high surface quality of the spiral phase plate throughout its area. This circumstance gives us grounds to believe that the vortex wave front to be reconstructed by the Hartmann-Shack sensor has to be close to the ideal LG_0^1 wave front.

5. WAVEFRONT SENSING OF OPTICAL VORTEX

The problem of phase reconstruction using the Shack-Hartmann technique was successfully solved for optical fields with smooth wavefronts [145, 146, 147]. In the simplest case, to obtain the phase $S(r)$ using the results of measurement of phase gradient projection $\nabla_\perp S_m(\mathbf{r})$ on the transverse plane it is possible to employ the numeric integration of the gradient over a contour Γ:

$$S(r) = S_0(r) + \int_\Gamma \nabla_\perp S_m(\rho)d\rho,$$

(9)

where r={x, y}. Since for the ordinary wave fields the phase distribution is the potential function, the values of S(r) do not depend formally on the configuration of the integration path. In the actual experiments, however, some errors are always present, so the potentiality of phase is violated and the results of phase reconstruction depend on the integration path [147]. To reduce the noise influence on the results it was proposed to consider the phase reconstruction as the minimization of a certain functional. The most commonly used functional is the criterion corresponding to the minimum of the weighted square of residual error of gradient of the reconstructed phase $\nabla_\perp S(r)$ and phase gradients $\nabla_\perp S_m(r)$, obtained in the measurements:

$$\int_D (W(r) \cdot (\nabla_\perp S(r) - \nabla_\perp S_m(r))^2 dr \to \min,$$

(10)

where W(r)={$W_x(x, y)$, $W_H(x, y)$} is the vector weighting function introduced to account for the reliability of $\nabla_\perp S_m(r)$ measurements. This method is known as the least mean square phase reconstruction.

The approaches to the solution of variation problem (10) are well known [146, 147] and actually mean the solution of the Poisson equation written with partial derivatives. Allowing for the weighting function W(r), it acquires the following form:

$$W_x(x, y)\left(\frac{\partial S}{\partial x} - \frac{\partial S_m}{\partial x}\right) + W_y(x, y)\left(\frac{\partial S}{\partial y} - \frac{\partial S_m}{\partial y}\right) = 0,$$

(11)

where $\partial S/\partial x$ and $\partial S/\partial y$ are gradients of reconstructed phase, $\partial S_m/\partial x$ and $\partial S_m/\partial y$ are measured gradients of the phase.

There are a wide variety of methods [145, 146, 147, 148] which can be used to solve the discrete variants of equation (11). For example, one can use the representation of (11) as a system of algebraic equations, the fast Fourier transform, or the Gauss-Zeidel iteration method applied to the multi-grid algorithm. This group of methods is equally well adopted for the application of centroid coordinates measured by the Shack-Hartmann sensor as input data:

$$\{\nabla_{\perp}S_m(r, z)\}_I = \frac{\iint V(r-r_0)I(r_0)\nabla_{\perp}S(r_0)dr_0}{\iint V(r-r_0)I(r_0)dr_0},$$

(12)

where the integration is performed over the square of the subaperture, V is a subaperture function, and $I(r_0)$ is intensity of the input beam.

The sensing of wavefront with screw phase dislocations by the least mean square method is not agreeable. With this technique (along with other methods based on the assumption that phase surface is a continuous function of coordinates) it is possible to reconstruct only a fraction of the entire phase function. As it turned out [68, 149], the differential properties of the vector field of phase gradients help to find some similarity between this field and the field of potential flow of a liquid penetrated by vortex strings. It is also possible to represent this vector field as a sum of potential and solenoid components:

$$\nabla_{\perp}S(\mathbf{r}) = \nabla_{\perp}S_p(\mathbf{r}) + \nabla_{\perp}S_c(\mathbf{r})$$

(13)

where $\nabla_{\perp}S_p$ is the gradient of potential phase component and $\nabla_{\perp}S_c$ is the gradient of vortex (solenoidal) component. By using only the ordinary methods of phase reconstruction it is possible to reproduce just the part of phase distribution that corresponds to potential component in (13).

However, if the quantity $\nabla_{\perp}S_c$ is considered as a rotor of vector potential H, namely, $\nabla_{\perp}S_c=\nabla\times H$, which is dependent only on the coordinates of optical vortices [68] then potential phase component in (13) can be found by the least mean square method [68, 147]. By means of a novel "hydrodynamic" approach to the properties of the vector filed of phase gradients a new group of methods was formed [72, 150, 151], employing the discovered coordinates of dislocations and reconstructing the potential phase component with the least mean square method. Within another technique [74, 152] reproduction of the scalar potential is also based on the least square method but the vortex component is calculated with Eq. (9) via a consistent rotor of vector potential. The method of matching the vortex component was proposed in [153] and is based on the following equation:

$$\nabla^2\left(Rot_{-\pi/2}\left(\nabla_{\perp}S_c\right)\right) = \left(\nabla\times\mathbf{H}\right)\cdot\mathbf{e}_z$$

(14)

where e_z is the unit vector of z axis and $Rot_{-\pi/2}(\nabla_\perp S_c)$ is operation of the rotation of each vector on $-\pi/2$ angle. This relation is a Poisson equation which allows one to find components of consistent vectors of vortex phase gradient. Now it is possible to take Eq. (9) and obtain a vortex phase component, assuming that the consistent gradients of vortex phase are measured without errors.

The searching for dislocation located positions, which is required in algorithms of phase reconstruction [72, 150, 151], is a sufficiently difficult problem. Because of the infinite phase gradients in the points of zero intensity, the application of methods based on solution of (13) [74, 152] is also not straightforward. Presently there is no such an algorithm, which guarantees the required fidelity of wavefront reconstruction in the presence of dislocations [64]. However, according to some estimations [154, 155, 156] the accurate detection of vortex coordinates and their topological charges insures the sensing of wavefronts with high precision. Therefore we expect a future improvement in reconstruction algorithms by involving more sophisticated methods into the consideration of gradient fields, insuring more accurate detection of dislocation positions and their topological charges.

Analysis shows that from the point of view of experimental realization, of the considered approaches of wavefront reconstruction the algorithm of D. Fried [74] is one of the best algorithms (with respect to accuracy, effectiveness and resistance to measurement noises) of recovery of phase surface $S(x, y)$ from its measured gradient $\nabla S_\perp$ distribution in the presence of optical vortices. Fried's algorithm (a *noise-variance-weighted complex exponential reconstructor*) consists of three parts: reduction or simplification, solving, and reconstruction. The algorithm designed for work in Hadjin geometry reconstructs the phase in the nodes of a quadratic grid with the dimensions $(2^N+1)\times(2^N+1)$, using the phase differences between these nodes. Obviously, to employ the algorithm we need $(2^N+1)\times2^N$ and $2^N\times(2^N+1)$ array of phase differences along x and y axes. The words "*complex exponential*" mean that the phase reconstruction problem is reformulated to a task of recovery of "phasors" u (the complex number with a unity absolute value and an argument that is equal to the phase of optical field) distribution in transverse section of the beam. Here, the analysis and transformation of differential complex vectors (differential phasors) $\Delta_x u \equiv \exp(i\Delta_x\varphi)$, $\Delta_y u \equiv \exp(i\Delta_y\varphi)$, corresponding to phase differences $\Delta_x\varphi$, $\Delta_y\varphi$ between different nodes of the computational grid, are used. The words "*noise-variance-weighted*" mean that the algorithm takes into account the distinctions of measurement variance of individual differential phasors, i.e. the influence ("weight") of differential phasors on the recovery result is inversely proportional to their variance. This feature of Fried's algorithm allows us to apply it to a computational grid of arbitrary dimension, not only to the $(2^N+1)\times(2^N+1)$ grid [74]; to take into account the average statistical inequality of measurement errors of phase gradient in different areas of the beam (for example, on the sub-apertures of the Hartmann-Shack sensor) if the repeated characterization of the same beam is performed; to consider a prior concept of the inequality of measurement errors of phase gradient in these areas if the measurement of the beam characteristics is single.

In Fried's algorithm the differential phasors are unit vectors. The operation of normalization of a complex vector is applied to provide for this requirement. However, the amplitudes of differential phasors and phasors, obtained under reduction and reconstruction, contain information about measurement errors of phase differences in the actual experiment. Based on this reason the algorithm in question has been modified [157, 158, 159]. The modification involves exclusion of the operation of complex vector normalization and allows an increase in algorithm accuracy.

The experimental setup for registration of an optical vortex wavefront consists of a system for formation of collimated laser beam, the Mach-Zehnder interferometer (as in the scheme in Figure 10), and the additionally induced the Hartmann-Shack wavefront sensor [160, 161]. It is shown in Figure 15. The system of formation of collimated beam includes a He-Ne laser 1 (λ=0.633 мкм) and collimator 2 composed of lenses with focal lengths 5 cm and 160 cm. The collimator forms the reference basic Gaussian beam with a diameter of 1 cm and the plane wave front. The Mach-Zehnder interferometer includes two optical plates 3 and two mirrors 4. The spiral phase plate 5 for formation of the optical vortex is interposed into one of the interferometer arms. The working surface of the phase plate, with a diameter of 2 cm completely covers the beam. After passing through the spiral phase plate the Gaussian beam turns into an optical vortex (the LG_0^1 mode) in the focal plane 8 of the lens 6, i.e. in far field, with high conversion coefficient. Focal plane 8 of the lens 6 with focal length 700 cm is transferred (for purposes of magnification) by an objective 8 in the optically conjugated plane 8'. The wavefront sensor consists of a lenslet array 9 situated in the plane 8'and a CCD camera 10.

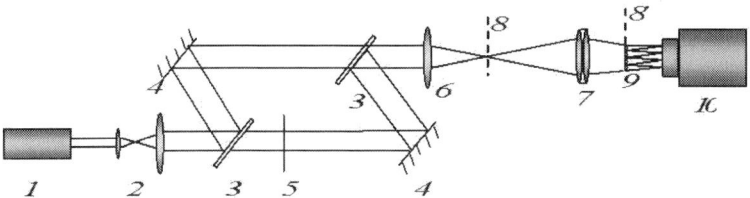

Figure 15. Experimental setup for wavefront sensing of optical vortex in far field: 1 – He-Ne laser; 2 – collimator; 3 – optical plate; 4 – plane mirror; 5 – spiral phase plate; 6 – the lens F=6 m; 7 – the objective; 8 and 8' - focal plane of lens 6 and its optically conjugated plane, respectively; 9 – lenslet array; 10 – CCD camera.

A technical feature of the Hartmann-Shack wavefront sensor used involves the employment of a raster of 8-level diffraction Fresnel lenses as the lenslet array (see Figure 16). The raster is fabricated from fused quartz by kinoform technology, similar to the aforesaid spiral phase plate, with the minimum size of microlens d=0.1 mm and diffraction efficiency up to 90% [162]. The accuracy of etching profile depth is not worse than 2%, the difference of the focal spot size from the theoretical size is ~1%. The spatial resolution of the wavefront sensor and its sensibility depend on the microlens geometry, the number of registered focal spots and their size with respect to the CCD camera pixel size.

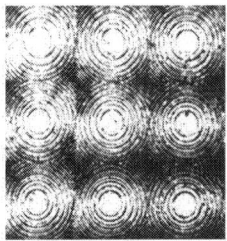 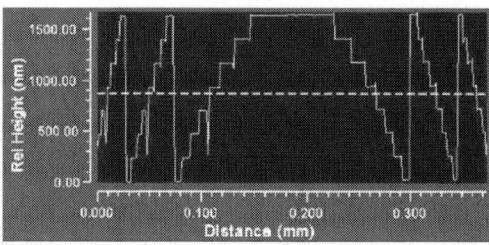

Figure 16. Photo image of a fragment of the lenslet array and image of surface profile of a microlens.

Under the registration of phase front the reference beam in the second arm of the interferometer is blocked. In the beginning the wavefront sensor is calibrated by a reference beam with plane phase front (the spiral phase plate is removed from the scheme). Then the spiral phase plate is inserted, and the picture of focal spots correspondent to singular phase front is registered. From the values of displacement of focal spots from initial positions, the local tilts of wave front on the sub-apertures of lenslet array are determined.

Experiments with a different number of registration spots on the hartmannogram have been carried out [160, 161]. When using a lenslet array with subaperture size d=0.3 mm, focal length f=25 mm and d=0.2 mm, f=15 mm, the picture from 8×8 and 16x16 focal spots on the CCD camera screen has been registered, respectively. The results of experimental measurements of wave front gradients are given in Figure 17 for measurement points 8×8, where the picture of displacements of focal spots in the hartmannogram is shown. The vortex center is situated between the sub-apertures of the array. Displacement of each spot is demonstrated by the arrow (line segment). The arrow origin corresponds to the reference spot position whereas the arrow end corresponds to spot position after the insertion of a spiral phase plate in the experimental scheme. In Figure 17 the results are also shown, which are obtained in calculation and are correspondent to the ideal LG_0^1 mode with high accuracy. It is seen from Figure 17 that the experimental and calculated pictures of spots' displacements agree with each other. Some local data difference is caused by the distinction between the phase and amplitude structure of the beam incident on the phase plate in the calculation and actual experiment, by the inaccurate location of the vortex in the optical axis assumed in calculations and by the inevitable noises of the measurement.

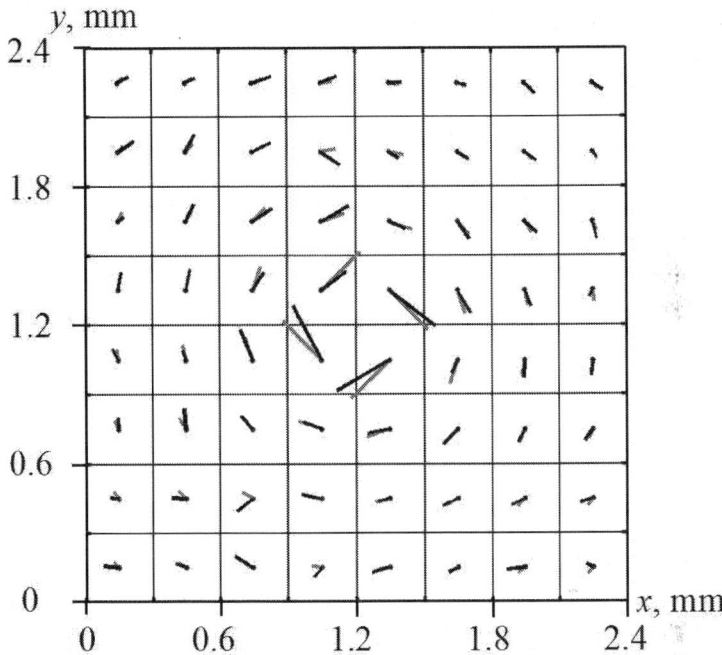

Figure 17. The picture of displacements of focal spots of the hartmannogram in experiment (black arrows) and calculation (grey arrows).

In Ref. [163] the vortex-like structure of displacements of spots in the hartmannogram was registered for the LG_0^1 and higher-order modes. As the primary information, the spot displacements can be used for deriving the Poynting vector skews (in fact, wavefront tilts), as it was made in [163], as well as for wavefront sensing that is more nontrivial. In this chapter we simply consider the reconstruction of singular phase surface by the Hartmann-Shack sensor and describe the realization of this operation with the new reconstruction technique.

In Figure 18 we present the wave front surface of optical vortex reconstructed by the Hartmann-Shack sensor [161, 164] with software incorporating the code of restoration of singular phase surfaces [157-159]. Comparison of experimental data with calculated results shows that the wave front surface is restored by the actual Hartmann Shack wavefront sensor with good quality despite the rather small size of the matrix of wave front tilts (spots in the hartmannogram). The reconstructed wave front has the characteristic spiral form with a break of the surface about 2π. Analysis shows that the accuracy of wave front reconstruction (of course, from the viewpoint of its proximity to the theoretical results) is not worse than $\lambda/20$. The accuracy of recovery of phase surface break increases at the measurement spots of 16x16. For comparison purposes in Figure 18 the result is demonstrated of the vortex wavefront reconstruction with the help of the standard least-squares restoration technique in the Hartmann-Shack sensor. It is seen that the conventional approach obviously fails.

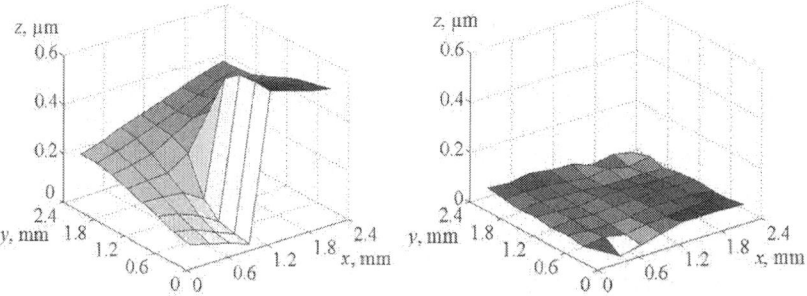

Figure 18. Experimental vortex phase surface reconstructed using modified Fried's (left) and conventional least-squares (right) procedure.

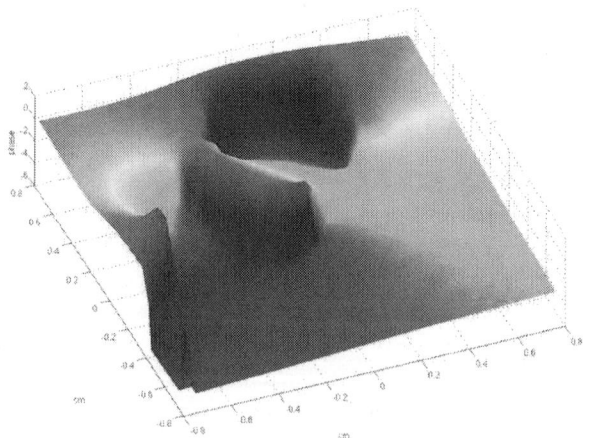

Figure 19. The phase surface fragment of the beam after the turbulent path reconstructed using the modified Fried's algorithm.

In Figure 19 we show the calculation results [165] of phase front reconstruction of the beam passed through the turbulent atmosphere in the case of $C_n^2=10^{-14}$ cm$^{-2/3}$ after 1 km distance propagation (see Figure 4). The modified Fried's algorithm embedded into the Hartmann-Shack sensor software correctly restores the complicated singular structure of the phase surface.

6. PHASE CORRECTION OF OPTICAL VORTEX

Next we consider the possibility to transform the wavefronts of the vortex beam by means of the closed-loop adaptive optical system with a wavefront sensor and a flexible deformable wavefront corrector. We can use the bimorph [166] as well as pusher-type [167, 168] piezoceramic-based adaptive mirrors as a wavefront corrector. In the experiments a flexible bimorph mirror [166] and the Hartmann-Shack wavefront sensor with a new reconstruction algorithm [157-159] are employed. An attempt is made to correct the laser beam carrying the optical vortex (namely, the Laguerre-Gaussian LG_0^1 mode), i.e., to remove its singularity. The dynamic effects are not considered, the goal is the estimation of the ability of the bimorph mirror to govern the spatial features of the optical vortex. It is very interesting to determine whether the phase correction leads to full elimination of the singularity [165].

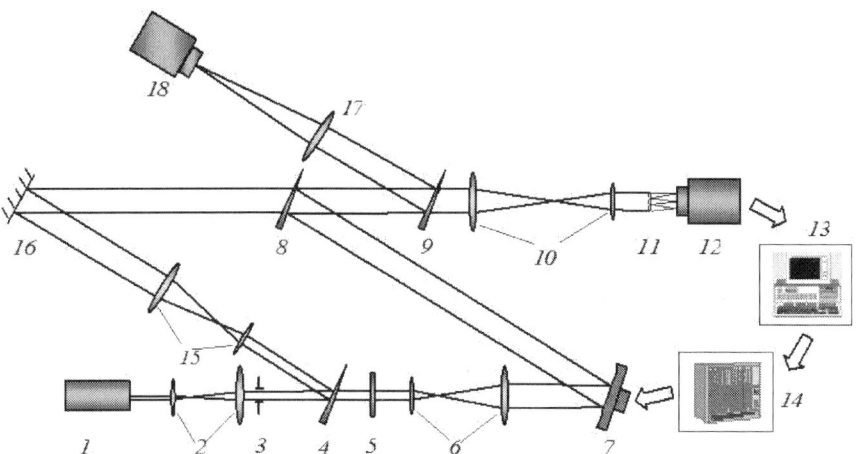

Figure 20. The close-loop adaptive system for optical vortex correction:1 – He-Ne laser; 2 – collimator; 3 – pinhole 10×10 mm; 4, 8, 9 – optical plates; 5 -spiral phase plate; 6, 10, 15 – telescopes; 7 – deformable adaptive mirror; 11 –lenslet array; 12, 18– CCD cameras; 13 –computer; 14 – control unit of adaptive mirror, 16 – plane mirror; 17 – lens.

A closed-loop adaptive system intended for performance of the necessary correction of vortex wavefront is shown in Figure 20 [169]. A reference laser beam is formed using a He-Ne laser 1, a collimator 2, and a square pinhole 3, which restricts the beam aperture to a size of 10×10 mm^2. Next the laser beam passes through a 32-level spiral phase plate 5 of a diameter of 2 cm, a fourfold telescope 6 and comes to an adaptive deformable mirror 7. It should be noted that the laser beam with the plane phase front that passes through the spiral phase plate maximally resembles the Laguerre-Gaussian LG_0^1 mode in the far field. In Figure 20 the wavefront corrector is situated in near rather than far field but at a relatively large distance from the spiral phase plate so that the proper vortex structure of the phase distribution is already formed in the wavefront corrector plane.

The wavefront corrector (the bimorph adaptive mirror) 7 [166] is shown in Figure 21. It is composed of a substrate of LK-105 glass with reflecting coating and two foursquare piezoceramic plates, each measuring 45x45 mm and 0.4 mm thick. The first piezoplate is rigidly glued to rear side of the substrate. It is complete, meaning it serves as one electrode, and is intended to compensate for the beam defocusing if need be. The second piezoplate destined to transform the vortex phase surface is glued to the first one. The 5x5=25 electrodes are patterned on the surface of the second piezoplate in the check geometry (close square packing). Each electrode has the shape of a square, with each side measuring 8.5 mm. The full thickness of the adaptive mirror is 4.5 mm. The wavefront corrector is fixed in a metal mounting with a square 45x45 mm window. The surface deformation of the adaptive mirror under the maximal voltage ±300 V applied to any one electrode reaches ±1.5 μm.

25	26	11	12	13
24	10	3	4	14
23	9	2	5	15
22	8	7	6	16
21	20	19	18	17

Figure 21. The deformable bimorph mirror and the scheme of arrangement of control elements on the second piezoplate.

The radiation beam reflected from the adaptive mirror 7 (see Figure 20) is directed by a plane mirror 8 through a reducing telescope 10 to a Hartmann-Shack sensor including a lenslet array 11 with d=0.2 mm, f=15 mm and a CCD camera 12. At the field size of 3.2 mm there are 16x16 spots in the hartmannogram on the CCD camera screen. The planes of adaptive mirror and lenslet array are optically conjugated so that the wavefront sensor reconstructs in fact the phase surface of the beam just in the corrector plane.

A beam part is derived by a dividing plate 9 to a CCD camera 18 for additional characterization (see Figure 20). In addition, the wavefront corrector 7, plates 4, 8 and rear mirror 16 form a Mach-Zehnder interferometer. On blocking the reference beam from the mirror 16, the CCD cameras 12 and 18 simultaneously register, respectively, the hartmannogram and intensity picture of the beam going from the adaptive mirror. Upon admission of the reference beam from the mirror 16, the CCD camera 18 registers the interference pattern of the beam going from the adaptive mirror with an obliquely incident reference beam. Screen of CCD camera 18 is situated at a focal distance from the lens 17 or in a plane of the adaptive mirror image (like the lenslet array) thus registering the intensity/interferogram of the beam in far or near field, respectively.

The wavefront has no singularity upon removal of the spiral phase plate 5 from the scheme in the Figure 20 and when switching off the wavefront corrector. The reference beam phase surface in the corrector plane is shown in Figure 22a. It is not an ideal plane (PV=0.33 µ) but it is certainly regular. Therefore the picture of diffraction at the square diaphragm 3 (see Figure 20) roughly takes place in far field in Figure 23a.

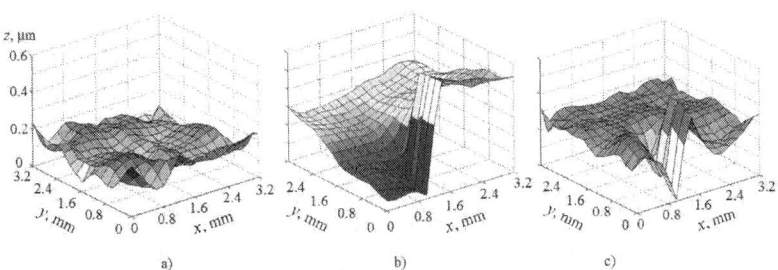

Figure 22. Experimental phase surface in near field: (a) reference beam and beam (b) before and (c) after correction.

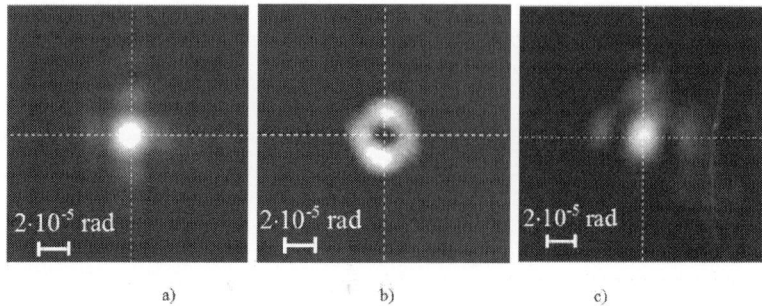

a)　　　　　　　　　b)　　　　　　　　　c)

Figure 23. Experimental far field intensity: (a) reference beam and beam (b) before and (c) after correction.

After inserting the spiral phase plate 5 and when switching off the adaptive mirror, the wavefront in near field in Figure 22b acquires the spiral form with λ-break (PV=0.63 μ) so the far-field intensity has a doughnut form (see Figure 23b). Note that the wavefront corrector software in the computer 13 is based on the singular reconstruction technique [157-159]; the conventional least-squares approach fails here. The vortex in far field is the LG_0^1 mode distorted by presence of other modes mainly because of the phase surface imperfection of the reference beam. Note that for the task of adjusting the vortex wavefront sensing technique it was necessary to form a close-to-ideal LG_0^1 mode as a "reference" optical vortex with maximally predetermined phase surface, to determine the sensing algorithm accuracy. Here, for the correction task, it is even more attractive to work with a distorted vortex.

In order to correct the vortex wavefront in the closed loop, the recovered phase surface in Figure 22b is decomposed on the response functions of control elements of the deformable mirror. The response function of a control element is the changing of the shape of the deformable mirror surface upon the energizing of this control element with zero voltages applied to the others actuators. The expansion coefficients on response functions are proportional to voltages to be applied from control unit 14 to appropriate elements of the deformable mirror. When applying control voltages to the adaptive mirror its surface is deformed to reproduce the measured vortex wavefront maximally and thus to obtain a wavefront close to a plane one upon reflection from the corrector. However, each superposition of the response functions of a flexible wavefront corrector is a smooth function, and the corrector is not able to exactly reproduce the phase discontinuity of a depth of 2π. The phase surface after the correction in Figure 22c is close to the reference one (Figure 22a) except for a narrow region at the break line (PV=0.5 μ). As the radiation from this part of the beam is scattered to larger angles and its portion in the beam is relatively small, the far field intensity picture after the correction in Figure 23c is much closer to the reference beam (Figure 23a) rather than the vortex before correction (Figure 23b). Thus, the doughnut-like vortex beam is focused into a beam with a bright axial spot and weaker background that radically increases the Strehl ratio and resolution of the optical system.

The beam interferograms in near field before and after correction are shown in Figure 24. Unlike the former, the latter contains no resolved singularities (at least, under the given fringe density). The vortices, however, may appear under beam propagation from the adaptive mirror plane as it was in the case of combined propagation of the vortex beam with a regular beam [170]. The experimental and calculated (at the reflection of an ideal LG_0^1 mode from the actual deformed adaptive mirror surface) interferograms of a corrected beam in far field are shown in Figure 25. Two off-axis vortices (denoted by light circles) of opposite topological charge are seen here. The first of them is initial vortex shifted from the axis whereas the second arises in the process of beam propagation from the far periphery of the beam (in fact from infinity, according to the terminology of [170]).

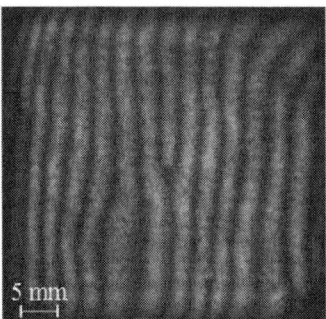 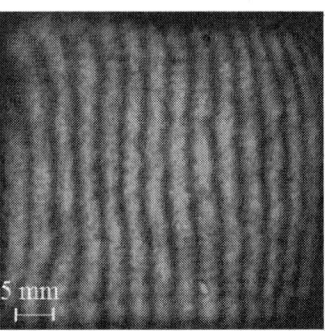

Figure 24. Experimental pattern of interference of the beam with an obliquely incident regular wave in the near field (left) before and (right) after correction.

Thus, the phase surface of the distorted LG_0^1 mode is corrected in the closed-loop adaptive optical system, including the bimorph piezoceramic mirror and the Hartmann-Shack wavefront sensor with the singular reconstruction technique. Experiments demonstrate the ability of the bimorph mirror to correct the optical vortex in a practical sense, namely, to focus the doughnut-like beam into a beam with a bright axial spot that considerably increases the Strehl ratio and optical system resolution. Since the phase break is not reproduced exactly on the flexible corrector surface, the off-axis vortices can appear in far field at the beam periphery.

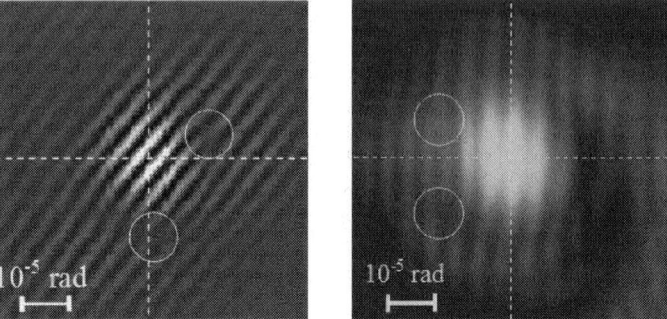

Figure 25. The pattern of interference of the corrected beam with an obliquely incident regular wave in the far field in (left) calculation and (right) experiment.

7. CONCLUSIONS

This chapter is dedicated to research of the possibility to control the phase front of a laser beam carrying an optical vortex by means of linear adaptive optics, namely, in the classic closed-loop adaptive system including a Hartmann-Shack wavefront sensor and a deformable mirror. On the one hand, the optical vortices appear randomly under beam propagation in the turbulent atmosphere, and the correction of singular phase front presents a considerable problem for tasks in atmospheric optics, astronomy, and optical communication. On the other hand, the controllable optical vortices have very attractive potential applications in optical data processing and many other scientific and practical fields where the regulation of singular phase is needed. This chapter discusses the main properties and applications of optical vortices, the problem of adaptive correction of singular phase in turbulent atmosphere, the issues of generating the "reference" laser vortex beam, its wavefront sensing and phase correction in the widespread adaptive optical system including a Hartmann-Shack wavefront sensor and a flexible deformable mirror.

The vortex beam is generated with help of a spiral phase plate made of fused quartz by kinoform technology. Provided that the optical quality of the spiral phase plate is good, such a means of vortex formation seems to be more preferable as compared with other considered methods of vortex generation with a well-determined phase surface. As a result, it becomes possible to obtain a singular beam very close to a Laguerre-Gaussian LG_0^1 mode with a well-determined singular phase structure that is necessary for checking the accuracy of subsequent wavefront reconstruction. The developed spiral phase plates are characterized by high laser damage resistance, the good surface profile accuracy and they facilitate formation of a high quality optical vortex.

The vortex phase surface measurement is carried out by a Hartmann-Shack wavefront sensor which is simpler in design and construction, more reliable and more widespread in various fields of adaptive optics when compared with other types of sensors. The commonly accepted Hartmann-Shack wavefront reconstruction is performed on the basis of the least-mean-square approach. This approach works well in the case of continuous phase distributions but is completely unsuitable for singular phase distributions. Therefore a new reconstruction technique has been developed for the reconstruction of singular phase surface, starting from the measured phase gradients. The measured shifts of focal spots in the hartmannogram are in good agreement with the calculation results. Using new software in the Hartmann-Shack sensor, the reconstruction of the "reference" vortex phase surface has been carried out to a high degree of accuracy.

The vortex laser beam (distorted LG_0^1 mode) is corrected in the closed-loop adaptive system including a Hartmann-Shack wavefront sensor with singular reconstruction technique and a flexible bimorph piezoceramic mirror with 5x5 actuators allocated in the check geometry. The mirror has high laser damage resistance meaning it can operate with powerful laser beams. The purpose of the correction is to eliminate the singularity of the beam to the highest degree possible. Experiments have demonstrated the ability of the bimorph mirror to correct the optical vortex in a practical sense. As a result of phase correction, the doughnut-like beam is focused into a beam with a bright axial spot that considerably increases the Strehl ratio and is important for practical applications. However, since the wavefront break cannot be reproduced exactly by a mirror with a flexible surface, the residual off-axis vortices can appear in far field at the beam periphery.

The investigations described above consolidate the actual birth of the experimental field of novel scientific branch – singular adaptive optics.

REFERENCES

1. M. A. Vorontsov, V. I. Shmalgauzen, 1985 Principles of adaptive optics
2. M. C. Roggemann, B. M. Welsh, 1996 Imaging through turbulence. Boca Raton, FL: CRC Press
3. R. K. Tyson, 1998 Principles of adaptive optics
4. J. W. Hardy, 1998 Adaptive optics for astronomical
5. F. Roddier, 1999 Adaptive optics in astronomy Cambridge: Cambridge University Press
6. R. K. Tyson, 2000 Introduction to adaptive optics
7. Optical Vortices. 1999 Vasnetsov M, Staliunas K. (eds.) Horizons in World Physics 228 New York: Nova Science
8. M. S. Soskin, M. V. Vasnetsov, 2001 Singular optics. Wolf E. (ed.) Progress in Optics, V.XLII. Amsterdam: Elsevier 219 276

9. A. Bekshaev, M. Soskin, M. Vasnetsov, 2009 Paraxial light beams with angular momentum Schulz M. (ed.) Progress in Optics Research. New York: Nova Science Publishers 1 75

10. M. Berry, 1981 Singularities in waves and rays. Balian R., Kleman M., Poirier J.-P. (ed.) Physics of Defects. Amsterdam: North-Holland 453 543

11. O. Bryngdahl, 1973 Radial- and circular-fringe interferograms. J. Opt. Soc. Am 63 9 1098 1104

12. J. F. Nye, M. V. Berry, 1974 Dislocations in wave trains. Proc. R. Soc. A 336 165 190

13. D. Rozas, C. T. Law, G. A. Jr Swartzlander, 1997 Propagation dynamics of optical vortices. J. Opt. Soc. Am. B 14 11 3054 3065

14. Z. S. Sacks, D. Rozas, G. A. Jr Swartzlander, 1998 Holographic formation of optical-vortex filaments. Jr. Opt. Soc. Am. B 15 8 2226

15. P. Coullet, L. Gil, F. Rocca, 1989 Optical vortices Optics Commun 73 5 403 408

16. A. E. Siegman, 1986 Lasers. Sausalito, CA: University Science Books

17. H. Kogelnik, T. Li, 1966 Laser beams and resonators. Applied Optics 5 10 1550 1567

18. L. Allen, M. W. Beijersbergen, R. J. C. Spreeuw, J. P. Woerdman, 1992 Orbital angular momentum of light and the transformation of Laguerre-Gaussian modes. Phys Rev. A 45 11 8185 8189

19. L. Allen, M. J. Padjett, M. Babiker, 1999 Orbital angular momentum of light Wolf E. (ed.) Progress in Optics 34 Amsterdam: Elsevier 291 370

20. D. A. B. Miller, 1998 Spatial channels for communicating with waves between volumes. Optics Letters 23 21 1645 1647

21. G. Gibson, J. Courtial, M. Padgett, M. Vasnetsov, V. Pas'ko, S. Barnett, S. Franke-Arnold, 2004 Free-space information transfer using light beams carrying orbital angular momentum. Optics Express 12 22 5448 5456

22. Z. Bouchal, R. Celechovsky, 2004 Mixed vortex states of light as information carriers New J. Phys 6 1 131 145

23. J. Scheuer, M. Orenstein, 1999 Optical vortices crystals: spontaneous generation in nonlinear semiconductor microcavities Science 285 5425 230 233

24. L. Mandel, E. Wolf, 1995 Optical Coherence and Quantum Optics New York: Cambridge University Press

25. A. Ashkin, 1992 Forces of a single-beam gradient laser trap on a dielectric sphere in the ray optics regime. Biophys. J 61 2 569 582

26. K. T. Gahagan, G. A. Jr Swartzlander, 1998 Trapping of low-index microparticles in an optical vortex. J. Opt. Soc. Am. B 15 2 524 534

27. K. T. Gahagan, G. A. Jr Swartzlander, 1999 Simultaneous trapping of low-index and high-index microparticles observed with an optical-vortex trap J. Opt. Soc. Am. B 16 4 533 539

28. J. E. Curtis, B. A. Koss, D. G. Grier, 2002 Dynamic holographic optical tweezers. Optics Commun 207 1-6 169 175

29. K. Ladavac, D. G. Grier, 2004 Microoptomechanical pump assembled and driven by holo- graphic optical vortex arrays. Optics Express 12 6 1144 1149

30. V. Daria, P. J. Rodrigo, J. Glueckstad, 2004 Dynamic array of dark optical traps Appl. Phys. Lett 84 3 323 325

31. S. N. Khonina, V. V. Kotlyar, M. V. Shinkaryev, V. A. Soifer, G. V. Uspleniev, 1992 J. Mod. Optics 39 5 1147 1154

32. G. A. Jr Swartzlander, 2001 Peering into darkness with a vortex spatial filter. Optics Letters 26 8 497 499

33. D. Rouan, P. Riaud, A. Boccaletti, Y. Clénet, A. Labeyrie, 2000 The four-quadrant phase-mask coronagraph. I. Principle. Pub. Astron. Soc. Pacific 112 1479 1486

34. A. Boccaletti, P. Riaud, P. Baudoz, J. Baudrand, D. Rouan, D. Gratadour, F. Lacombe, A. Lagrange-M, 2004 The four-quadrant phase-mask coronagraph. IV. First light at the very large telescope. Pub. Astron. Soc. Pacific 116 1061 1071

35. G. Foo, D. M. Palacios, G. A. Jr Swartzlander, 2005 Optical vortex coronagraph. Optics Letters 30 24 3308 3310

36. J. H. Lee, G. Foo, E. G. Johnson, G. A. Jr Swartzlander, 2006 Experimental verification of an optical vortex coronagraph. Phys. Rev. Lett 97 5 053901 1

37. J. A. Davis, D. E. Mc Namara, D. M. Cottrell, 2000 Image processing with the radial Hilbert transform: theory and experiments. Optics Letters 25 2 99 101

38. K. G. Larkin, D. J. Bone, M. A. Oldfield, 2001 Natural demodulation of two-dimensional fringe patterns. I. General background of the spiral phase quadrature transform. J. Opt. Soc. Am. A 18 8 1862 1870

39. J. Masajada, A. Popiołek-Masajada, D. M. Wieliczka, 2002 The interferometric system using optical vortices as phase markers. Optics Commun 207 1 85 93

40. P. Senthilkumaran, 2003 Optical phase singularities in detection of laser beam collimation. Applied Optics 42 31 6314 6320

41. V. Westphal, S. W. Hell, 2005 Nanoscale Resolution in the Focal Plane of an Optical Microscope. Phys. Rev. Lett 94 14 143903 1

42. M. D. Levenson, T. J. Ebihara, G. Dai, Y. Morikawa, N. Hayashi, S. M. Tan, 2004 Optical vortex mask via levels. J. Microlithogr. Microfabr. Microsyst 3 2 293 304

43. R. Menon, H. I. Smith, 2006 Absorbance-modulation optical lithography. J. Opt. Soc. Am. A 23 9 2290 2294

44. A. Mair, A. Vaziri, G. Weihs, A. Zeilinger, 2001 Entanglement of the orbital angular momentum states of photons. Nature 412 7 313 316

45. H. H. Arnaut, G. A. Barbosa, 2000 Orbital and intrinsic angular momentum of single photons and entangled pairs of photons generated by parametric down-conversion. Phys. Rev. Lett 85 2 286 289

46. S. Franke-Arnold, S. M. Barnett, M. J. Padgett, L. Allen, 2002 Two-photon entanglement of orbital angular momentum states Phys. Rev. A 65 3 033823

47. J. R. Abo-Shaeer, C. Raman, J. M. Vogels, W. Ketterle, 2001 Observation of vortex lattices in Bose-Einstein condensates. Science 292 5516 476 479

48. K. Dholakia, N. B. Simpson, M. J. Padgett, L. Allen, 1996 Second-harmonic generation and the orbital angular momentum of light Phys. Rev. A 54 5 R3742 3745

49. F. A. Starikov, G. G. Kochemasov, 2001 Novel phenomena at stimulated Brillouin scattering of vortex laser beams. Optics Commun 193 1-6 207 215

50. F. A. Starikov, G. G. Kochemasov, 2001 Investigation of stimulated Brillouin scattering of vortex laser beams. Proc. SPIE 4403 217

51. F. A. Starikov, 2007 Stimulated Brillouin scattering of Laguerre-Gaussian laser modes: new phenomena. Gaponov-Grekhov AV., Nekorkin VI. (ed). Nonlinear waves 2006. N.Novgorod: IAP RAS 206 221

52. F. A. Starikov, Yu. V. Dolgopolov, A. V. Kopalkin, et al. 2006 About the correction of laser beams with phase front vortex. J. Phys. IV 133 683 685

53. F. A. Starikov, Yu. V. Dolgopolov, A. V. Kopalkin, et al. 2008 New phenomena at stimulated Brillouin scattering of Laguerre-Gaussian laser modes: theory, calculation, and experiments. Proc. SPIE 70090E 1 11

54.   A. V. Kopalkin, V. A. Bogachev, Yu. V. Dolgopolov, et al. 2011 Conjugation and transformation of the wave front by stimulated Brillouin scattering of vortex Laguerre-Gaussian laser modes. Quantum Electronics 41 11 1023 1026

55.   N. B. Baranova, B. Zel'dovich, Mamaev. A. V. Ya., N. V. Pilipetskii, V. V. Shkunov, 1981 Dislocations of the wavefront of a speckle-inhomogeneous field (theory and experiment). Sov. Phys. JETP Lett 33 4 195 199

56.   N. B. Baranova, B. Ya. Zel'dovich, A. V. Mamaev, N. V. Pilipetskii, V. V. Shkunov, 1982 Dislocation density on wavefront of a speckle-structure light field. Sov. Phys. JETP 56 5 983 988

57.   N. B. Baranova, A. V. Mamaev, N. V. Pilipetskii, V. V. Shkunov, B. Ya. Zel'dovich, 1983 Wavefront dislocations: topological limitations for adaptive systems with phase conjugation. J. Opt. Soc. Am. A 73 5 525 528

58.   V. K. Ladagin, 1985 About the numerical integration of a quasi-optical equation. Questions of Atomic Science and Technology. Ser. Methods and codes of numerical solution of tasks of mathematical physics 1 19 26

59.   M. D. Feit, J. A. Jr Fleck, 1988 Beam nonparaxiality, filament formation, and beam breakup in the self-focusing of optical beams. J. Opt. Soc. Am. B 7 3 633 640

60.   V. P. Kandidov, 1996 Monte Carlo method in nonlinear statistical optics. Physics Usp 39 12 1243 1272

61.   J. W. Goodman, 2000 Statistical optics. New York: Wiley

62.   D. L. Fried, J. L. Vaughn, 1992 Branch cuts in the phase function. Applied Optics 31 15 2865 2882

63.   A. Primmerman, R. Pries, R. A. Humphreys, B. G. Zollars, H. T. Barclay, J. Herrmann, 1995 Atmospheric-compensation experiments in strong-scintillation conditions. Applied Optics 34 12 081 088

64.   J. D. Barchers, D. L. Fried, D. J. Link, 2002 Evaluation of the performance of Hartmann sensors in strong scintillation. Applied Optics 41 6 1012 1021

65.   F. Yu. Kanev, V. P. Lukin, N. A. Makenova, 2002 Analysis of adaptive correction efficiency with account of limitations induced by Shack-Hartmann sensor. Proc. SPIE 5026 190 197

66.   J. C. Ricklin, F. M. Davidson, 1998 Atmospheric turbulence effects on a partially coherent Gaussian beam: implication for free-space laser communication. Applied Optics 37 21 4553 4561

67. M. Levine, E. A. Martinsen, A. Wirth, A. Jankevich, M. Toledo-Quinones, F. Landers, Th. L. Bruno, 1998 Horizontal line-of-sight turbulence over near-ground paths and implication for adaptive optics

68. D. L. Fried, 1998 Branch point problem in adaptive optics

69. E. O. Le Bigot, W. J. Wild, E. J. Kibblewhite, 1998 Reconstructions of discontinuous light phase functions. Optics Letters 23 1 10 12

70. H. Takijo, T. Takahashi, 1988 Least-squares phase estimation from the phase difference. J. Opt. Soc. Am. A 5 3 416 425

71. V. P. Aksenov, V. A. Banakh, O. V. Tikhomirova, 1998 Potential and vortex features of optical speckle field and visualization of wave-front singularities. Applied Optics 37 21 4536 4540

72. W. W. Arrasmith, 1999 Branch-point-tolerant least-squares phase reconstructor. J. Opt. Soc. Am. A 16 7 1864 1872

73. G. A. Tyler, 2000 Reconstruction and assessment of the least-squares and slope discrepancy components of the phase. J. Opt. Soc. Am. A 17 10 1828 1839

74. D. L. Fried, 2001 Adaptive optics wave function reconstruction and phase unwrapping when branch points are present Optics Commun 200 1 43 72

75. V. P. Aksenov, O. V. Tikhomirova, 2002 Theory of singular-phase reconstruction for an optical speckle field in the turbulent atmosphere. J. Opt. Soc. Am. A 19 2 345 355

76. C. Rockstuhl, A. A. Ivanovskyy, M. S. Soskin, et al. 2004 High-resolution measurement of phase singularities produced by computer-generated holograms Optics Commun 242 1-3 163 169

77. K. L. Baker, E. A. Stappaerts, D. Gavel, et al. 2004 High-speed horizontal-path atmospheric turbulence

78. J. Notaras, C. Paterson, 2007 Demonstration of closed-loop adaptive optics

79. J. Hartmann, 1904 Objetivuntersuchungen. Z. Instrum 1 1 33 97

80. R. B. Shack, B. C. Platt, 1971 Production and use of a lenticular Hartmann screen. J. Opt. Soc. Am 6 5 656 662

81. J. W. Hardy, J. E. Lefebvre, C. L. Koliopoulos, 1977 Real-time atmospheric compensation. J. Opt. Soc. Am 67 3 360 369

82. D. G. Sandler, L. Cuellar, M. Lefebvre, et al. 1994 Shearing interferometry for laser-guide-star atmospheric correction at large D/r0. J. Opt. Soc. Am. A 11 2 858 873

83. F. Roddier, 1988 Curvature sensing and compensation: a new concept in adaptive optics

84. G. Rousset, 1999 Wave-front sensors. Roddier F. (ed.) Adaptive optics in astronomyCambridge: Cambridge University Press 91 130

85. R. J. Dorn, 2001 A CCD based curvature wavefront sensor for adaptive optics

86. R. Ragazzoni, 1996 Pupil plane wavefront sensing with an oscillating prism Journal of Modern Optics 43 2 289 293

87. R. Ragazzoni, A. Ghedina, A. Baruffolo, E. Marchetti, et al. 2000 Testing the pyramid wavefront sensor on the sky. Proc. SPIE 4007 423 429

88. M. Vorontsov, J. Riker, G. Carhart, V. S. Rao Gudimetla, L. Beresnev, T. Weyrauch, L. C. Jr Roberts, 2009 Deep turbulence effects compensation experiments with a cascaded adaptive optics

89. V. Yu. Bazhenov, M. V. Vasnetsov, M. S. Soskin, 1990 Laser beams with screw wavefront dislocations. Sov. Phys. JETP Lett 52 8 429 431

90. M. Brambilla, F. Battipede, L. A. Lugiato, V. Penna, F. Prati, C. Tamm, C. O. Weiss, 1991 Transverse Laser Patterns. I. Phase singularity crystals. Phys. Rev. A 43 9 5090 5113

91. E. Abramochkin, V. Volostnikov, 1991 Beam transformations and nontransformed beams Optics Commun 83 1, 2 123 135

92. L. E. Grin', P. V. Korolenko, N. N. Fedotov, 1992 About the generation of laser beams with screw wavefront structure. Optics and Spectroscopy 73 5 1007 1010

93. V. Y.u Bazhenov, M. S. Soskin, M. V. Vasnetsov, 1992 Screw dislocations in light wavefronts. J. Mod. Optics 39 5 985 990

94. M. W. Beijersbergen, L. Allen, H. E. L. O. van der Veen, J. P. Woerdman, 1993 Astigmatic laser mode converters and transfer of orbital angular momentum Optics Commun 96 1-3 123 132

95. M. W. Beijersbergen, R. P. C. Coerwinkel, M. Kristensen, J. P. Woerdman, 1994 Helical- wavefront laser beams produced with a spiral phase plate. Optics Commun 112 5-6 321 327

96. K. Dholakia, N. B. Simpson, M. J. Padgett, L. Allen, 1996 Second harmonic generation and the orbital angular momentum of light. Phys. Rev. A 54 5 R3742 R3745

97. R. Oron, Y. Danziger, N. Davidson, A. Friesem, E. Hasman, 1999 Laser mode discrimination with intra-cavity spiral phase elements Optics Commun 169 1-6 115

98. A. Wada, Y. Miyamoto, T. Ohtani, N. Nishihara, M. Takeda, 2001 Effects of astigmatic aberration in holographic generation of Laguerre-Gaussian beam. Proc. SPIE 4416 376 379

99. Y. Miyamoto, M. Masuda, A. Wada, M. Takeda, 2001 Electron-beam lithography fabrication of phase holograms to generate Laguerre-Gaussian beams. Proc. SPIE 3740 232 235

100. D. W. Zhang, -C. Yuan X., 2002 Optical doughnut for optical tweezers. Optics Letters 27 15 1351 1353

101. A. A. Malyutin, 2004 On a method for obtaining laser beams with a phase singularity. Quantum Electronics 34 3 255 260

102. Y. Izdebskaya, V. Shvedov, A. Volyar, 2005 Generation of higher-order optical vortices by a dielectric wedge. Optics Letters 30 18 2472 2474

103. S. Vyas, P. Senthilkumaran, 2007 Interferometric optical vortex array generator Applied Optics 46 15 2893 2898

104. V. V. Kotlyar, A. A. Kovalev, 2008 Fraunhofer diffraction of the plane wave by a multilevel (quantized) spiral phase plate. Optics Letters 33 2 189 191

105. F. T. Arecchi, S. Boccaletti, G. Giacomelli, G. P. Puccioni, P. L. Ramazza, S. Residori, 1992 Patterns, space-time chaos and topological defects in nonlinear optics. Physica D: Nonlinear Phenomena 61 1-4 25 39

106. G. Indebetouw, D. R. Korwan, 1994 Model of vortices nucleation in a photorefractive phase-conjugate resonator J. Mod. Optics 41 5 941 950

107. M. S. Soskin, M. V. Vasnetsov, 1998 Nonlinear singular optics. Pure and Applied Optics 7 2 301 311

108. A. Berzanskis, A. Matijosius, A. Piskarskas, V. Smilgevicius, A. Stabinis, 1997 Conversion of topological charge of optical vortices in a parametric frequency converter Optics Commun 140 4-6 273 276

109. J. Yin, Y. Zhu, W. Wang, Y. Wang, W. Jhe, 1998 Optical potential for atom guidance in a dark hollow beam. J. Opt. Soc. Am. B 15 1 25 33

110. B. Darsht, Zel'dovich. B. Ya., Kataevskaya. I. V. Ya., N. D. Kundikova, 1995 Formation of a single wavefront dislocation. Zh. Eksp. Teor. Fiz 107 5 1464 1472

111. T. A. Fadeeva, S. A. Reshetnikoff, A. V. Volyar, 1998 Guided optical vortices and their angular momentum in low-mode fibers. Proc. SPIE 3487 59 70

112. A. Sobolev, T. Cherezova, V. Samarkin, A. Kydryashov, 2007 Bimorph flexible mirror for vortex beam formation. Proc. SPIE 63462A 1 8

113. R. K. Tyson, M. Scipioni, J. Viegas, 2008 Generation of an optical vortex with a segmented deformable mirror Applied Optics 47 33 6300 6306

114. J. M. Vaughan, D. V. Willets, 1979 Interference properties of a light beam having a helical wave surface. Optics Commun 30 3 263 270

115. N. N. Rozanov, 1993 About the formation of radiation with wavefront dislocations. Optics and Spectroscopy 75 4 861 867

116. R. Oron, N. Davidson, A. A. Friesem, E. Hasman, 2000 Efficient formation of pure helical laser beams. Optics Commun 182 1-3 205 208

117. E. Abramochkin, N. Losevsky, V. Volostnikov, 1997 Generation of spiral-type laser beams. Optics Commun 141 1-2 59 64

118. C. Tamm, C. O. Weiss, 1990 Bistability and optical switching of spatial patterns in laser. J. Opt. Soc. Am. B 7 6 1034 1038

119. M. Padgett, J. Arlt, N. Simpson, L. Allen, 1996 An experiment to observe the intensity and phase structure of Laguerre-Gaussian laser modes Am. J. Phys 64 1 77 82

120. D. V. Petrov, F. Canal, L. Torner, 1997 A simple method to generate optical beams with a screw phase dislocation Optics Commun 143 4-6 265

121. M. J. Snadden, A. S. Bell, R. B. M. Clarke, E. Riis, D. H. Mc Intyre, 1997 Doughnut mode magneto-optical trap. J. Opt. Soc. Am. B 14 3 544 552

122. Y. Yoshikawa, H. Sasada, 2002 Versatile of optical vortices based on paraxial mode expansion. J. Opt. Soc. Am. A 19 10 2127 2133

123. J. Courtial, M. J. Padjett, 1999 Performance of a cylindrical lens mode converter for producing Laguerre-Gaussian laser modes. Optics Commun 159 1-3 13 18

124. V. Kh. Bagdasarov, S. V. Garnov, N. N. Denisov, A. A. Malyutin, Yu. V. Dolgopolov, A. V. Kopalkin, F. A. Starikov, 2009 Laser system emitting 100 mJ in Laguerre-Gaussian modes. Quantum Electronics 39 9 785 788

125. A. A. Malyutin, 2006 Tunable astigmatic π/2 converter of laser modes with a fixed distance between input and output planes. Quantum Electronics 36 1 76 78

126. D. Gabor, 1948 A new microscopic principle. Nature 161 4098 777 778

127. E. Leith, J. Upatnieks, 1961 New technique in wavefront reconstruction. J. Opt. Soc. Am 51 11 1469 1473

128. N. R. Heckenberg, R. Mc Duff, CP Smith, A. G. White, 1992 Generation of optical phase singularities by computer-generated holograms Optics Letters 17 3 221 223

129. F. A. Starikov, V. V. Atuchin, Yu. V. Dolgopolov, et al. 2004 Generation of optical vortex for an adaptive optical system for phase correction of laser beams with wave front dislocations. Proc. SPIE 5572 400 408

130.   F. A. Starikov, V. V. Atuchin, Yu. V. Dolgopolov, et al. 2005 Development of an adaptive optical system for phase correction of laser beams with wave front dislocations: generation of an optical vortex. Proc. SPIE 5777 784 787

131.   F. A. Starikov, G. G. Kochemasov, 2005 ISTC Projects from RFNC-VNIIEF devoted to improving laser beam quality. Springer Proceedings in Physics 102 291 301

132.   Z. S. Sacks, D. Rozas, G. A. Jr Swartzlander, 1998 Holographic formation of optical-vortex filaments. J. Opt. Soc. Am. B 15 8 2226 2234

133.   G. H. Kim, J. H. Jeon, K. H. Ko, H. J. Moon, J. H. Lee, J. S. Chang, 1997 Optical vortices produced with a nonspiral phase plate. Applied Optics 36 33 8614 8621

134.   V. G. Shvedov, Ya. V. Izdebskaya, A. N. Alekseev, A. V. Volyar, 2002 The formation of optical vortices in the course of light diffraction on a dielectric wedge. Technical Physics Letters 28 3 256 260

135.   S. S. R. Oemrawsingh, J. A. W. van Houwelingen, E. R. Eliel, J. P. Woerdman, E. J. K. Verstegen, J. G. Kloosterboer, G. W. Hooft, 2004 Production and characterization of spiral phase plates for optical wavelengths. Applied Optics 43 3 688 694

136.   D. Ganic, X. Gan, M. Gu, 2002 Generation of doughnut laser beams by use of a liquid-crystal cell with a conversion efficiency near 100%. Optics Letters 27 15 1351 353

137.   J. E. Curtis, D. G. Grier, 2003 Structure of Optical Vortices. Phys. Rev. Lett 90 13 133901

138.   A. I. Fishman, 1999 Phase optical elements - kinoforms. Soros Educ. J 12 76 83

139.   T. Kamimura, S. Akamatsu, H. Horibe, et al. 2004 Enhancement of surface-damage resistance by removing subsurface damage in fused silica and its dependence on wavelength Jap. J. Appl. Phys 43 9 L1229 L1231

140.   J. W. Sung, H. Hockel, J. D. Brown, E. G. Johnson, 2006 Development of two-dimensional phase grating mask for fabrication of an analog-resist profile. Applied Optics 45 1 33 43

141.   G. A. Jr Swartzlander, 2006 Achromatic optical vortex lens. Optics Letters 31 13 2042 2044

142.   V. V. Atuchin, S. L. Permyakov, I. S. Soldatenkov, F. A. Starikov, 2006 Kinoform generator of vortex laser beams. Proc. SPIE 6054 1 4

143. Yu. I. Malakhov, 2006 ISTC projects devoted to improving laser beam quality. Proc. SPIE 6346 1 8

144. Yu. I. Malakhov, V. V. Atuchin, A. V. Kudryashov, F. A. Starikov, 2009 Optical components of adaptive systems for improving laser beam quality. Proc. SPIE 7131 1 5

145. W. H. Southwell, 1980 Wave-front estimation from wave-front slope measurements. J. Opt. Soc. Am 70 8 998 1006

146. M. A. Vorontsov, A. V. Koryabin, V. I. Shmalgausen, 1988 Controlled optical systems. Moscow: Nauka

147. D. C. Ghiglia, M. D. Pritt, 1998 Two-dimensional phase unwrapping: theory, algorithms, and software. New-York: Wiley

148. W. Zou, Z. Zhang, 2000 Generalized wave-front reconstruction algorithm

149. V. Aksenov, V. Banakh, O. Tikhomirova, 1998 Potential and vortex features of optical speckle fields and visualization of wave-front singularities Applied Optics 37 21 4536 4540

150. M. C. Roggemann, A. C. Koivunen, 2000 Branch-point reconstruction in laser beam projection through turbulence with finite-degree-of-freedom phase-only wave-front correction. J. Opt. Soc. Am. A 17 1 53 62

151. V. P. Aksenov, O. V. Tikhomirova, 2002 Theory of singular-phase reconstruction for an optical speckle field in the turbulent atmosphere. J. Opt. Soc. Am. A 19 2 345 355

152. G. A. Tyler, 2000 Reconstruction and assessment of the least-squares and slope discrepancy components of the phase. J. Opt. Soc. Am. A 17 10 1828 1839

153. E.-O. Le Bigot, W. J. Wild, 1999 Theory of branch-point detection and its implementation. J. Opt. Soc. Am. A 16 7 1724 1729

154. V. P. Aksenov, I. V. Izmailov, F. Yu. Kanev, F. A. Starikov., 2005 Localization of optical vortices and reconstruction of wavefront with screw dislocations. Proc. SPIE 5894 1 11

155. V. P. Aksenov, I. V. Izmailov, F. Yu. Kanev, 2005 Algorithms of a singular wavefront reconstruction. Proc. SPIE 6018 1 11

156. V. P. Aksenov, I. V. Izmailov, F. Yu. Kanev, F. A. Starikov, 2006 Screening of singular points of vector field of phase gradient, localization of optical vortices and reconstruction of wavefront with screw dislocations. Proc. SPIE 6162 1 12

157. V. P. Aksenov, I. V. Izmailov, F. Yu. Kanev, F. A. Starikov, 2006 Performance of a wavefront sensor in the presence of singular point. Proc. SPIE 634133 1 6

158. V. P. Aksenov, I. V. Izmailov, F. Yu. Kanev, F. A. Starikov, 2007 Singular wavefront reconstruction with the tilts measured by Shack-Hartmann sensor. Proc. SPIE 63463 1 8

159. V. P. Aksenov, I. V. Izmailov, F. Yu. Kanev, F. A. Starikov., 2008 Algorithms for the reconstruction of the singular wavefront of laser radiation: analysis and improvement of accuracy. Quantum Electronics 38 673 677

160. F. A. Starikov, V. V. Atuchin, G. G. Kochemasov, et al. 2005 Wave front registration of an optical vortex generated with the help of spiral phase plates. Proc. SPIE 589 1 11

161. F. A. Starikov, V. P. Aksenov, I. V. Izmailov, et al. 2007 Wave front sensing of an optical vortex. Proc. SPIE 634 1 8

162. V. V. Atuchin, I. S. Soldatenkov, A. V. Kirpichnikov, et al. 2004 Multilevel kinoform microlens arrays in fused silica for high-power laser optics. Proc. SPIE 5481 43 46

163. J. Leach, S. Keen, M. Padgett, C. Saunter, G. D. Love, 2006 Direct measurement of the skew angle of the Poynting vector in helically phased beam. Optics Express 14 25 11919 11923

164. F. A. Starikov, G. G. Kochemasov, S. M. Kulikov, et al. 2007 Wave front reconstruction of an optical vortex by Hartmann-Shack sensor. Optics Letters 32 16 2291 2293

165. F. A. Starikov, V. P. Aksenov, V. V. Atuchin, et al. 2009 Correction of vortex laser beams in a closed-loop adaptive system with bimorph mirror. Proc. SPIE 7131 1 8

166. F. A. Starikov, V. P. Aksenov, V. V. Atuchin, et al. 2007 Wave front sensing of an optical vortex and its correction in the close-loop adaptive system with bimorph mirror. Proc. SPIE 6747 1 8

167. S. Yu. Bokalo, S. G. Garanin, S. V. Grigorovich, et al. 2007 Deformable mirror based on piezoelectric actuators for the adaptive system of the Iskra-6 facility. Quantum Electronics 37 8 691 696

168. S. G. Garanin, A. N. Manachinsky, F. A. Starikov, S. V. Khokhlov, 2012 Phase correction of laser radiation with the use of adaptive optical systems

169. F. A. Starikov, G. G. Kochemasov, M. O. Koltygin, et al. 2009 Correction of vortex laser beam in a closed-loop adaptive system with bimorph mirror. Optics Letters 34 15 2264 2266

170.   M. S. Soskin, V. N. Gorshkov, M. V. Vasnetsov, J. T. Malos, N. R. Heckenberg, 1997 Topological charge and angular momentum of light beams carrying optical vortices. Phys. Rev. A 56 5 4064 4075

Chapter 3

Adaptive Optics Technology for High-Resolution Retinal Imaging

Marco Lombardo [1,*], Sebastiano Serrao [1], Nicholas Devaney [2], Mariacristina Parravano [1] and Giuseppe Lombardo [3,4]

[1] Fondazione G.B. Bietti IRCCS, Via Livenza 3, 00198 Rome, Italy

[2] Applied Optics Group, School of Physics, National University of Ireland, Galway, Ireland

[3] CNR-IPCF Unit of Support of Cosenza, c/o University of Calabria, Ponte P. Bucci Cubo 31/C, 87036 Rende, Italy

[4] Vision Engineering, Via Adda 7, 00198 Rome, Italy

ABSTRACT

Adaptive optics (AO) is a technology used to improve the performance of optical systems by reducing the effects of optical aberrations. The direct visualization of the photoreceptor cells, capillaries and nerve fiber bundles represents the major benefit of adding AO to retinal imaging. Adaptive optics is opening a new frontier for clinical research in ophthalmology, providing new information on the early pathological changes of the retinal microstructures in various retinal diseases. We have reviewed AO technology for retinal imaging, providing information on the core components of an AO retinal camera. The most commonly used wavefront sensing and correcting elements are discussed. Furthermore, we discuss current applications of AO imaging to a population of healthy adults and to the most frequent causes of blindness, including diabetic retinopathy, age-related macular degeneration and glaucoma. We conclude our work with a discussion on future clinical prospects for AO retinal imaging.

Keywords: adaptive optical systems; optical sensors; biomedical imaging techniques; eye

1. INTRODUCTION

The human eye functions as an optical system whose purpose is to bring the outside world into focus on the retina, thereby allowing us to see. However, the optics and how they are aligned are not perfect and the consequence is that incoming light rays deviate from the desired path that reaches the foveal center or foveola. These deviations are defined as optical aberrations [1]. Aberrations in the eye optics not only blur images formed on the retina, thus impairing vision, but also blur images taken of the retina by ophthalmic imaging cameras.

Ocular aberrations can be classified into low-order (LOA) and high-order aberrations (HOA). Low-order aberrations, such as defocus and astigmatism, are the predominant optical aberrations and they account for approximately 90% to the overall wavefront aberration (WA) of the eye. Although HOA make a small contribution (on average ≤ 10%) to the total variance of the eye WA, their effect on image quality is well known as well as the fact that their correction can significantly improve visual performance and retinal imaging [2–4].

Diagnosis of retinal diseases at an early stage is crucial for the treatment and avoidance of serious visual loss. Across the developed world, the major causes of vision loss can be attributed to age-related macular degeneration, diabetic retinopathy and glaucoma [5]. Diagnosis usually occurs once damage has already happened to some extent and this is due to the relatively poor resolution of current retinal imaging techniques which limits their ability to detect abnormalities of the retinal microstructures, including photoreceptors cells, capillaries and nerve fiber bundles. Imaging modalities able to detect and monitor pathological variations of the retinal microstructures at a pre-clinical stage of disease represent the basis for designing new diagnostic and treatment protocols to preserve the normal integrity and function of the retina. The need for sensitive and accurate diagnostic tools is increasing as a growing number of new treatment options become available to ophthalmologists, including gene therapy, nano-targeted drug delivery devices and micro-pulsed lasers.

Optical imaging is the preferred method to non-invasively investigate the retina and to perform efficient retinal therapy. Retinal cameras for wide field clinical imaging are generally designed without considering aberration correction beyond defocus, the resolution at which images could be recovered from the retina in vivo has been limited to the macroscopic scale. In order to bring the lateral resolution of ophthalmoscopes to the microscopic scale, it is necessary to compensate not only for defocus, but also astigmatism and higher order aberrations. Adaptive optics (AO) is a technology used to improve the performance of optical systems by minimizing aberrations. It has been developed for astronomical telescopes in order to remove the effect of atmospheric turbulence from astrophysics objects and only in recent years has been extended to ophthalmology [6,7]. When applied to the optical system of the eye, AO technology can provide substantial improvements in the sharpness of retinal images that are normally degraded by ocular aberrations [8]. AO retinal imaging can theoretically improve the lateral resolution to 2 μm and therefore provide information about the retinal microstructures that cannot be obtained with current retinal imaging techniques.

The scope of this review is to describe some of the technical aspects that have made AO retinal imaging in human eyes possible and to discuss applications and future prospects for noninvasive clinical imaging of retinal diseases with microscale level of resolution.

2. THE OPTICAL SYSTEM OF THE HUMAN EYE

The eye's optical system consists of three main components: the cornea, the crystalline lens and in between them the iris (Figure 1). The cornea, the outermost optical element, is responsible for about 2/3 of the optical power and aberrations of the eye. The iris controls the amount of light coming into the retina by regulating the pupil diameter. As in any optical system, the size of the pupil has important consequences for image formation: a smaller pupil increases the depth of focus and minimizes the effects of high-order aberrations. Conversely, the magnitude of aberrations increases with pupil dilation leading to a decrease in both visual performance and optical quality of the retinal image [1]. The crystalline lens accounts for about 1/3 of the optical power of the eye but it is capable of changing its focusing properties: controlled changes in the shape and thickness of the crystalline lens allow the eye to accommodate, the process by which the eye focuses on near objects. Even in the normal eye, departures from ideal focus (i.e., aberrations) exist and degrade the eye's optical performance [8–10]. LOA are the predominant optical aberrations (90% of the overall WA of the eye): defocus (positive or negative; i.e., hyperopia and myopia respectively) is the dominant aberration, followed by astigmatism (orthogonal or oblique). It is well known that the human eye suffers from HOA that cannot yet be accurately corrected and that they greatly diminish the overall optical quality of the eye, though their contribution to the overall WA of the eye is ≤10% [3,8]. The presence of HOA, beyond defocus and astigmatism, has been known by researchers since the 19th century, but only in the 1990s have wavefront sensors been developed to allow routine estimation of the eye's WA. The development of ocular wavefront sensors has allowed rapid, accurate and objective measurements of wave aberrations and made large population studies possible [2,9–11]. Several authors [12–15] have measured the distribution and contribution of both LOA and HOA to the overall WA of the eye: between HOA, the magnitude of 3rd order coma-like aberrations (vertical coma, horizontal coma, oblique trefoil and horizontal trefoil) and spherical aberration is higher than other higher aberration modes [1]. The eye's WA is not static but fluctuates over time: the eye's focus exhibits fluctuations about its mean value for steady-state accommodation with amplitudes ranging between 0.03 and 0.5 diopters (D). In addition, a general tendency for spherical aberration to change in a negative direction with increase in accommodation (− 0.04 μm/D for accommodative levels of 1.0 to 6.0 D) has been measured, while the other HOA are not significantly influenced by accommodation [16,17]. The largest source of temporal short-term instability (seconds and minutes) of HOA is then due to the micro-fluctuations in the accommodation of the lens: the anterior curvature increases centrally and flattens peripherally during accommodation, while at the same time, the lens thickness increases and the equatorial diameter decreases. These factors may contribute to the change in the measured aberrations. Another source of fluctuations is local changes in the tear film thickness over the cornea, due to evaporation and/or blinking [1,18]. If considering a long period of time (over the course of the day and between successive days), the WA of the eye has been demonstrated to be sufficiently stable, with no significant changes in the magnitude and contributions of HOA [1,17]. An AO ophthalmic device can measure and correct for the fluctuations of the eye's WA, thus improving the resolution of images taken from the retina of patients.

3. ADAPTIVE OPTICS TECHNOLOGY FOR RETINAL IMAGING

The history of adaptive optics for ophthalmic imaging is just over 15 years old. AO was first used by Dreher et al. in 1989 [19], but the correction was limited to only second order optical aberrations of the eye. In 1997, AO technology was successfully applied to high resolution imaging in the human eye by Liang et al.[20]. Since that time AO technology has advanced dramatically, including the integration of AO into different imaging modalities and improved optical design [21]. Almost all existing ophthalmic modalities have incorporated AO to enhance quality and resolution of the retinal images: flood illumination fundus imaging, confocal scanning laser ophthalmoscopy and ophthalmic optical coherence tomography, with each offering different benefits [6,7,22]. Current AO flood-illumination cameras incorporate a superluminescent diode to illuminate the retina, used in conjunction with a high-speed scientific-grade CCD. These improvements permit faster modulation (higher frame rates) and higher efficiency (shorter exposure durations) than former prototypes. Confocal scanning laser ophthalmoscopy (cSLO) and ophthalmic optical coherence tomography (OCT) are typically realized by raster scanning a focused point source across the retina. The cSLO uses the same optical path for scanning the laser spot on the retina and for delivering the returning light to the detector. A confocal pinhole just in front of the light detector collects most of the light from the retina while rejecting light originating from unwanted planes away from the image retinal plane. AO-OCT provides a high-resolution cross-sectional view through the living retina, comparable to a histological section. Adaptive optics provides three technical benefits for OCT that improve the visualization and detection of microscopic structures in the retina: (1) increased lateral resolution, (2) reduced speckle size (granular artifacts), and (3) increased sensitivity to weak reflections.

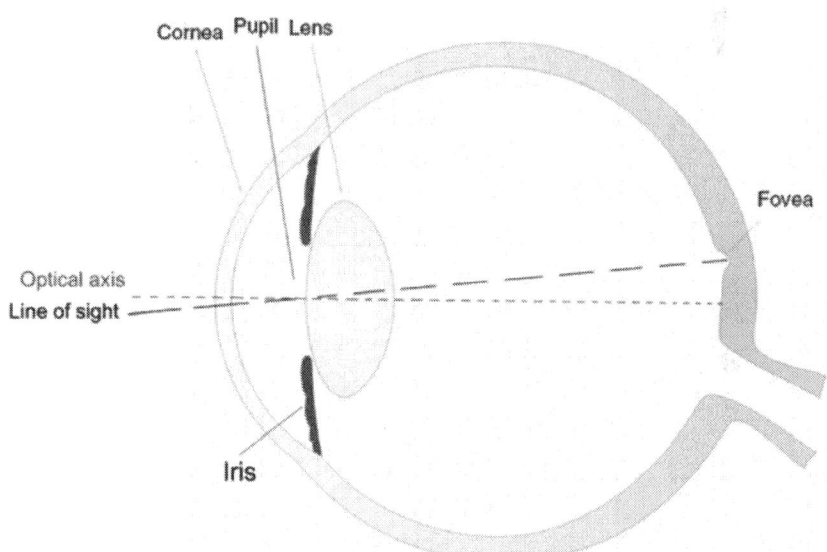

Figure 1. The optical system of the human eye consists of three main components, i.e., the cornea, the crystalline lens and the iris. The iris controls the amount of light coming into the retina by regulating the diameter of the pupil. Therefore, the pupil of the eye acts as the aperture of the system. The optical axis of the eye (dotted grey line) is defined as the line joining the centers of curvature of all the optical surfaces. However, the appropriate and convenient axis that should be used for describing the optical system of the eye is the line of sight (dashed black line), which is defined as the ray that passes through the center of the entrance pupil and strikes the center of the fovea (i.e., the foveola).

Adaptive optics by itself does not provide a retinal image, rather an AO subsystem must be incorporated into an imaging device. A typical AO retinal imaging camera has three principal components: a wavefront sensor, a corrective element and a control system. The wavefront sensor and corrector measure and correct the eye's wave aberrations respectively. The AO controller, programmed with a computer, controls the interaction between the wavefront sensor and the corrector element; it interprets the wavefront sensor data and computes the appropriate wavefront corrector drive signals. AO systems operating in closed-loop (Figure 2) place the wavefront sensor after the wavefront corrector. In this configuration, the measured wavefront is the error signal that gets fed back to the controller to further reduce the residual aberrations in the next iteration, theoretically correcting the retinal images up to the diffraction limit.

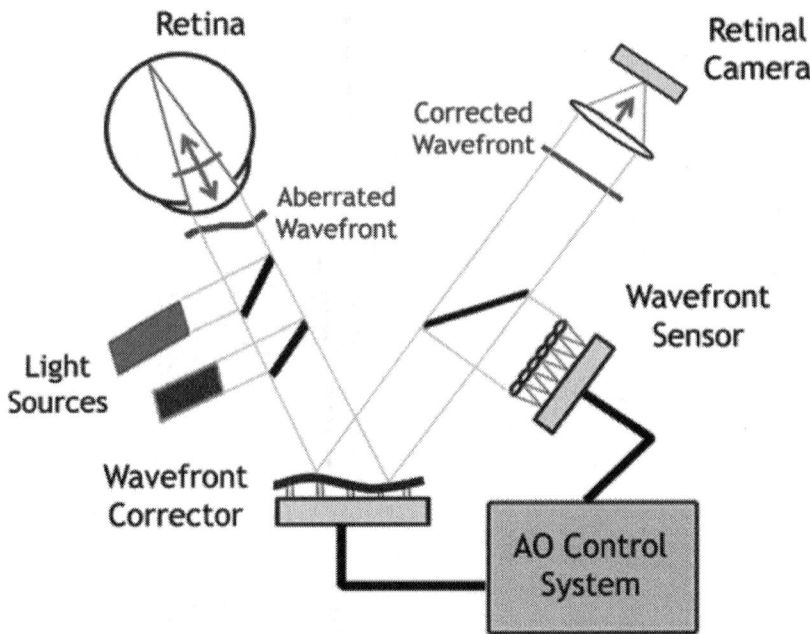

Figure 2. Basic layout of an adaptive optics system for retinal imaging. The system measures the ocular aberrations with a wavefront sensor and corrects for them with a wavefront corrector to achieve high lateral resolution imaging. Two light sources are generally used by an AO system: one is used to measure and correct the wavefront aberration of the eye; the second source is used to illuminate the retinal field being imaged. The AO compensated retinal image is captured by a high-resolution imaging camera.

The benefit of AO for noninvasive retinal imaging has been clearly shown with previous high-resolution images of cone photoreceptors in normal subjects [20] and with the discovery of differences in the pattern of the cone mosaic in colour blind eyes with respect to controls [23]. Since then, the use of AO retinal imaging to detect and monitor retinal abnormalities has clearly been shown in studies where high resolution images of the cone mosaic were acquired in normal eyes and eyes with retinal diseases [24]. The early diagnosis of retinal diseases and the monitoring of treatment efficacy at a cellular level provides promising clinical applications of AO technology.

Continuing efforts in this fascinating area of research now promote the design and development of innovative systems and devices for improving resolution and contrast of AO retinal images [21]. Although different methods have been proposed for measuring aberrations, the Hartmann-Shack wavefront sensor is still most widely used. Regarding the aberration correctors, many different devices have been proposed and demonstrated for ophthalmic applications. In this review, emphasis is placed on principles rather than details of individual instruments, most of which can be found in the specialized literature [25].

4. WAVEFRONT SENSING TECHNOLOGY

Several wavefront sensing techniques have been developed for the measurement of the wavefront error in human eyes. Most wavefront sensors are based on the same principle, which is an indirect measurement of the first or second derivatives of the wavefront at the location of the pupil plane of the eye and the reconstruction of the complete wavefront by integrating these derivatives. In general, the apparatus directs a small beam of light into the eye which then backscatters off the retina. The scattered light leaving the eye gets aberrated by the eye's optics before being recorded by the imaging components of the wavefront sensor.

The Shack-Hartmann (S-H) type wavefront sensor is the most common method for measuring ocular aberrations. The device was introduced in clinical ophthalmology in 1994 [26,27] and uses an array of micro-lenslets each of which observes the out-going beam at a different pupil location. The resulting light, aberrated by the eye's optics, is focused as an array of spots onto a detector. The best estimate of each spot's position and how far it deviates from a reference is calculated. The result is proportional to the average wavefront slope at that particular location of the pupil. Detailed information on the principle of S-H wavefront sensing can be found in previous work [25]. Limitations of the method include the quantity of light needed, the dynamic range and the fact that analysis of data from a S-H sensor does not consider the quality of the individual spots formed by the lenslet array [25]. However, experience has shown that the quality of spot images can vary greatly over the pupil of a human eye. If the wavefront shape within a single lenslet varies significantly, the spot pattern formed by that lenslet will be blurred making it more difficult to estimate its location. In the context of wavefront sensing, the term dynamic range represents the maximum wavefront slope that can be reliably measured. The dynamic range of the S-H sensor is strictly related to the optical parameters of the S-H microlenses: the lenslet spacing (or number of lenslet across the pupil) and the focal length of the lenslet array. Several methods have been used to increase the dynamic range of the S-H sensor, including a precompensation of LOA, the use of a larger lenslet diameter and/or a shorter focal length lenslet, restriction of the illuminating beam diameter, the magnification of the pupil at the lenslet array etc.[25,28–32].

Due to the increasing interest and application of wavefront sensing techniques in the ophthalmic community in recent years, innovative methodologies have been designed and developed. Curvature sensing, pyramid sensing and interferometry are the most promising emerging wavefront sensing techniques.

Curvature sensing technology is based on phase-diversity. It depends on comparisons between phases in adjacent areas in the image and objective plane of an optical system [33–37]. Phase diversity is normally implemented by recording two images at different focal planes, and then reconstructing the wavefront from these images. Aberrations in the wavefront in the measurement plane will alter the local intensities of the wavefront as it propagates. A convex distortion will cause the wavefront to converge and hence become more intense, the contrary will be the case for a concave distortion. The change in intensity is a measure of the local wavefront curvature and may be used to reconstruct the wavefront. The Curvature Sensor (C-S) was originally developed for astronomical observation based on the technique developed by Roddier in 1988 [38]. His original idea was to directly couple a curvature sensor element and a bimorph deformable mirror, without the need for intermediate calculations, in an AO astronomical telescope. Experimental work by Diaz-Douton et al.[39] has demonstrated the feasibility of a curvature sensor for ocular wavefront measurement. The advantages of the C-S are its relatively high dynamic range and low cost, while the disadvantages are related to the prolonged time of computing and the fact that a large defocus is needed to measure the wavefront with higher resolution, thus reducing the sensitivity of the sensor. This means that the C-S might not be as accurate as expected for measuring HOA. On the other hand, this problem could be overcome by designing a C-S in which the defocusing distance can be adjusted to the individual eye [25].

The pyramid sensor (P-S) was developed by Ragazzoni and implemented for the first time in an astronomical AO system [40,41]. In this system, a transparent pyramid splits the stellar image into four parts. Each beam forms an image of the telescope pupil on the same detector. A four-faceted refractive pyramid is placed in the optical path with its apex aligned to the optical axis facing the incoming beam. The wavefront gradients along two orthogonal directions are retrieved from the intensity distribution among the four pupil images resulting from this operation. The first application of a P-S sensor in ophthalmology was demonstrated by Iglesias et al.[42] in 2002, who used this method to measure the WA of artificial and normal eyes. They also pointed out the necessity to deal with spurious reflections from the anterior cornea. Chamot et al.[43] developed an ophthalmic AO system with a P-S in the measurement arm of the device and a piezoelectric deformable mirror (DM) as wavefront corrector element. The DM was optically conjugated to a steering mirror positioned in the wavefront sensor arm immediately before the tip of the diffractive pyramid element; this modulates the position of the image on the pyramid tip in a circular manner and facilitates quantitative measurement of the wavefront slopes. Tests were performed in artificial and human eyes achieving results similar to those obtained by other AO systems implemented with a S-H sensor, further demonstrating the feasibility and accuracy of the P-S sensing technique in an AO system for ophthalmic applications. One of the main advantages of a P-S is the easy adaptability of the system to the variations in the range of the aberrations one can expect in the human eye optics. The dynamic range of the sensor can be therefore quickly modified. Moreover, at small modulation amplitudes the sensitivity of a pyramidal sensor can be higher than that of a S-H sensor.

A variety of interferometry techniques have been suggested for ocular wavefront sensing, including shearing interferometry [44–49] and Talbot interferometry [50–52]. The common advantages of all the interferometric techniques is their relatively simple and inexpensive design if compared to other opto-electronic systems, as well as their accuracy, high spatial resolution and large dynamic range. The disadvantages are the sensitivity to vibration, changes in polarization of the beam coming back out of the eye and the complex reconstruction of the phase error. All these factors limit widespread application of this technology to human eyes. Among the interferometric techniques, Talbot interferometry has gained the most interest for application in vision science [25]. It is constructed with two gratings in which moiré fringes are generated by superimposing the Fourier image of the first grating on the second. The two gratings should have the same period. If the phase object is placed in front of the first grating, the light deflected by the object yields the shifted Fourier images and the resultant moiré fringes show the deflection mapping [50]. On the other hand, only one periodic grating can be used for phase distortion analysis, by exploiting the Talbot effect or self-imaging phenomenon [52]. The Talbot image can be directly detected by a detector placed at the Talbot distance from the periodic pattern, as described in previous work [25]. Distortion of the fringe pattern reflects the local tilt of the wavefront. This pattern can be observed only at a very short distance from the grating due to diffraction and the pattern disappears as distance increases. Diffraction patterns can be observed again at specific periodic distances from the grating. These are called Talbot images. Sekine et al.[53] used a two-dimensional grating for sensing the optical wavefront with the CCD placed in the plane of the Talbot image of the first order to maximize the contrast of the grating image. They obtained Talbot images from both artificial and human eyes and were able to successfully reconstruct wavefront shapes, with no discernible differences in comparison with those obtained by a S-H sensor. Warden et al.[54] demonstrated accurate results using a Talbot wavefront sensor which has recently become available commercially. A series of measurements was taken in model eyes demonstrating the high accuracy of the Talbot sensor in comparison with two commercially available S-H sensors, especially for HOA.

5. WAVEFRONT CORRECTING TECHNOLOGY

Wavefront correction has been performed by different techniques, depending on the source of aberrations, the field of application, and the instrumentation available. The correctors fall within two broad categories: (1) piston segmented devices [e.g., piston-tip-tilt, liquid crystal spatial light modulators (LC-SLM) and segmented mirrors] and (2) continuous surface mirrors (e.g., including deformable mirrors, membrane micro-mirrors, and bimorphs). The correction methods range from phase conjugation and computer-generated holograms to deformable mirrors, which nowadays are the most popular devices for adaptive optics.

Wavefront phase errors can be corrected by introducing an optical path difference in the beam by either varying the refractive index of the phase corrector (refractive devices) or by introducing a variable geometrical path difference (reflective devices i.e., deformable mirrors). The most common refractive devices use liquid crystals, the refractive index of which can be electrically controlled. Liquid crystal (LC) molecules are elongated and are polarised such that intermolecular forces between the crystals keep them relatively aligned. The basic form of a LC device consists of a layer of LC sandwiched between two pieces of optical quality glass. The orientation of the LCs is fixed by cutting microscopic grooves on the inner surfaces of the glass plates; the LCs line up in the direction of the grooves. Transparent electrodes are deposited on the glass surfaces, and they may be as small as 10 μm. LC devices can therefore provide many thousands of correction elements. These devices offer piston-like correction since there is practically no continuity requirement for the refractive index between pixels. They come in different varieties: (1) ferroelectric devices can produce phase changes of either 0 or pi radians and can operate at high frequencies; (2) nematic devices can provide continuous phase changes but are slower than ferroelectric devices. The first use of LC displays (LCDs) for dynamic ocular wavefront correction was by Prieto et al.[55]. More recently, ocular wavefront correction has been demonstrated using Liquid Crystal on Silicon devices (LCoS) [56]. In this device the LC is deposited on an array of silicon pixels and operated in reflective mode. The voltage applied to the pixels controls the refractive index, and the advantage is that higher speeds can be obtained. LC devices require the use of linearly polarised light [56]; Kong et al.[57], however, recently described an open loop ophthalmic AO system in which the uncorrected polarisation component is used by the wavefront sensor. A limitation of LC devices is the relatively limited dynamic range when using the device to correct ocular aberrations.

Deformable mirrors (DM) consist of a reflective faceplate acted on by a set of actuators. The faceplate can be continuous or segmented. While it is easier to manufacture segmented mirrors with a very large number of actuators, they have the disadvantage that the gaps between segments introduce greater spurious effects in the image (due to diffraction) than continuous mirrors. Many different types of actuators have been used and most of them have been applied to ocular AO systems [58]. The most common types of actuators are piezoelectric, electrostatic or magnetic.

In piezoelectric materials, an applied electric field gives rise to a change in shape. The most commonly used material is lead zirconate titanate (PZT). These mirrors were originally developed for use in astronomical AO [59] and were used in the first ophthalmic AO system [20]. They can be made to feature quite a large stroke, a linear actuation-to-voltage response, and low amounts of hysteresis. The typical size of a PZT actuator is 25 mm, so deformable mirrors using these actuators tend to be relatively large. Bimorph mirrors consist of two bonded piezoelectric ceramic wafers that are oppositely polarised parallel to their axis [60]. An array of electrodes is deposited between the wafers; applying a voltage to an electrode results in one wafer expanding locally and laterally while the other wafer contracts inducing a spherical bending. These mirrors are a natural choice for AO systems using a curvature wavefront sensor. However they can also be used with Shack-Hartmann (or any other) wavefront sensors, and were used in this way in an ocular AO system [61].

Electrostatic actuators are usually used in membrane-type mirrors or micromirrors [62]. Membrane mirrors consist of a reflective membrane which is deformed by means of electrostatic forces due to an array of electrodes placed a small distance behind the membrane. The local membrane curvature is proportional to the square of the signal voltage. The influence function of the actuators, i.e., the mirror shape when a single actuator has a voltage applied, is broader than those of other mirrors. They are very suitable for low-order correction, and the introduction of commercial models made low-cost adaptive optics possible for the first time. The stroke is limited by the fact that the actuators operate in 'pull-only' fashion in the usual configuration. Bonora and Poletto [63] introduced a 'push-pull' version having a transparent electrode in front of the mirror in order to extend the stroke. In another interesting development, Bonora et al.[64] proposed a "photocontrolled deformable mirror" (PCDM) in which a LCD is used to generate a light distribution on a photoconductive layer placed behind a membrane mirror. The control is therefore by light, and can be very high order or reduced to the order required. Electrostatic actuators are also applied in micromirrors (MEMS). In MEMS the electrodes do not act directly on the membrane, but rather on an intermediate membrane [65]. The mirror is attached to this membrane through posts, and the result is to localise the actuator influence functions. This can therefore provide higher order correction. Currently, there are MEMS by Boston Micromachines (Cambridge, MA, USA) with 32, 140 and 1020 actuators and the actuator size ranges from 300 to 500 μm and the corresponding stroke is from 1.5 to 5.5 μm or by IRIS AO, Inc. (Berkeley, CA, USA), with 111 to 489 actuators with inscribed apertures from 3.5 to 7.7 mm respectively and stroke from 5 to 8 μm.

Magnetic mirrors usually use voice coils to act on small magnets attached to the rear of a membrane mirror. These devices are therefore controlled by current rather than voltage. The main advantage is that a large stroke can be achieved, much larger than for other devices. A 52-Element device (Imagine Eyes, Orsay, France) was demonstrated in an ophthalmic AO system by Fernandez et al.[66] and more recently by Lombardo et al.[67]. The temporal response of this kind of mirror can be an issue, but it can be taken into account in the control system [68]. The ALPAO company (Montbonnot St. Martin, France) manufactures different versions of magnetic mirror having 37 to 277 actuators with actuator spacing of 1.5 or 2.5 mm and overall stroke up to 60 μm. They claim settling times of order 1 ms for their "hi-speed" range.

A novel magnetic fluid mirror is receiving attention [69]. In this device the surface of a magnetic fluid is coated with a reflective film and acted on by coils placed under the fluid volume. The deformation depends on the square of the magnetic field, but can be linearised by the addition of a strong, uniform magnetic field. The stroke can be large, and the response can be made fast by using high-viscosity ferro-fluids. However, these liquid mirrors must conserve volume which means that superposition of influence functions is not possible. This and the fact that they can evaporate are the main disadvantages of liquid mirrors. A liquid mirror which is electrostatically deformed has also been demonstrated recently [70].

The performance of wavefront correctors depends on a number of parameters, including the number and configuration of the actuators, the stroke, temporal response, linearity, hysteresis etc. Devaney et al.[71] compared eight different commercially available mirrors for correcting both ocular and atmospheric wave aberration. The sample included piezoelectric, membrane, bimorph and magnetic mirrors. Influence functions of all the mirrors were measured using interferometry. Wavefronts were simulated to have statistics corresponding to either ocular or atmospheric aberrations, with the correction achieved for each mirror determined by least-squares fitting the influence functions of each mirror to the wavefront. The number of mirror modes corrected and the size of the optical pupil projected on the actuator geometry were optimized for each device. In general, it was found that better correction can be obtained when there is a ring of actuators just outside the pupil. The optimal number of modes to correct depends on the mirror stroke and geometry. It was found that the mirrors with higher stroke (the magnetic and bimorph devices) should provide the best performance in terms of residual root-mean-square (RMS) wavefront error or Strehl ratio.

6. APPLICATIONS OF ADAPTIVE OPTICS RETINAL IMAGING

Adaptive optics retinal imaging has demonstrated the capability to image the living human retina at microscopic resolution. The improvement in retinal image contrast and resolution allows the direct observation of retinal microstructures giving the researcher the opportunity to analyze their integrity and/or pathological abnormalities. An AO ophthalmoscope provides en face images of the retinal layers showing photoreceptors, retinal vessels and nerve fiber bundles (Figure 3). Advances in image processing methods are required in order to maximize the value of the retinal data acquired and to make the AO retinal imaging technology accessible to the clinical ophthalmic community. Accurate automated routines are indeed mandatory when large quantities of data need to be analyzed.

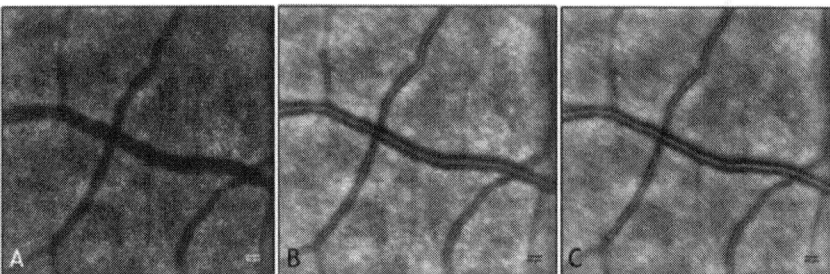

Figure 3. AO retinal images with adaptive compensation of the photoreceptor layer (**A**), vasculature (**B**) and retinal nerve fiber layer (**C**). Scale bars represent 50 µm. The blood vessels form a three-dimensional network across the inner retinal layers. All the images shown in this review have been acquired using a flood-illumination AO retinal camera (rtx1, Imagine Eyes, Orsay, France).

Currently, the majority of studies have focused on the generation and analysis of AO images of photoreceptor cells, including cones and rods [67,72–75]. Over the last few years, efforts have been made to develop reliable methods to measure the cone density as a function of retinal eccentricity in populations of

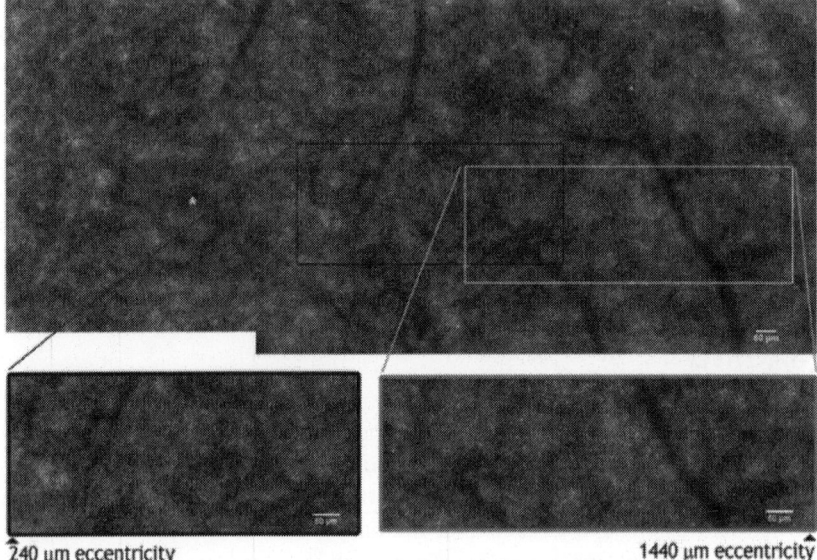

240 μm eccentricity   1440 μm eccentricity

Figure 4.AO montage of the photoreceptor mosaic in a 46 year old subject. The asterisk shows the foveal center. The black and grey boxes enclose two high-magnification images of the photoreceptor layer from 240–1,440 μm eccentricities from the foveal center. The center-to-center distance between cones has been shown to increase with greater eccentricities from the fovea; the cone density, accordingly, declines with increasing eccentricity from the fovea. Scale bars represent 50 μm.

healthy adults [76–81]. In general, data on populations of healthy eyes are fundamental for characterizing the density and the spacing distribution as well as the brightness of healthy photoreceptor cells in vivo (Figure 4). This will allow measurement of the normal ranges which can be compared to pathological photoreceptors, even in the early stages of retinal diseases. An increasing number of studies [67,82–90] are showing the distribution of cone photoreceptor density and spacing in adults. The in vivo measurements of cone density [67,82–85] have shown good agreement with histologic data from cadaver eyes [86–90]. Using an AO scanning laser ophthalmoscope (AOSLO), researchers [82–85] found the cone density to drop on average from 120,000 cones/mm^2 at 0.1 mm to 20,000 cones/mm^2 at 1.0 mm from the foveal center. In general, the cone density values at the same eccentricities of the nasal and temporal retina within 2 mm eccentricity from the fovea were found to be within 10% of each other both in ex vivo and in vivo studies [84–87,91]; a 10% higher density of cones along the horizontal than the vertical meridian was, in general, found. A synopsis of the average cone density from previous AO studies in populations of healthy subjects is shown in Table 1. The parafoveal cone density showed a moderate to high inter-individual variation with coefficient of variation (defined as the ratio of the standard deviation to the mean) values ranging between 12% and 20% [82–86,91]. Discrepancies between studies could be due to different factors, such as the inclusion of subjects with different ages or eyes with different axial lengths and refractive corrections, the instrument used for biometry, the different model eye used to estimate the retinal image size, the use of foveal center or foveal fixation as reference point to define retinal eccentricities and the sampling window area used to count cones.

Table 1. Cone density estimates in histology and AO retinal imaging studies taken at increasing eccentricities from the foveal center.

Work	Subjects (N.); age (range, years); AxL (range, mm); sampling window area (µm or pixels); model eye	Cone density (average, cones/mm²) as a function of retinal eccentricity (range, µm) along the horizontal meridian			
		230–360 µm	400–540 µm	720–890 µm	1,000–1,350 µm
Curcio et al. [86]	7; 27–44 yrs; AxL not reported; variable sampling windows; anatomical schematic eye	Nasal/Temp: 60,000–55,000	Nasal/Temp: 40,000	Nasal/Temp: 26,000	Nasal/Temp: 20,000
Li et al. [83]	18; 23–43 yrs; 22.9–28.3 mm; adaptive windows (adjusted to contain 150 cones); Gullstrand schematic eye model	All meridians: 60,000–45,000	Not reported	Not reported	Not reported
Chui et al. [82]	11; 21–31 yrs; 22.8–27.5 mm; 150 × 150 pixels window; standard reduced eye model	Not reported	Nasal/Temp: 41,000	Nasal/Temp: 27,000	Nasal/Temp: 15,000
Chui et al. [84]	4; 24–54 yrs; AxL not reported; 22 × 22 µm window; model eye not reported	Not reported	Temp: 30,000	Not reported	Temp: 15,000
Song et al. [85]	10; 22–35 yrs; 22.1–26.1 mm; 50 × 50 µm window; Indiana model eye	Nasal: 59,700–50,000 Temp: 59,200–50,500	Nasal: 43,700–37,800 Temp: 41,200–37,300	Nasal: 29,100–24,200 Temp: 28,100–24,100	Nasal: 19,100–16,800 Temp: 19,900–16,300
Lombardo et al. [91]	12 (24 eyes); 24–38 yrs; 22.6–26.6 mm; 50 × 50 µm window; Gullstrand schematic eye model	Nasal/Temp: 49,400	Nasal/Temp: 38,500	Nasal/Temp: 30,000	Not reported

AO retinal imaging in healthy subjects also makes it possible to better understand the sampling limit of resolution of the cone mosaic in vivo. Analysis of the spatial distribution of the cone photoreceptors provides new information on the physical aspects of visual sampling of the human eye. In a recent work, our group [91] found that the mean Nyquist limit sampling of resolution of the cone mosaic (N_c) was 33 ± 2 cycles/degree (c/deg) at 260 μm eccentricity, declining to 26 ± 2 c/deg at 600 μm eccentricity in a population of twelve young adults (age range: 24–38 years; 24 eyes; axial length of the eye (AxL) range: 22.61–26.63 mm). Authors previously found comparable results for young adults [83,84,92]. The N_c was calculated to be 34 c/deg at 1° eccentricity (approximately 270 μm) by Chui et al.[84]. Coletta and Watson [92] estimated N_c to range between 50 and 42 c/deg at the fovea and between 24 to 22 c/deg at 4° eccentricity (approximately 1.1 mm) in a population of subjects with spherical equivalent error (SEr) ranging between 0 D and −14 D. In a population of 18 healthy subjects (age range: 23–43 years; 18 eyes: AxL 22.86–28.31 mm), Li et al.[83] found that cone density tended to decrease with increasing axial length at eccentricities between 100 and 300 μm from the foveal center, with no statistically significantly differences between emmetropes and myopes at the fovea. In a population of 11 healthy subjects (age range 21–31 years; 11 eyes), cone density in moderate myopes (up to−7.50 D) was found to be significantly lower than in emmetropes within 2.00 mm eccentricity from the fovea [84]. The spatial vision of the cone mosaic reduces with increasing axial length [91]: the lower Nyquist limit sampling of resolution of the cone mosaic in myopes than emmetropes has been postulated to be caused by retinal stretching, occurring at the posterior pole of myopic eyes, due to the eye's increased axial length [84,91,92]. A higher difference between myopes and emmetropes has been demonstrated when the acuity limits were expressed in retinal units (c/mm) rather than in angular units (c/deg). The discrepancy between resolution and N_c expressed in retinal acuity was primarily considered a result of psychophysical and neural factors, including the neural sampling rate (i.e., the retinal ganglion cells receptive field density) [93–95].

AO retinal imaging has also been used to study the waveguide and reflectance properties of cone photoreceptors in vivo[96–104]. Variation of brightness between adjacent cones was observed, even when imaging the retina of healthy subjects. Differences in reflectance between adjacent cones were seen both at the boundaries of retinal vessels and in areas devoid of vessels. While intra-retinal scattering was considered as the primary source of the higher reflectance of cones that reside beneath the vessels in comparison with adjacent cones, this phenomenon could not explain the variation in brightness between adjacent cones in areas devoid of vessels. Various hypotheses have been made to find the possible factors that influence the intensity variation of the cone mosaic: (1) the differences in reflectivity could be caused by molecular differences within the cones that are due to phototransduction [96]; (2) the reflectance variation could be based on the cone outer segment length [97–100,102–104]: fluctuations in reflectivity indeed could be due to changes in the outer segment length, related to disk shedding. (3) The cause of the variation could also be related to the pointing direction of cones and the angle of incidence of the light entering the pupil [105–108]. (4) Finally, technical factors should be considered to contribute to differences in reflectance between adjacent areas of cones, including the light source (i.e., laser or SLED emitter) and/or the wavelength of light used to illuminate the retina, the axial resolution of the imaging system (confocal or flood-illumination) and the imaging process [100,101,104]. The relative contribution of the sources of reflection within the cones was shown to be not constant with time [98,100,102,103] and the photopigment density has been found to decrease with increasing eccentricity from the fovea up to 2° eccentricity [109].

7. ADAPTIVE OPTICS RETINAL IMAGING OF RETINAL DISEASES

In addition to the above basic applications, ophthalmic AO systems are translating into clinical applications that are rapidly expanding. It is expected that AO technology will soon translate into clinical applications. The most promising application of AO retinal imaging is the detection of early signs of retinal diseases. AO retinal imaging indeed provided new information about the pathological changes of the retinal microstructure in several diseases, including inherited or acquired neuro-retinal degenerations [110–112] and vaso-occlusive diseases [113,114]. Because the rate of disease progression is typically slow for most retinal degenerations, it is estimated that patients show a clinical loss of visual function years after the onset of the disease. If AO imaging can provide high-resolution measurements of the cone photoreceptor structure, micro-vasculature and nerve fiber bundles in patients, then it may open a new frontier for the early diagnosis of retinal diseases and for monitoring the efficacy of therapies at a cellular level. Researchers [7,85,110] have demonstrated that AO devices allow imaging of exactly the same retinal area over days or months in order to follow disease progression with microscopic accuracy.

Across the developed world, the major causes of vision loss are attributed to diabetic retinopathy (DR), age-related macular degeneration (AMD) and glaucoma [5]. Diagnosis usually occurs once the damage has already happened to some extent, because of the relatively low resolution of current retinal imaging techniques when used to detect abnormalities of the retinal microstructures and the relatively poor sensitivity of functional testing. Considering that structural damage of these microstructures precedes their functional impairment, the detection of pathological variations of photoreceptors, capillaries and nerve fiber bundles at the pre-clinical stage of the disease can be beneficial for early treatment and avoidance of serious visual loss. Furthermore, a better understanding of the microscopic alterations in retinal tissue may provide further insight into the mechanisms of disease progression and be helpful to identify new approaches for therapeutic intervention.

Diabetic retinopathy (DR) is a frequently occurring complication of diabetes mellitus, that is a metabolic disease in which the patient has a high serum glucose level. According to the World Health Organization, diabetes mellitus is responsible for about 12% of new cases of blindness between the ages of 45 and 74 years in the developed world. With the incidence of diabetes throughout the world projected to rise from 150 million to approximately 300 million by the year 2025, DR represents a major threat to the global population and will likely present ever-increasing burdens on the health care delivery system [115,116]. DR can be classified as non proliferative (NPDR) or proliferative (PDR). NPDR is further graded as mild, moderate and severe according to the Early Treatment Diabetic Retinopathy Study (ETDRS) severity scale [117].

The earliest clinical pathological changes of DR occur in the microvascular structures [118,119]. According to the most commonly accepted patho-physiological model (i.e., the microvascular theory) [120], DR consists of a microangiopathy that induces pathological changes of the vascular structures and the blood rheological properties as a consequence of chronic hyperglycaemia. It has been also postulated that DR is a multifactorial disease involving the retinal neuronal cells (neurodegenerative theory) [121–124]. The neurodegenerative changes are apoptosis of several populations of retinal cells, including photoreceptors, bipolar and ganglion cells. The functional and structural impairment of neural tissue has been theorized to participate in the generation of the earliest morphological alterations of the vascular structures [121–125]. The question of whether the photoreceptor loss could be initially determined by a neuronal or vascular breakdown during diabetes mellitus remains unsolved. Work is needed to understand whether the neurodegenerative changes of retinal cells may precede the vascular damages or the two processes are simultaneous. The exact nature of their interdependence is complex and still not known.

Noninvasive detection of the pathological signs of DR is usually performed by dilated fundus examination, colour fundus photography and more sophisticated retinal imaging techniques, such as Spectral Domain Optical Coherence Tomography (SD-OCT). Fluorescein angiography (FA) can be useful to assess the integrity of the blood retinal barrier as the amount of fluorescein leakage is related to the dysfunction of the retinal vascular endothelium. Even if FA represents an important diagnostic tool and improves the accuracy of laser treatment of DR complications (e.g., diabetic macular oedema and new vitreo-retinal vessels) [126,127], on the other hand, it requires injection of a fluorescent dye agent that can lead to unintended systemic complications. AO retinal imaging could be useful to detect pathological changes early in the course of DR, such as microaneurysms, micro-haemorrhages and loss of photoreceptors [113,114,128–130] (Figure 5). The detection of the early retinal changes related to DR might be important in the management of the patient requiring a better glyco-metabolic control. Authors [128,129] described image processing and analysis algorithms to extract the capillary vessel information and provided excellent visualization of the parafoveal capillary network in healthy eyes using AOSLO devices. In a recent study [114], the capillary network was evaluated in patients with type 2 diabetes. Researchers found a higher capillary dropout and a higher tortuosity of the arterovenous channels in patients with diabetes and no diabetic retinopathy than in healthy controls. Tam et al.[113] analyzed the retinal microvasculature in a patient with type 1 diabetes and severe NPDR over a 16 months follow-up period. Longitudinal assessment of the capillaries showed microaneurysm formation and disappearance as well as the formation of tiny capillary bends similar in appearance to intraretinal microvascular abnormalities. In vivo imaging of the capillary network has been also shown using an AO-SD-OCT [131–133]. The 3D information provided by OCT represents a major advantage compared to en face imaging techniques. The AO-SD-OCT can provide a theoretical spot volume of 3 μm^3, capable of reconstructing the entire retinal capillary network of the inner retina. On the other hand, OCT cannot distinguish the lumen from the vascular walls of larger vessels [134]. In a preliminary investigation on large retinal arterioles in patients with type 1 diabetes and NPDR, we acquired images of the vessel walls using an AO flood-illumination retinal camera (Figure 6 and Movie 1 (movie showing a retinal artery, as seen in Figure 6). The blood vessel shows periodic twitching, probably corresponding to the cardiac cycle of the patient, as previously shown by Zhong et al.[135], showing the capability of AO ophthalmoscopy to evaluate them for monitoring both arterial lumen and walls in vaso-occlusive retinal diseases. Both axial and en face AO imaging techniques could therefore be complementary to noninvasively analyze the retinal vessels at high resolution. The detection of pre-clinical abnormalities of retinal microcirculation in patients with diabetes could represent a valuable advantage of AO retinal imaging in comparison with current noninvasive imaging modalities.

Non-necrotic photoreceptor loss, in addition to microangiopathy, has been considered to be responsible for the vision loss associated to DR [121–124]; on the other hand, a loss of cone photoreceptors has never been clinically demonstrated. An objective of our current work is to evaluate the photoreceptor mosaic geometry and reflectance: the procedure has the potential to provide additional and valuable information about the cellular changes of retinal pathologies. In a recent study [136] we found that, in a population of eleven patients with type 1 diabetes, the parafoveal cone photoreceptors showed a higher variation in intensity than in healthy controls at the same retinal eccentricity (Figure 7). This phenomenon was particularly evident near the areas of intraretinal focal oedema (Figure 8). The regional differences in image intensity with areas of cones brighter than adjacent areas were also previously seen in other retinal diseases, including cone-rod dystrophy [110]. The significance of this phenomenon is not yet clear. In general, there are various hypotheses on the intensity variation of cones in AO imaging, as discussed in the previous section [97–104]. Fluctuations in reflectivity may be caused by molecular changes within the cones that are due to phototransduction or to changes in outer segment length, related to disk shedding. It has been also theorized that areas of cones darker than adjacent cones may reflect a disruption of the waveguide properties of the cones themselves [105,109] or to light interactions between the end of the OS tip and the pigments in macrophages or retinal pigment epithelial (RPE) cells [102,103]. Technical factors, however, should be considered to contribute to the spatial variation of reflectance between adjacent areas of cones, as previously discussed [100,102,104].

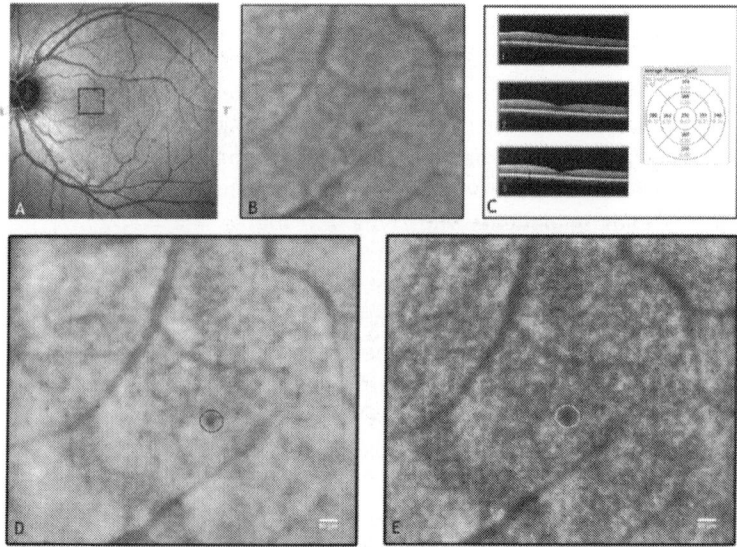

Figure 5. (A) Wide-field SLO image of the left eye in a 38-year old female patient with type 1 diabetes and mild non proliferative diabetic retinopathy. **(B)** High-magnification SLO image of a region of interest (encircled in A, 0.88 × 1.53 mm) showing a micro-haemorrhage, though with low resolution. **(C)** The SD-OCT horizontal scans of the central retina, corresponding to the green lines superimposed to the SLO image in A show a preserved retinal microstructure. The retinal thickness map is also shown. **(D and E)** The same region of interest imaged by an AO system with adaptive compensation of the vessels and photoreceptor layer, respectively. Scale bars represent 50 μm. In **(D)**, the small arterioles and the capillaries create a web between larger vessels; the spot haemorrhage (encircled) shows distinct margins. In **(E)**, the AO image of the cone mosaic. The haemorrhage projects a dense shadow, with defocused margins, onto the mosaic completely masking the underlying photoreceptors.

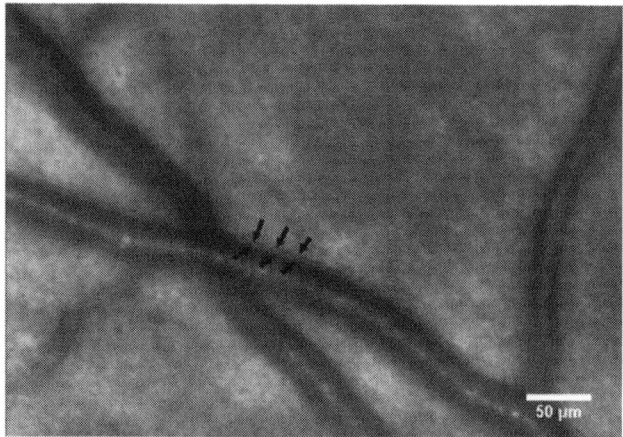

Figure 6. AO image showing a retinal artery. In en face retinal imaging, the lumen of a blood vessel appears as a streak of variable diameter and morphology, depending on the vessel order; however, it always shows the same pattern, consisting of a central high-intensity channel and two peripheral darker channels (likely due to the curved vessel wall). The artery wall appears as a grey line outside the peripheral lumen vessel. The arrows define the inner and outer wall borders. Scale bar: 50 µm.

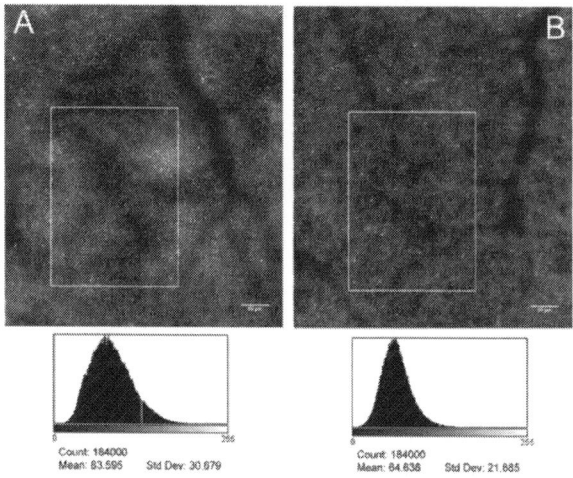

Figure 7. AO images, both centered at coordinates: x = 1.5° temporal and y = 1.8° superior from the fovea, of the photoreceptor mosaic in the left eye of a patient with type 1 diabetes without clinical signs of diabetic retinopathy (**A**, 36 years old female, noDR eye) and a healthy control (**B**, 37 years old female). Scale bars represent 50 µm. In the lower row, the image histograms (x-axis: 0–255 grey level; y-axis: pixel intensity level) of the selected areas are shown: both a higher average and a higher standard variation of the intensity level distribution of pixels was found in the noDR eye than in control. Biological and technical factors can contribute to variations in brightness between adjacent domains of photoreceptors.

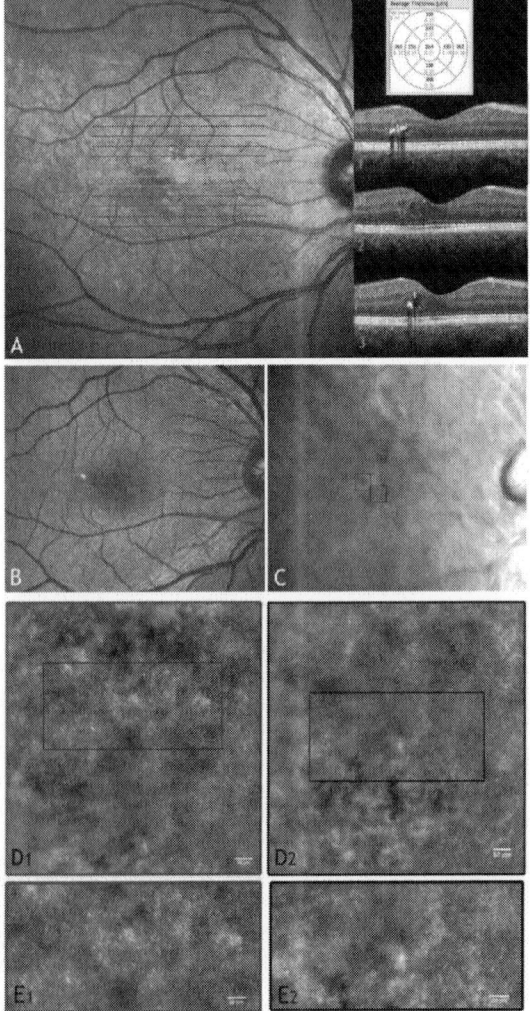

Figure 8. (**A**) Wide-field SLO image and OCT scans of the right eye in a 49 year old patient with mild non proliferative diabetic retinopathy showing hard exudates and a focal macular oedema. (**B** and **C**) Wide-field digital images (retromode modality by F10, Nidek, Japan) of the posterior pole showing the locations of the retinal exudate and the micro-cystic oedema (black box), respectively. (**D1** and **D2**) Adaptive optics images of the photoreceptor layer acquired within the regions of interest enclosed in C (scale bars represent 50 µm). In panel D2, cones are highly resolved only in part probably due to increased scattering from oedematous inner retinal layers. (**E1** and **E2**) High-magnification images of the photoreceptor layer shown in D1 and D2 respectively (scale bars: 50 µm). High variations in brightness between adjacent domains of photoreceptors can be seen in regions close to retinal oedema (E1). Intraretinal oedema reduces high-resolution imaging of photoreceptors (E2).

Age Related Macular Degeneration (AMD) is the leading cause of blindness in the elderly across the developed world. It represents a deterioration of the macular area and usually affects older adults. A recent study found that the late stage AMD (the most disabling form of the pathology) is present in approximately 5% of the over 65 's and 12% of the over 80 's [137]. AMD is a multifactorial disease, involving ocular, systemic and genetic risk factors. The ocular risk factors include darker iris pigmentation and hyperopic refraction, while systemic risk factors include cigarette smoking, obesity, sunlight exposure and cardiovascular diseases [138,139]. Genes influence several biological pathways related to AMD, including the immune processes, mechanisms involving collagen and glycosaminoglycans synthesis and angiogenesis. All these factors have been associated with the onset, progression and bilateral involvement of early, intermediate, and advanced states of AMD [140–146]. Genetic susceptibility can be influenced by the environmental factors: taken together, both factors are highly predictive of the onset, progression and response to treatments [147]. Several patho-biological pathways have been implicated in the pathogenesis of AMD: these include senescence, shown by a lipofuscin accumulation in the RPE cells, choroidal ischemia and oxidative damage [148,149].

There are two clinical types of AMD, the "dry" and "wet" form. In the early stages of AMD, which is asymptomatic, insoluble extracellular aggregates called drusen accumulate in the retina. The late stage of dry AMD, which is also known as geographic atrophy (GA), is characterized by scattered or confluent areas of degeneration of RPE cells and the overlying retinal photoreceptors, which rely on the RPE for trophic support. The other late stage form of AMD, the wet form (10–15%), is typified by choroidal neo-vascularization (CNV), where newly immature blood vessels grow toward the outer retina from the underlying choroid leaking fluid below or within the retina [150,151].

Retinal imaging for the management of AMD includes fluorescein angiography. In the wet AMD form, leakage of dye (hyperfluorescence) is noted and classified by location (subfoveal, juxtafoveal, or extrafoveal) and by type (classic, occult, or mixed). Indocyanine green angiography (ICG) uses an intravenous dye with different characteristics from fluorescein (e.g., less melanin absorbance). It improves identification and characterisation of neovascular variants of AMD (e.g., the polypoidal choroidal vasculopathy). SD-OCT enables high-resolution in vivo cross-sectional or volumetric tomographic visualisation of the retinal micro-architecture. It allows visualisation of the cross-sectional outline of the neovascular choroidal complex, but its internal structure cannot be well resolved and the neovascular components cannot be distinguished from the fibrous components, haemorrhages or dense exudates within the lesion. With the advent of the anti-VEGF therapy, SD-OCT imaging is widely used for the early diagnosis of CNV and for the treatment and re-treatment management.

In the early, asymptomatic, AMD stages (i.e., presence of drusen), the ability to predict the rate of progression is currently limited. By monitoring drusen over time, en face AO imaging (Figure 9), also combined with AO-SD-OCT imaging, can theoretically detect both their progression, in terms of size, and analyze their direct effect on the overlying photoreceptor mosaic [152–154]. Godara et al.[152] showed, for a 45-year old female, that both the cone density and cone arrangement of the mosaic overlying the drusen were within normal limits. AO imaging revealed a regular photoreceptor mosaic with areas of hyper-reflectivity coinciding with the location of the drusen. The increased reflectivity associated with the drusen could be attributed to increased scatter from the RPE (due to decreased melanin or accumulation of some other waste material), to loss of outer segment pigment or loss of the photoreceptor outer segment. Boretsky et al.[154] identified several small drusen deposits that were not observed with standard wide field imaging techniques in early AMD. They also investigated large coalescent drusen and areas of geographic atrophy in advanced stage dry AMD, showing significant decrease in visible photoreceptor density. A sensitive, non-invasive, imaging tool could help to better recognize the earliest retinal changes and to identify patients who could progress rapidly and may benefit from a more intensive observation and management. Furthermore in the near future, an early diagnosis of the macular disease could have an important role in the evaluation of the effectiveness of new prevention strategies.

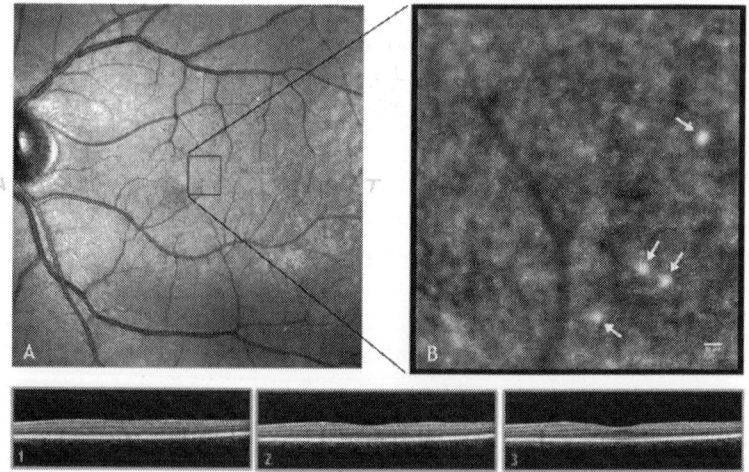

Figure 9. (A) Wide-field SLO image of the left eye in a 49 year old male showing hard drusen located near the fovea. The SD-OCT horizontal scans (1, 2 and 3) of both the photoreceptor and retinal pigment epithelial (RPE) layers have almost normal appearances. **(B)** AO image of the region of interest showing drusen (white arrows). The cone photoreceptors above drusen are well resolved, showing a higher brightness than adjacent cones. Similar findings have been shown using AOSLO (References [152–154]). Scale bar: 50 μm.

Glaucoma is the leading cause of irreversible, preventable blindness worldwide [155]. It has been estimated that over 11 million glaucoma sufferers worldwide are bilaterally blind from the disease [156]. Primary open angle glaucoma (POAG) is a chronic disease characterized by progressive loss of retinal ganglion cells, usually associated with ocular hypertension, that leads to structural damage of the inner retinal layers, as shown by progressive regional or diffuse thinning of the retinal nerve fiber layer (RNFL) [157]. Axonal tissue loss in the RNFL has been reported to be one of the earliest detectable glaucomatous changes, preceding morphologic changes of the optic nerve head (ONH), followed by functional loss, as shown by progressive visual field (VF) defects. The temporal sequence of glaucomatous structural/functional damage suggests that looking for structural changes at the ONH/RNFL level should theoretically allow an earlier diagnosis than detection of functional defects [158,159].

Many imaging modalities have been used to analyze RNFL loss in glaucomatous eyes. Colour and red-free fundus photography [160] represent the standard approaches, but the changes of RNFL are not detectable until there is more than 50% nerve fiber loss. Scanning laser polarimetry, scanning laser ophthalmoscopy and SD-OCT are imaging modalities that allow a quantitative analysis of the ONH and RNFL. Commercially available OCT devices, however, cannot provide sufficiently clear images of individual nerve fiber bundles to identify the specific structural abnormality that underlies the pathogenesis of glaucoma [161,162]. Adding AO to imaging systems such as flood-illuminated ophthalmoscopes, SLO equipment or OCT has recently allowed researchers to identify individual nerve fiber bundles [163–165], providing high-resolution images of both the RNFL and the ONH (Figure 10). Takayama et al.[163], using an AOSLO, measured the individual nerve fiber bundles width in a population of twenty healthy adults. In all the eyes, the AOSLO images showed hyperreflective bundles, representing the nerve fiber bundles, in the RNFL. Dark lines among the hyperreflective bundles were considered to represent Müller cell septa. The width of nerve fiber bundles, at distances from the edge of the optic disc ranging between 1.00 and 6.00 mm, was 22 ± 6 µm. There were no significant differences among the bundle widths at these distances along the same meridian. The hyperreflective bundles on the temporal and nasal sides of the optic disc were, however, narrower (on average 20 µm width) than those above and below the optic disc (on average 30 µm width). In the central retina, the hyperreflective bundles nasal to the fovea were narrower than those above or below the fovea.

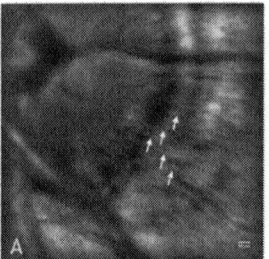 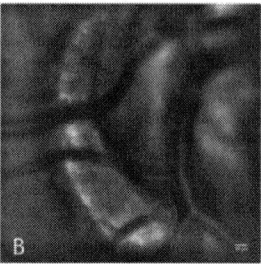 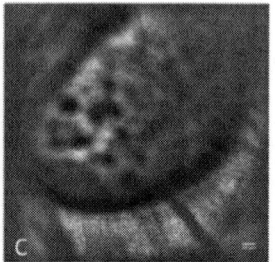

Figure 10. (A) The optic nerve head of the left eye in a healthy subject (26 years old female). The nerve fiber bundles can be highly resolved (some are indicated by white arrows) via AO imaging. In (**B**) and (**C**), the optic nerve head in two subjects suffering from advanced glaucoma showing a large papillary excavation. The nerve fiber bundles in advanced glaucoma cannot be resolved as usual in healthy eyes. In C, the lamina cribosa can be seen through the large papillary excavation. A, B and C: scale bars represent 50 μm.

Kocaoglu et al.[165] used an AO-OCT to obtain images of the RFNL in four healthy subjects. The imaging sessions were confined to three locations: retinal eccentricities of 6 degree superior and inferior to the fovea, and 3 degree nasal to the fovea. The authors showed that the nerve fiber bundles reflect noticeably more light than the surrounding tissue, a factor of approximately two times more. As they approach the fovea, the nerve fiber bundles become thin (both in width and depth) and separate. Bundles at 3 degrees demonstrate a larger aspect ratio (width to thickness) than those at 6 degrees with average width and thickness ranging from 30–50 μm and 10–15 μm, respectively.

Currently, there are no studies on glaucomatous eyes or eyes with ocular hypertension. AO imaging could indeed be useful to evaluate the morphological characteristics of the RNFL in patients with ocular hypertension in order to identify possible risk factors implicated in the ONH damage progression. AO could also help to recognize early glaucomatous damages and to identify patients who could progress rapidly and also may benefit from more intensive observation and management. Moreover, AO imaging could have an important role in the evaluation of neuroprotection strategies.

8. CONCLUSIONS

The optical system of the eye imposes a limit for retinal imaging. If the eye's wavefront aberrations are completely corrected across the pupil, significant improvement in the eye's optical quality is gained. Such improvement has been shown to be valuable in recent clinical applications of adaptive optics retinal imaging [110–114,152,154].

Over the last ten years, advances in understanding of the eye's aberrations have moved the wave aberration theory from an academic concept to an engineering level that is central to improving ophthalmic technologies. In the first experiment using a fundus camera equipped with adaptive optics, Liang et al.[20] were able to image individual cones in the living human retina. Since this seminal paper, adaptive optics for retinal imaging applications has entered the field of clinical research. The ability to image the photoreceptor layer, the retinal micro-vasculature and the nerve fiber bundles in vivo now provides the opportunity to better understand the pathological processes leading to visual impairment [72–90,129–131,163–170]. The future of AO retinal imaging promises early detection of degenerative retinal diseases and monitoring the efficacy of treatments at a cellular level. In retinal diseases, early detection and treatment are essential to prevent the occurrence of serious damage and visual loss. It has been shown that it is possible to take images, with cellular resolution, in exactly the same retinal area over days, months and years [85,110,171]. The ability to longitudinally track disease progression serves as the foundation for an imaging-based approach to track treatment response with greater sensitivity and on a much shorter time scale than current outcome measures such as visual acuity and visual field sensitivity can allow. As novel treatments to slow disease progression in both inherited and acquired retinal degenerations are developed, it will be critical to evaluate the effect treatments have on individual photoreceptor cells. It is expected that AO high-resolution imaging tools will allow clinicians to track retinal disease and the efficacy of therapy with great accuracy, helping to accelerate the search for new strategies of secondary prevention to avoid serious visual loss.

High cost and system complexity currently hinder the wide adoption of AO technology in clinical ophthalmology. Most AO retinal cameras have been designed and constructed for the best imaging performance possible, with the exclusion of all other factors, including size, cost, complexity, ease-of-use, time required to obtain, process, and analyze the retinal images, etc. This is delaying the transition of AO from the research lab to the clinic. Nevertheless, important progress has been made in this regard during recent years [6,172]: with reports of significant performance improvement of AO methods and systems for a growing number of ophthalmic applications, demand for a "commercial" AO instrument will increase and its cost should probably decrease. A compact, simplified AO instrument that can be used by ophthalmologists will facilitate the introduction of this technology into clinical practice and the development of new methodologies to detect and treat retinal diseases. Four companies have developed an AO prototype as a clinical viable tool at the time of this review (Boston Micromachines Corporation; Canon, Inc.; Imagine Eyes; and Physical Sciences, Inc.).

Another current drawback of AO imaging is the time required to obtain, process and analyze the retinal images: the continuous advances in the development of automated and reliable methods to evaluate the retinal microstructures, including cell photoreceptors, vessels and nerve fiber bundles [82–85,113,114,128,129,131,163–165] is however expected to resolve this issue soon. Accurate automated routines are indeed mandatory when large quantities of data need to be analyzed. A number of research groups are evaluating methods of analysis and interpretation of AO retinal images, such as cone density and spacing and packing regularity of the cone mosaic. Functional features of the cone mosaic can be used to capture additional information that cannot be described by the above metrics, such as the variation of cell brightness between adjacent domains of healthy and abnormal cones [153]. Continuing efforts to develop new image analysis metrics will increase the clinical utility of AO retinal imaging.

REFERENCES

1. Lombardo, M.; Lombardo, G. Wave aberration of human eyes and new descriptors of image optical quality of the eye. J. Cataract Refract. Surg.**2010**, 36, 313–331.

2. Thibos, L.N.; Hong, X.; Bradley, A.; Cheng, X. Statistical variation of aberration structure and image quality in a normal population of healthy eyes. JOSA A**2002**, 19, 2329–2348.

3. Charman, W.N.; Chateau, N. The prospects for super-acuity: Limits to visual performance after correction of monochromatic ocular aberration. Ophthalmic Physl. Opt.**2003**, 23, 479–493.

4. Guirao, A.; Porter, J.; Williams, D.R.; Cox, I.G. Calculated impact of higher-order monochromatic aberrations on retinal image quality in a population of human eyes. JOSA A**2002**, 19, 1–9.

5. Resnikoff, S.; Pascolini, D.; Etya'ale, D.; Kocur, I.; Pararajasegaram, R.; Pokharel, G.P.; Mariotti, S.P. Global data on visual impairment in the year 2002. Bull. WHO**2004**, 82, 844–851.

6. Williams, D.R. Imaging single cells in the living retina. Vis. Res.**2011**, 51, 1379–1396.

7. Godara, P.; Dubis, A.M.; Roorda, A.; Duncan, J.L.; Carroll, J. Adaptive optics retinal imaging: Emerging clinical applications. Optom. Vis. Sci.**2010**, 87, 930–941.

8. Williams, D.R.; Yoon, G.Y.; Porter, J.; Guirao, A.; Hofer, H.; Cox, I. Visual benefit of correcting higher order aberrations of the eye. J. Refract. Surg.**2000**, 16, S554–S559.

9. Castejon-Mochón, J.F.; López-Gil, N.; Benito, A.; Artal, P. Ocular wavefront aberration statistics in a normal young population. Vis. Res.**2002**, 42, 1611–1617.

10. Porter, J.; Guirao, A.; Cox, I.G.; Williams, D.R. Monochromatic aberrations of the human eye in a large population. JOSA A**2001**, 18, 1793–1803.

11. Thibos, L.N. The prospects for perfect vision. J. Refract. Surg.**2000**, 16, S540–S546.

12. Thibos, L.N.; Bradley, A.; Hong, X. A statistical model of the aberration structure of normal, well-corrected eyes. Ophthalmic Physl. Opt.**2002**, 22, 427–433.

13. Salmon, T.O.; Van de Pol, C. Normal-Eye zernike coefficients and root-mean-square wavefront errors. J. Cataract Refract. Surg.**2006**, 32, 2064–2074.

14. Guirao, A.; Porter, J.; Williams, D.R.; Cox, I.G. Calculated impact of higher-order monochromatic aberrations on retinal image quality in a population of human eyes. JOSA A**2002**, 19, 1–9.

15. Wang, Y.; Zhao, K.; Jin, Y.; Zuo, T. Changes of higher order aberration with various pupil sizes in myopic eyes. J. Refract. Surg.**2003**, 19, S270–S274.

16. Cheng, H.; Barnett, J.K.; Vilupuru, A.S.; Marsack, J.D.; Kasthurirangan, S.; Applegate, R.A.; Roorda, A. A population study on changes in wave aberrations with accommodation. J. Vis.**2004**, 4, 272–280.

17. Hofer, H.; Artal, P.; Singer, B.; Aragon, J.L.; Williams, D.R. Dynamics of the eye's wave aberration. JOSA A**2001**, 18, 497–506.

18. Li, K.; Yoon, G. Changes in aberration and retinal image quality due to tear film dynamics. Opt. Express**2006**, 14, 12552–12559.

19. Dreher, A.W.; Bille, J.F.; Weinreb, R.N. Active optical depth resolution improvement of the laser tomographic scanner. Appl. Opt.**1989**, 28, 804–808.

20. Liang, J.; Miller, D.T.; Williams, D.R. Supernormal and high-resolution retinal imaging through adaptive optics. JOSA A**1997**, 14, 2884–2892.

21. Dubra, A.; Sulai, Y. Reflective afocal broadband adaptive optics scanning ophthalmoscope. Biomed. Opt. Express**2011**, 2, 1757–1768.

22. Miller, D.T.; Kocaoglu, O.P.; Wang, Q.; Lee, S. Adaptive optics and the eye (super resolution OCT). Eye**2011**, 25, 321–330.

23. Carroll, J.; Neitz, M.; Hofer, H.; Neitz, J.; Williams, D.R. Functional photoreceptor loss revealed with adaptive optics: An alternate cause of color blindness. PNAS**2004**, 101, 8461–8466.

24. Godara, P.; Dubis, A.M.; Roorda, A.; Duncan, J.L.; Carroll, J. Adaptive optics retinal imaging: Emerging clinical applications. Optom. Vis. Sci.**2010**, 87, 930–941.

25. Lombardo, M.; Lombardo, G. New methods and techniques for sensing the wave aberration of human eyes. Clin. Exp. Optom.**2009**, 92, 176–186.

26. Liang, J.; Grimm, B.; Goelz, S.; Billie, J.F. Objective measurement of wave aberrations of the human eye with the use of a Hartmann-Shack wave-front sensor. JOSA A**1994**, 11, 1949–1957.

27. Platt, R.; Shack, R. History and principles of Shack-Hartmann wavefront sensing. J. Refract. Surg.**2001**, 17, S573–S577.

28. Pfund, J.; Lindlein, N.; Schwider, J. Dynamic range expansion of a shack-hartmann sensor by use of a modified unwrapping algorithm. Opt. Lett.**2000**, 39, 561–567.

29. Miller, J.M.; Anwaruddin, R.; Straub, J.; Schwiegerling, J. Higher order aberrations in normal, dilated, intraocular lens, and laser in situ keratomileusis corneas. J. Refract. Surg.**2002**, 18, S579–S583.

30. Kuroda, T.; Fujikado, T.; Maeda, N.; Oshika, T.; Hirohara, Y.; Mihashi, T. Wavefront analysis of higher-order aberrations in patients with cataract. J. Cataract Refract. Surg.**2002**, 28, 438–44.

31. Marsack, J.; Milner, T.; Rylander, G.; Leach, N.; Roorda, A. Applying wavefront sensors and corneal topography to keratoconus. Biomed. Sci. Instrum.**2002**, 38, 471–476.

32. Yoon, G.; Pantanelli, S.; Nagy, L.J. Large-Dynamic-Range shack-hartmann wavefront sensor for highly aberrated eyes. J. Biomed. Opt.**2006**, 11, 30502.

33. Gonsalves, R.A. Phase retrieval and diversity in adaptive optics. Opt. Eng.**1982**, 21, 829–832.

34. Fienup, J.R. Phase retrieval algorithms: A comparison. Appl. Opt.**1982**, 21, 2758–2769.

35. Teague, M.R. Deterministic phase retrieval: A Green's function solution. JOSA A**1983**, 73, 1434–1441.

36. Gureyev, T.E.; Roberts, A.; Nugent, A. Phase retrieval with the transport-of-intensity equation: Matrix solution with use of Zernike polynomials. JOSA A**1995**, 12, 1932–1941.

37. Gureyev, T.E.; Nugent, A. Phase retrieval with the transport-of-intensity equation. II. orthogonal series solution for nonuniform illumination. JOSA A**1996**, 13, 1670–1682.

38. Roddier, F. Curvature sensing and compensation: A new concept in adaptive optics. Appl. Opt.**1988**, 27, 1223–1225.

39. Diaz-Douton, F.; Pujol, J.; Arjona, M.; Luque, S.O. Curvature sensor for ocular wavefront measurement. Opt. Lett.**2006**, 31, 2245–2247.

40. Ragazzoni, R. Pupil plane wavefront sensing with an oscillating prism 1996. J. Mod. Opt.**1996**, 43, 289–293.

41. Ragazzoni, R.; Farinato, J. Sensitivity of a pyramidic wave front sensor in closed loop adaptive optics. Astron. Astrophys**1999**, 350, L23–L26.

42. Iglesias, I.; Ragazzoni, R.; Julien, Y.; Artal, P. Extended source pyramid wave-front sensor for the human eye. Opt. Express**2002**, 10, 419–428.

43. Chamot, S.R.; Dainty, C.; Esposito, S. Adaptive optics for ophthalmic applications using a pyramid wavefront sensor. Opt. Express**2006**, 2, 518–526.

44. Leibbrandt, G.W.R.; Harbers, G.; Kunst, P.J. Wavefront analysis with high accuracy by use of a double-grating lateral shearing interferometer. Appl. Opt.**1996**, 35, 6151–6161.

45. Gundlach, A.; Huntley, J.M.; Manzke, B.; Schwider, J. Speckle shearing interferometry using a diffractive optical beamsplitter. Opt. Eng.**1997**, 36, 1488–1493.

46. Griffin, D.W. Phase-Shifting shearing interferometer. Opt. Lett.**2001**, 26, 140–141.

47. Harbers, G.; Kunst, P.J.; Leibbrandt, W.R. Analysis of lateral shearing interferograms by use of Zernike polynomials. Appl. Opt.**1996**, 35, 6162–6172.

48. Karp, J.H.; Chan, T.K.; Ford, J.E. Integrated diffractive shearing interferometry for adaptive wavefront sensing. Appl. Opt.**2008**, 35, 6666–6674.

49. Chanteloup, J.C. Multiple-Wave lateral shearing interferometry for wavefront sensing. Appl. Opt.**2005**, 44, 1559–1571.

50. Siegel, C.; Loewenthal, F.; Balmer, J.E. A wavefront sensor based on the fractional talbot effect. Opt. Commun.**2001**, 194, 265–275.

51. Nakano, Y.; Murata, K. Measurements of phase objects using the Talbot effect and moiré techniques. Appl. Opt.**1984**, 23, 2296–2299.

52. Salama, N.H.; Patrignani, D.; De Pasquale, L.; Sicre, E.E. Wavefront sensor using the talbot effect. Opt. Laser Technol.**1999**, 31, 269–272.

53. Sekine, R.; Shibuya, T.; Ukai, K.; Komatsu, S.; Hattori, M.; Mihashi, T.; Nakazawa, N.; Hirohara, Y. Measurement of wavefront aberration of human eye using Talbot image of two-dimensional grating. Opt. Rev.**2006**, 13, 207–211.

54. Warden, L.; Liu, Y.; Binder, P.S.; Dreher, A.W.; Sverdrup, L. Performance of a new binocular wavefront aberrometer based on a self-imaging diffractive sensor. J. Refract. Surg.**2008**, 24, 188–196.

55. Prieto, P.; Fernández, E.; Manzanera, S.; Artal, P. Adaptive optics with a programmable phase modulator: Applications in the human eye. Opt. Exp.**2004**, 12, 4059–4071.

56. Fernández, E.; Prieto, P.; Artal, P. Adaptive optics binocular visual simulator to study stereopsis in the presence of aberrations. JOSA A**2010**, 27, A48–A55.

57. Kong, N.; Li, C.; Xia, M.; Li, D.; Qi, Y.; Xuan, L. Optimization of the open-loop liquid crystal adaptive optics retinal imaging system. J. Biomed. Opt.**2012**.

58. Hampson, K. Topical review: Adaptive optics and vision. J. Mod. Opt.**2008**, 55, 3425–3467.

59. Hardy, J. Adaptive Optics for Astronomical Telescopes (Oxford Series in Optical and Imaging Sciences); Oxford University Press: New York, NY, USA, 1998.

60. Horsley, D.; Park, H.; Laut, S.; Wernet, J. Characterisation for vision science applications of a bimorph deformable mirror using phase-shifting interferometry. Proc. SPIE**2005**, 5688, 133–144.

61. Chen, D.; Jones, S.; Silva, D.; Olivier, S. High-Resolution adaptive optics scanning laser ophthalmoscope with dual deformable mirrors. JOSA A**2007**, 24, 1305–1312.

62. Vdovin, G.; Sarro, P. Flexible mirror micromachined in silicon. Appl. Opt.**1995**, 34, 2968–2972.

63. Bonora, S.; Poletto, L. Push-Pull membrane mirrors for adaptive optics. Opt. Express**2006**, 14, 11935–11944.

64. Bonora, S.; Coburn, D.; Bortolozzo, U.; Dainty, C.; Residori, S. High resolution wavefront correction with photocontrolled deformable mirror. Opt. Express**2012**, 20, 5178–5188.

65. Bifano, T.; Perreault, J.; Bierden, P.; Dimas, C. Micromachined deformable mirrors for adaptive optics. Proc. SPIE**2002**, 4825, 10–13.

66. Fernández, E.; Vabre, L.; Hermann, B.; Unterhuber, A.; Považay, B.; Drexler, W. Adaptive optics with a magnetic deformable mirror: applications in the human eye. Opt. Express**2006**, 14, 8900–8917.

67. Lombardo, M.; Serrao, S.; Ducoli, P.; Lombardo, G. Adaptive optics photoreceptor imaging. Ophthalmology**2012**, 119.

68. Ödlund, E.; Raynaud, H.F.; Kulcsár, C.; Harms, F.; Levecq, X.; Martins, F.; Chateau, N.; Podoleanu, A. Control of an electromagnetic deformable mirror using high speed dynamics characterization and identification. Appl. Opt.**2010**, 49, G120–G128.

69. Iqbal, A.; Wu, Z.; Amara, F. Closed-Loop control of magnetic fluid deformable mirrors. Opt. Express**2009**, 17, 18597–18970.

70. Vdovin, G. Closed-loop adaptive optical system with a liquid mirror. Opt. Lett.**2009**, 34, 524–526.

71. Devaney, N.; Dalimier, E.; Farrell, T.; Coburn, D.; Mackey, R.; Mackey, D.; Laurent, F.; Daly, E.; Dainty, C. Correction of ocular and atmospheric wavefronts: A comparison of the performance of various deformable mirrors. Appl. Opt.**2008**, 47, 6550–6562.

72. Alpern, M.; Ching, C.C.; Kitahara, K. The directional sensitivity of retinal rods. J. Physiol.**1983**, 343, 577–592.

73. Carroll, J.; Choi, S.S.; Williams, D.R. In vivo imaging of the photoreceptor mosaic of a rod monochromat. Vis. Res.**2008**, 48, 2564–2568.

74. Dubra, A.; Sulai, Y.; Norris, J.L.; Cooper, R.F.; Dubis, A.M.; Williams, D.R.; Carroll, J. Noninvasive imaging of the human rod photoreceptor mosaic using a confocal adaptive optics scanning ophthalmoscope. Biomed. Opt. Express**2011**, 2, 1864–1876.

75. Doble, N.; Choi, S.S.; Codona, J.L.; Christou, J.; Enoch, J.M.; Williams, D.R. In vivo imaging of the human rod photoreceptor mosaic. Opt. Lett.**2011**, 36, 31–33.

76. Garrioch, R.; Langlo, C.; Dubis, A.M.; Cooper, R.F.; Dubra, A.; Carroll, J. Repeatability on in vivo cone density and spacing measurements. Optom. Vis. Sci.**2012**, 89, 632–643.

77. Li, K.Y.; Roorda, A. Automated identification of cone photoreceptors in adaptive optics retinal images. JOSA A**2007**, 24, 1358–1363.

78. Xue, B.; Choi, S.S.; Doble, N.; Werner, J.S. Photoreceptor counting and montaging of en-face retinal images from an adaptive optics fundus camera. JOSA A**2007**, 24, 1364–1372.

79. Wojtas, D.H.; Wu, B.; Ahnelt, P.K.; Bones, P.J.; Millane, R.P. Automated analysis of differential interference contrast microscopy images of the foveal cone mosaic. JOSA A**2008**, 25, 1181–1189.

80. Rodieck, R.W. The density recovery profile: A method for the analysis of points in the plane applicable to retinal studies. Vis. Neurosci.**1991**, 6, 95–111.

81. Brostow, W.; Dussault, J.P.; Fox, B.L. Construction of Voronoi polyhedra. J. Comput. Phys.**1978**, 29, 81–92.

82. Chui, T.Y.P.; Song, H.; Burns, S. Individual variations in human cone photoreceptor packing density: Variations with refractive error. Invest. Ophthalmol. Vis. Sci.**2008**, 49, 4679–4687.

83. Li, K.Y.; Tiruveedhula, P.; Roorda, A. Intersubject variability of foveal cone photoreceptor density in relation to eye length. Invest. Ophthalmol. Vis. Sci.**2010**, 51, 6858–6867.

84. Chui, T.Y.P.; Song, H.; Burns, S. Adaptive-Optics imaging of human cone photoreceptor distribution. JOSA A**2008**, 25, 3021–3029.

85. Song, H.; Chui, T.Y.P.; Zhong, Z.; Elsner, A.E.; Burns, S.A. Variation of cone photoreceptor packing density with retinal eccentricity and age. Invest. Ophthalmol. Vis. Sci.**2011**, 52, 7376–7384.

86. Curcio, C.A.; Sloan, K.R.; Kalina, R.E.; Hendrickson, A.E. Human photoreceptor topography. J. Comp. Neurol.**1990**, 292, 497–523.

87. Curcio, C.A.; Sloan, K.R. Packing geometry of human cone photoreceptors: variation with eccentricity and evidence of local anisotropy. Vis. Neurosci.**1992**, 9, 169–180.

88. Curcio, C.A.; Sloan, K.R.; Packer, O.; Hendrickson, A.E.; Kalina, R.E. Distribution of cones in human and monkey retina: Individual variability and radial asymmetry. Science**1987**, 236, 579–582.

89. Østerberg, G.A. Topography of the layer of rods and cones in the human retina. Acta Ophthalmol.**1935**, 13, 1–97.

90. Jonas, J.B.; Schneider, U.; Naumann, G.O.H. Count and density of human retinal photoreceptors. Graef. Arch. Clin. Exp. Ophthal.**1992**, 230, 505–510.

91. Lombardo, M.; Serrao, S.; Ducoli, P.; Lombardo, G. Variations in the image optical quality of the eye and the sampling limit of resolution of the cone mosaic with axial length in young adults. J. Cataract Refract. Surg.**2012**, 38, 1147–1155.

92. Coletta, N.J.; Watson, T. Effect of myopia on visual acuity measured with laser interference fringes. Vis. Res.**2006**, 46, 636–651.

93. Rossi, E.A.; Roorda, A. The relationship between visual resolution and cone spacing in the human fovea. Nat. Neurosci.**2010**, 13, 156–157.

94. Sjöstrand, J.; Olsson, V.; Popovic, Z.; Conradi, N. Quantitative estimations of foveal and extra-foveal retinal circuitry in humans. Vis. Res.**1999**, 39, 2987–2998.

95. Drasdo, N.; Millican, C.L.; Katholi, C.R.; Curcio, C.A. The length of Henle fibers in the human retina and a model of ganglion receptive field density in the visual field. Vis. Res.**2007**, 47, 2901–2911.

96. Pallikaris, A.; Williams, D.R.; Hofer, H. The reflectance of single cones in the living human eye. Invest. Ophthalmol. Vis. Sci.**2003**, 44, 4580–4592.

97. Ravi, S.J.; Besecker, J.R.; Derby, J.C.; Kocaoglu, O.P.; Cense, B.; Gao, W.; Wang, Q.; Miller, D.T. Imaging outer segment renewal in living human cone photoreceptors. Opt. Express**2010**, 18, 5257–5270.

98. Ravi, S.J.; Rha, J.; Zhang, Y.; Cense, B.; Gao, W.; Miller, D.T. In vivo functional imaging of human cone photoreceptors. Opt. Express**2007**, 15, 16141–16160.

99. Cooper, R.F.; Dubis, A.M.; Pavaskar, A.; Rha, J.; Dubra, A.; Carroll, J. Spatial and temporal variation of rod photoreceptor reflectance in the human retina. Biomed. Opt. Express**2011**, 2, 2577–2589.

100. Rha, J.; Schroeder, B.; Godara, P.; Carroll, J. Variable optical activation of human cone photoreceptors visualized using a short coherence light source. Opt. Lett.**2009**, 34, 3782–3784.

101. Choi, S.S.; Doble, N.; Lin, J.; Christou, J.; Williams, D.R. Effect of wavelength on in vivo images of the human cone mosaic. JOSA A**2005**, 22, 2598–2605.

102. Kocaoglu, O.P.; Lee, S.; Jonnal, R.S.; Wang, Q.; Herde, A.E.; Derby, J.C.; Gao, W.; Miller, D.T. Imaging cone photoreceptors in three dimensions and in time using ultrahigh resolution optical coherence tomography with adaptive optics. Biomed. Opt. Express**2011**, 2, 748–763.

103. Pircher, M.; Kroisamer, J.S.; Felberer, F.; Sattmann, H.; Göttzinger, E.; Hitzenberger, C.K. Temporal changes of human cone photoreceptors observed in vivo with SLO/OCT. Biomed. Opt. Express**2010**, 2, 100–112.

104. Rha, J.; Jonnal, R.S.; Thorn, K.E.; Qu, J.; Zhang, Y.; Miller, D.T. Adaptive optics flood-illumination camera for high-speed retinal imaging. Opt. Express**2006**, 14, 4552–4569.

105. Roorda, A.; Williams, D.R. Optical fiber properties of individual human cones. J. Vis.**2002**, 35, 607–614.

106. Burns, S.A.; Wu, S.; He, J.C.; Elsner, A.E. Variations in photoreceptor directionality across the central retina. JOSA A**1997**, 14, 2033–2040.

107. He, J.C.; Marcos, S.; Burns, S.A. Comparison of cone directionality determined by psychophysical and reflectometric techniques. JOSA A**1999**, 16, 2363–2369.

108. Rativa, D.; Vohnsen, B. Analysis of individual cone-photoreceptor directionality using scanning laser ophthalmoscopy. Biomed. Opt. Express**2011**, 2, 1423–1431.

109. Marcos, S.; Tornow, R.P.; Elsner, A.E.; Navarro, R. Foveal cone spacing and cone photopigment density difference: objective measurements in the same subjects. Vis. Res.**1997**, 37, 1909–1915.

110. Duncan, J.L.; Zhang, Y.; Gandhi, J.; Nakanishi, C.; Otham, M.; Brahnam, K.E.H.; Swaroop, A.; Roorda, A. High-Resolution imaging with adaptive optics in patients with inherited retinal degeneration. Invest. Ophthalmol. Vis. Sci.**2007**, 48, 3283–3291.

111. Chen, Y.; Ratnam, K.; Sundquist, S.M.; Lujan, B.; Ayyagari, R.; Gudiseva, V.H.; Roorda, A.; Duncan, J.L. Cone photoreceptor abnormalities correlate with vision loss in patients with Stargardt disease. Invest. Ophthalmol. Vis. Sci.**2011**, 52, 3281–3292.

112. Choi, S.S.; Zawadzki, R.J.; Lim, M.C.; Brandt, J.D.; Keltner, J.L.; Doble, N.; Werner, J.S. Evidence of outer retinal changes in glaucoma patients as revealed by ultrahigh-resolution in vivo retinal imaging. Br. J. Ophthalmol.**2011**, 95, 131–141.

113. Tam, J.; Dhamdhere, K.P.; Tiruveedhula, P.; Lujan, B.J.; Johnson, R.N.; Bearse, M.A.; Adams, A.J., Jr.; Roorda, A. Subclinical capillary changes in non-proliferative diabetic retinopathy. Optom. Vis. Sci.**2012**, 89, E692–E703.

114. Tam, J.; Dhamdhere, K.P.; Tiruveedhula, P.; Manzanera, S.; Barez, S.; Bearse, M.A., Jr.; Adams, J.A.; Roorda, A. Disruption of the retinal parafoveal capillary network in type 2 diabetes before the onset of diabetic retinopathy. Invest. Ophthalmol. Vis. Sci.**2012**, 52, 9257–9266.

115. Klein, R.; Knudtson, M.D.; Lee, K.E.; Gangnon, R.; Klein, B.E. The wisconsin epidemiologic study of diabetic retinopathy xxiii: The twenty-five-year incidence of macular edema in persons with type 1 diabetes. Ophthalmology**2009**, 116, 497–503.

116. Scully, T. Diabetes in numbers. Nature**2012**, 485, S2–S3.

117. Early Treatment Diabetic Retinopathy Study Research Group. Fundus photographic risk factors for progression of diabetic retinopathy. ETDRS Report Number 12. Ophthalmology**1991**, 98, 823–833.

118. Moore, J.; Bagley, S.; Ireland, G.; McLeod, D.; Boulton, M.E. Three dimensional analysis of microaneurysms in the human diabetic retina. J. Anat.**1999**, 194, 89–110.

119. Kern, T.S.; Engerman, R.L. Vascular lesions in diabetes are distributed non-uniformly within the retina. Exp. Eye Res.**1995**, 60, 545–549.

120. Cunha-Vaz, J.G. Pathophysiology of diabetic retinopathy. Br. J. Ophthalmol.**1978**, 62, 351–355.

121. Barber, A.J. A new view of diabetic retinopathy: a neurodegenerative disease of the eye. Prog. Neuro-Psych. Biol. Psych.**2003**, 27, 283–290.

122. Verma, A.; Rani, P.K.; Raman, R.; Pal, S.S.; Laxmi, G.; Gupta, M.; Sahu, C.; Vaitheeswaran, S.T. Is neuronal dysfunction on early sign of

diabetic retinopathy? Microperimetry and Spectral Domain Optical Coherence Tomography (SD-OCT) study in individuals with diabetes, but no diabetic retinopathy. Eye**2009**, 23, 1824–1830.

123. Fletcher, E.L.; Phipps, J.A.; Wilkinson-Berka, J.L. Dysfunction of retinal neurons and glia during diabetes. Clin. Exp. Optom.**2005**, 88, 132–145.

124. Lieth, E.; Gardner, T.W.; Barber, A.J.; Antonetti, D.A. Retinal neurodegeneration: Early pathology in diabetes. Clin. Exp. Ophthalmol.**2000**, 28, 3–8.

125. Van Dijk, H.W.; Kok, P.H.; Garvin, M.; Sonka, M.; De Vries, J.H.; Michels, R.P.; Van Velthoven, M.E.; Schlingemann, R.O.; Verbraak, F.D.; Abràmoff, M.D. Selective loss of inner retinal layer thickness in type 1 diabetic patients with minimal diabetic retinopathy. Invest. Ophthalmol. Vis. Sci.**2009**, 50, 3404–3409.

126. Kylstra, J.A.; Brown, J.C.; Jaffe, G.J.; Cox, T.A.; Gallemore, R.; Greven, C.M.; Hall, J.G.; Eifrig, D.E. The importance of fluorescein angiography in planning laser treatment of diabetic macular edema. Ophthalmology**1999**, 106, 2068–2073.

127. Mendis, K.R.; Balaratnasingam, C.; Yu, P.; Barry, C.J.; McAllister, I.L.; Cringle, S.J.; Yu, D.Y. Correlation of histological and clinical images to determine the diagnostic value of fluorescein angiography for studying retinal capillary detail. Invest. Ophthalmol. Vis. Sci.**2010**, 51, 5864–5869.

128. Popovic, Z.; Knutsson, P.; Thaung, J.; Petersen, M.O.; Sjostrand, J. Noninvasive imaging of human foveal capillary network using dual-conjugate adaptive optics. Invest. Ophthalmol. Vis. Sci.**2011**, 52, 2649–2655.

129. Tam, J.; Martin, J.A.; Roorda, A. Noninvasive visualization and analysis of parafoveal capillaries in humans. Invest. Ophthalmol. Vis. Sci.**2010**, 51, 1691–1698.

130. Uji, A.; Hangai, M.; Ooto, S.; Takayama, K.; Arakawa, N.; Imamura, H.; Nozato, K.; Yoshimura, N. The source of moving particles in parafoveal capillaries detected by adaptive optics scanning laser ophthalmoscopy. Invest. Ophthalmol. Vis. Sci.**2012**, 53, 171–178.

131. Wang, Q.; Kocaoglu, O.P.; Cense, B.; Bruestle, J.; Jonnal, R.S.; Gao, W.; Miller, D.T. Imaging retinal capillaries using ultrahigh-resolution optical coherence tomography and adaptive optics. Invest. Ophthalmol. Vis. Sci.**2011**, 52, 6292–6299.

132. Hammer, D.X.; Iftimia, N.V.; Ferguson, R.D.; Bigelow, C.E.; Ustun, T.E.; Barnaby, A.M.; Fultun, A.B. Foveal fine structure in retinopathy of

prematurity: An adaptive optics fourier domain optical coherence tomography study. Invest. Ophthalmol. Vis. Sci.**2008**, 49, 2061–2070.

133. Schnoll, T.; Singh, A.S.G.; Blatter, C.; Schriefl, S.; Ahlers-Erfurth, U.S.; Leitgeb, R.A. Imaging of the parafoveal capillary network and its integrity analysis using fractal dimension. Biomed. Opt. Expess**2011**, 2, 1159–1168.

134. Fischer, M.D.; Huber, G.; Feng, Y.; Tanimoto, N.; Mühlfriedel, R.; Beck, S.C.; Tröger, E.; Kernstock, C.; Preising, M.N.; Lorenz, B.; Hammes, H.P.; Seeliger, M.W. In vivo assessment of retinal vascular wall dimensions. Invest. Ophthalmol. Vis. Sci.**2010**, 51, 5254–5259.

135. Zhong, Z.; Petrig, B.L.; Qi, X.; Burns, S. In vivo measurement of erythrocyte velocity and retinal blood flow using adaptive optics scanning laser ophthalmoscopy. Opt. Expr.**2008**, 16, 12746–12755.

136. Parravano, M.; Lombardo, M.; Lombardo, G.; Boccassini, B.; Lioi, S.; Varano, M. In Vivo investigation of the retinal microscopy in patients with type 1 Diabetes Mellitus. Invest. Ophthalmol. Vis. Sci.**2012**, 53. E-Abstract: 5657.

137. Smith, W.; Assink, J.; Klein, R.; Mitchell, P.; Klaver, C.C.; Klein, B.E.; Hofman, A.; Jensen, S.; Wang, J.J.; De Jong, P.T. Risk factors for age-related macular degeneration: Pooled findings from three continents. Ophthalmology**2001**, 108, 697–704.

138. Choudhury, F.; Varma, R.; McKean-Cowdin, R.; Klein, R.; Azen, S.P.; Los angeles latino eye study group. Risk factors for four-year incidence and progression of age-related macular degeneration: The los angeles latino eye study. Amer. J. Ophthalmol.**2011**, 152, 385–395.

139. Chakravarthy, U.; Wong, T.Y.; Fletcher, A.; Piault, E.; Evans, C.; Zlateva, G.; Buggage, R.; Pleil, A.; Mitchell, P. Clinical risk factors for age-related macular degeneration: A systematic review and meta-analysis. BMC Ophthalmol.**2010**, 13, 10–31.

140. Klein, R.J.; Zeiss, C.; Chew, E.Y.; Tsai, J.Y.; Sackler, R.S.; Haynes, C.; Henning, A.K.; SanGiovanni, J.P.; Mane, S.M.; Mayne, S.T.; et al. Complement factor H polymorphism in age-related macular degeneration. Science**2005**, 15, 385–389.

141. Maller, J.B.; Fagerness, J.A.; Reynolds, R.C.; Neale, B.M.; Daly, M.J.; Seddon, J.M. Variation in complement factor 3 is associated with risk of age-related macular degeneration. Nat. Genet.**2007**, 39, 1200–1201.

142. Reynolds, R.; Rosner, B.; Seddon, J.M. Serum lipid biomarkers and hepatic lipase gene associations with age-related macular degeneration. Ophthalmology**2010**, 117, 1989–1995.

143. Sobrin, L.; Ripke, S.; Yu, Y.; Fagerness, J.; Bhangale, T.R.; Tan, P.L.; Souied, E.H.; Buitendijk, G.H.S.; Merriam, J.E.; Richardson, A.J. Heritability and genome-wide association study to assess genetic differences between advanced age-related macular degeneration subtypes. Ophthalmology**2012**, 119, 1874–1885.

144. Neale, B.M.; Fagerness, J.; Reynolds, R.; Sobrin, L.; Parker, M.; Raychaudhuri, S.; Tan, P.L.; Oh, E.C.; Merriam, J.E.; Souied, E. Genome-Wide association study of advanced age-related macular degeneration identifies a role of the hepatic lipase gene (LIPC). PNAS**2010**, 20, 7395–7400.

145. McKay, G.J.; Patterson, C.C.; Chakravarthy, U.; Dasari, S.; Klaver, C.C.; Vingerling, J.R.; Ho, L.; De Jong, P.T.V.M.; Fletcher, A.E.; Young, I.S. Evidence of association of APOE with age-related macular degeneration: A pooled analysis of 15 studies. Hum Mutat.**2011**, 32, 1407–1416.

146. Yu, Y.; Bhangale, T.R.; Fagerness, J.; Ripke, S.; Thorleifsson, G.; Tan, P.L.; Souied, E.H.; Richardson, A.J.; Merriam, J.E.; Buitendijk, G.H.S. Common variants near FRK/COL10A1 and VEGFA are associated with advanced age-related macular degeneration. Hum. Mol. Genet.**2011**, 20, 3699–3709.

147. Seddon, J.M.; Reynolds, R.; Yu, Y.; Daly, M.J.; Rosner, B. Risk models for progression to advanced age-related macular degeneration using demographic, environmental, genetic, and ocular factors. Ophthalmology**2011**, 118, 2203–2211.

148. Ding, X.; Patel, M.; Chan, C.C. Molecular pathology of age-related macular degeneration. Prog. Retin. Eye Res.**2009**, 28, 1–18.

149. Grisanti, S.; Tatar, O. The role of vascular endothelial growth factor and other endogenous interplayers in age-related macular degeneration. Prog. Retin. Eye Res.**2008**, 27, 372–390.

150. Ambati, J.; Fowler, B.J. Mechanisms of age-related macular degeneration. Neuron**2012**, 12, 26–39.

151. Lim, L.S.; Mitchell, P.; Seddon, J.M.; Holz, F.G.; Wong, T.Y. Age-Related macular degeneration. Lancet**2012**, 5, 1728–1738.

152. Godara, P.; Siebe, C.; Rha, J.; Michaelides, M.; Carroll, J. Assessing the photoreceptor mosaic over drusen using adaptive optics and SD-OCT. Ophthalmic Surg. Lasers Imaging**2010**, 41, S104–S108.

153. Godara, P.; Wagner-Schuman, M.; Rha, J.; Connor, T.B., Jr.; Stepien, K.E.; Carroll, J. Imaging the photoreceptor mosaic with adaptive optics: Beyond counting cones. Advan. Exp. Med. Biol.**2012**, 723, 451–458.

154. Boretsky, A.; Khan, F.; Burnett, G.; Hammer, D.X.; Ferguson, R.D.; Van Kuijk, F.; Motamedi, M. In vivo imaging of photoreceptor disruption associated with age-related macular degeneration: A pilot study. Laser Surg. Med.**2012**, 44, 603–610.

155. Kotecha, A.; Fernandes, S.; Bunce, C.; Franks, W.A. Avoidable sight loss from glaucoma: Is it unavoidable? Br. J. Ophthalmol.**2012**, 96, 816–820.

156. Quigley, H.A.; Broman, A.T. The number of people with glaucoma worldwide in 2010 and 2020. Br. J. Ophthalmol.**2006**, 90, 262–167.

157. Quigley, H.A. Glaucoma. Lancet**2011**, 377, 1367–1377.

158. Alencar, L.M.; Zangwill, L.M.; Weinreb, R.N.; Bowd, C.; Sample, P.A.; Girkin, C.A.; Liebmann, J.M.; Medeiros, F.A. A comparison of rates of change in neuroretinal rim area and retinal nerve fiber layer thickness in progressive glaucoma. Invest. Ophthalmol. Vis. Sci.**2010**, 51, 3531–3539.

159. Sakamoto, A.; Hangai, M.; Nukada, M.; Nakanishim, H.; Morim, S.; Koteram, Y.; Inoue, R.; Yoshimura, N. Three-dimensional imaging of macular retinal nerve fiber layer in glaucoma using spectral-domain optical coherence tomography. Invest. Ophthalmol. Vis. Sci.**2010**, 51, 5062–5070.

160. Quigley, H.A.; Reacher, M.; Katz, J.; Strahlman, E.; Gilbert, D.; Scott, R. Quantitative grading of nerve fiber layer photographs. Ophthalmology**1993**, 100, 1800–1807.

161. Mansouri, K.; Leite, M.T.; Medeiros, F.A.; Leung, C.K.; Weinreb, R.N. Assessment of rates of structural change in glaucoma using imaging technologies. Eye**2011**, 25, 269–277.

162. Lim, T.C.; Chattopadhyay, S.; Acharya, U.R. A survey and comparative study on the instruments for glaucoma detection. Med. Eng. Phys.**2012**, 34, 129–139.

163. Takayama, K.; Ooto, S.; Hangai, M.; Arakawa, N.; Oshima, S.; Shibata, N.; Hanebuchi, M.; Inoue, T.; Yoshimura, N. High-Resolution imaging of the retinal nerve fiber layer in normal eyes using adaptive optics scanning laser ophthalmoscopy. PLoS ONE**2012**.

164. Huang, G.; Qi, X.; Chui, T.Y.; Zhong, Z.; Burns, S.A. A clinical planning module for adaptive optics SLO imaging. Optom. Vis. Sci.**2012**, 89, 593–601.

165. Kocaoglu, O.P.; Cense, B.; Jonnal, R.S.; Wang, Q.; Lee, S.; Gao, W.; Miller, D.T. Imaging retinal nerve fiber bundles using optical coherence tomography with adaptive optics. Vis. Res.**2011**, 51, 1835–1844.

166. Merino, D.; Duncan, J.L.; Tiruveedhula, P.; Roorda, A. Observation of cone and rod photoreceptors in normal subjects and patients using a new generation adaptive optics scanning laser ophthalmoscope. Biomed. Opt. Express**2011**, 2, 2189–2201.

167. Zawadzki, R.J.; Jones, S.M.; Pilli, S.; Balderas-Mata, S.; Kim, D.Y.; Olivier, S.S.; Werner, J.S. Integrated adaptive optics optical coherence tomography and adaptive optics scanning laser ophthalmoscope system for simultaneous cellular resolution in vivo retinal imaging. Biomed. Opt. Express**2011**, 2, 1674–1686.

168. Tam, J.; Tiruveedhula, P.; Roorda, A. Characterization of single-file flow through human retinal parafoveal capillaries using an adaptive optics scanning laser ophthalmoscope. Biomed. Opt. Express**2011**, 2, 781–793.

169. Chui, T.Y.C.; Van Nasdale, D.A.; Burns, S.A. The use of forward scatter to improve retinal vascular imaging with an adaptive optics scanning laser ophthalmoscope. Biomed. Opt. Express**2012**, 3, 2537–2549.

170. Lombardo, M.; Lombardo, G.; Schiano, L.D.; Ducoli, P.; Stirpe, M.; Serrao, S. Interocular symmetry of parafoveal photoreceptor cone density distribution. Retina**2013**. in press.

171. Rha, J.; Dubis, A.M.; Wagner-Schuman, M.; Tait, D.M.; Godara, P.; Schroeder, B.; Stepien, K.; Carroll, J. Spectral domain optical coherence tomography and adaptive optics: imaging photoreceptor layer morphology to interpret preclinical phenotypes. Advan. Exp. Med. Biol.**2010**, 664, 309–316.

172. Seyedahmadi, B.J.; Vavvas, D. In vivo high-resolution retinal imaging using adaptive optics. Semin. Ophthalmol.**2010**, 25, 186–191.

Chapter 4

Devices and Techniques for Sensorless Adaptive Optics

S. Bonora[1], R.J. Zawadzki[2], G. Naletto[1, 3], U. Bortolozzo[4] and S. Residori[4]

[1] *CNR-IFN, Laboratory for UV and X-Ray and Optical Research, Padova, Italy*
[2] *VSRI, Department of Ophthalmology and Vision Science, University of California Davis, Sacramento, CA, USA*
[3] *Department of Information Engineering, University of Padova, Padova, Italy*
[4] *INLN, Université de Nice-Sophia Antipolis, CNRS, France*

1. INTRODUCTION

Minimizing the aberrations is the basic concern of all the optical system designers. For this purpose, a large amount of work has been carried out and plenty of literature can be found on the subject. Until the last twenty years, the large majority of the optical design was related to "static" optical systems, where several opto-mechanical parameters, such as refractive index, shape, curvatures, etc. are slowly time dependent. In these systems, simple mechanisms can be adopted to change the relative position of one or more optical elements (for example, the secondary mirror of many astronomical telescopes), or slightly modify their shape and curvature (as in some synchrotron beamlines, where some optical surfaces are mechanically bent) to compensate defocusing. In the last years, a new type of optical systems, that we may call "dynamical", have heavily occupied the interest of optical designers, opening the possibility of working also in situations where the system environment varies rather quickly with time, either in a controlled or not-controlled way. For this class of optical systems adaptive optics (AO) with a closed loop control system has to be implemented. The correction of dynamical systems was predicted by Babcock in the 1953 [1] and, then, the first prototypes were realized in the early 70s with the purpose of satellite surveillance and launching high power laser beams trough the atmosphere [2]. The most known scientific applications of closed loop correction by means of

AO is the acquisition of astronomical images in ground-based telescopes [3] and *in-vivo* imaging of cone photoreceptor mosaic by AO enhanced Fundus Cameras [4]. In astronomy, to remove the so called "seeing effect", the star light twinkling due to local dynamic variations of the atmospheric density in the air column above the telescope, it is necessary to have the real time knowledge of the wavefront of the observed object. This can be realized, for instance, by means of a Shack-Hartmann wavefront sensing device coupled to a dedicated fast algorithm which returns the mathematical description of the wavefront aberration, typically through a Zernike series decomposition [5]. Then, this information is suitably coded and passed to an AO, as a fast deformable mirror located along the optical path, that adapts its shape to compensate the time dependent aberrations. Similarly in vision science [6, 7] or retinal imaging [8-15], static and dynamic aberrations created by variation in shape of eye refractive elements and eye movements are measured by wavefront sensor, usually Schack-Hartmann and corrected by wavefront corrector, in most cases a deformable mirror. Other applications which make use of AO systems are for example: free space optical communication systems [16, 17], microscopy [18-20] or beam shaping in laser applications [21]. It is, however, rather obvious that not all AO applications have similar needs, and in particular that in some cases systems simpler than the astronomical ones can be realized. For example, in some cases there is no need to have the real time information about the aberrated wavefront: either because the aberration variation is slow [22] or because there is a specific phase which remains for a limited amount of time, as for example when correcting low order ocular aberrations "eyeglass prescription"for patient in ophthalmic diagnostics, or in optical devices in which the environmental conditions are not initially defined but the system remains stable in time [23-25]. In all these cases it can be convenient to have a simpler AO system, able to correct only the slow variations of the wavefront aberrations.

In the above mentioned cases the wavefront correction can be operated with a strong reduction in the hardware complexity, in particular by using a sensorless approach. Several techniques have been developed which use these simpler AO systems. They are generally based on the optimization of some merit function that depends on the optical system under consideration.

The algorithms for the sensorless correction can be divided into two main classes: the stochastic and the image-based ones. In the first class, the system is optimized starting from a random set and, then, applying an iterative selection of the best solutions. These algorithms have the advantage of not requiring any preliminary information about the system but they take a lot of time for converging. Many algorithms using this approach have been written and exploited successfully in different fields. Among them the most popular are: genetic algorithms [18, 26, 24], simulated annealing [13], simplex or ant colonies [27]. These approaches have the drawback of requiring a rather long computation time, or many iterations before converging, taking up to several minutes before reaching the desired system optimization.

Other sensorless techniques can be realized by analyzing some specific known feature, either intrinsic to the system or artificially introduced. An example of the latter case can be found in [28-29]. With respect to classical AO systems, the sensorless approach offers the advantage of not needing the wavefront sensor: this reduces the cost of the instrument and avoids all the problems related to maintaining the performance of such a device once installed and aligned. However, the absence of the wavefront sensor implies also some limitations, for instance, a much longer time before reaching an optimal image quality, or a final image not perfectly optimized. Clearly, the required final result and the available resources are the key elements driving the choice towards one system or another. In section 2, we will explain in detail the genetic algorithm and the ant colonies optimization process, while providing a few examples of their application in optical experimental setups.

The image-based algorithms will be explained in section 3, together with a few examples of recently reported successful applications in optical experiments. New devices useful to generate the bias aberrations will also be presented.

2. STOCHASTIC ALGORITHMS FOR SENSORLESS CORRECTION

2.1. Genetic algorithm

A genetic algorithm [30] searches the solution of a problem by simulating the evolution process. Starting from a population of possible solutions, it saves some of the strongest elements, that are the only ones selected to survive, and, thus, are able to reproduce themselves giving rise to the next generations. In general, the inferior individuals can survive and reproduce with a smaller probability.

This strategy allows solving a large class of problems without any initial hypothesis or preliminary knowledge. Its effectiveness was demonstrated in many experimental setups, as will be discussed in the following paragraphs.

The main steps of a genetic algorithm are depicted in Table 1 and in Fig. 1.

Table 1. Main steps required by a genetic algorithm.

Starting random Population
1_selection function
2_reproduction function
3_evaluate population
4_repeat from step 1

The initial population is chosen randomly in the whole set of possible solutions. The selection function can be either probabilistic or deterministic. In the probabilistic case, the strongest elements have more chances of being selected and of reproducing to the next generation. This decreases the possibility of falling in a "local" maximum solution.

The reproduction function creates new individuals from the old population. There are two kinds of functions: crossover and mutations.

CrossOver functions: they mix the genes of the two parents by slightly modifying them and by obtaining two sons.

Example: EuristicXOver:

From the parents $V_a^{(k-1)}$ and $V_b^{(k-1)}$, the children $V_a^{(k)}$ and $V_b^{(k)}$ are generated by the following rule:

$$V_a^{(k)} = V_a^{(k-1)} + r\left(V_b^{(k-1)} - V_a^{(k-1)}\right)$$

$$V_b^{(k)} = V_b^{(k-1)}$$

Mutations functions: the genes of the parent are randomly modified.

Example: Uniform Mutation:

The mutation take an element $V_{cj}^{(k-1)}$ and mutate it in a new one by the rule:

$$V_{cj}^{(k)} = \begin{cases} V_{cj}^{(k-1)} + w(k)\left(1 - V_{cj}^{(k-1)}\right) & if \ rand > 0.5 \\ V_{cj}^{(k-1)} + w(k)V_{cj}^{(k-1)} & if \ rand < 0.5 \end{cases}$$

where w(k) is weight function which decreases with the iteration k.

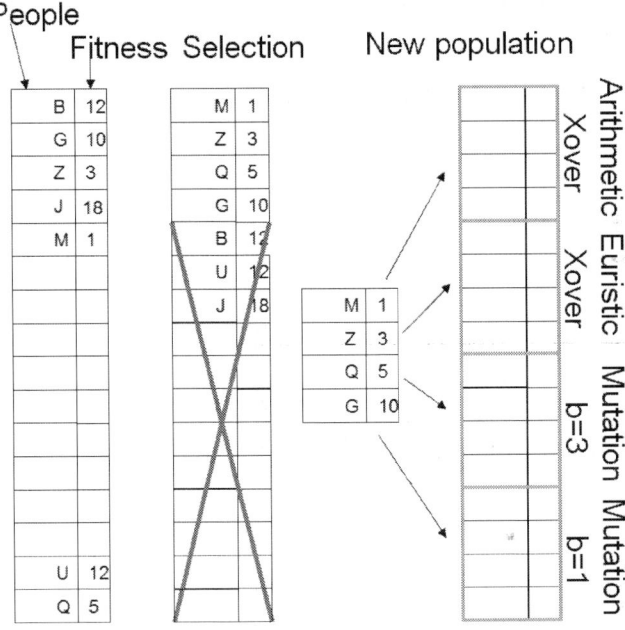

Figure 1. Diagram representing the genetic algorithm principle. The algorithm starts from a random population and then each individual is measured and the population is sorted according to its fitness. Then, some of the best individuals are selected for the generation of the next population.

2.1.1. Application example: Laser focalization

The intensity of a laser in its focal spot is largely dependent on the quality of the focal point, and this effect is even stronger in nonlinear optics. Often, in laser systems it is not simple to reach an optimal alignment, so that AO devices can be very useful in these cases.

For example, in ref. [24] it was demonstrated how an AO sensorless optimization based on a genetic algorithm can largely enhance the XUV high-order harmonics (HH) generated by the interaction of an ultrafast laser and a gas jet.

The AO system was composed by an electrostatic deformable mirror (Okotech) placed before the interaction chamber as illustrated in Fig. 2. The feedback for the genetic algorithm was the photon flux at the shortest wavelengths acquired placing a photomultiplier tube at the XUV spectrometer output.

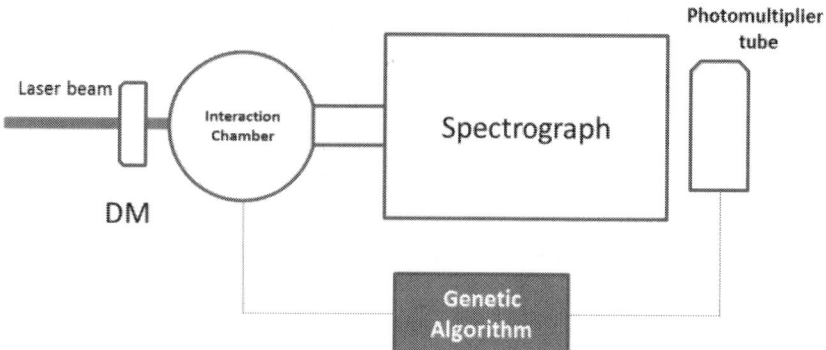

Figure 2. Experimental setup for the optimization of a laser focalization used for high order harmonics generation in ultrafast nonlinear optics. The pulsed laser beam interacts with a gas jet in the interaction chamber. The photomultiplier tube collects the signal from the spectrograph and feeds the genetic algorithm that drives the deformable mirror DM.

The laser pulse was generated by a Ti:S CPA laser system with a hollow-fiber to realize the compression of the pulse duration. The typical values used in the experiment are 6 fs of duration, 200 μJ of pulse energy, at 1 kHz repetition rate (all the experimental details are described in Villoresi et al. 2004). The focusing of the laser pulses on the gas jet, after the modifications introduced by the Deformable Mirror (DM), is obtained by means of a 250 mm focal length spherical mirror. The spectrometer that analyzes the HHs beam is based on a flat varied-line-spacing grazing-incidence grating with two toroidal mirrors.

The real-time acquisition of the spectral intensity is realized by the combination of a solar-blind open microchannel-plate (MCP) with MgF_2 photocathode and a phosphor screen placed on the spectrometer focal plane, which converts the HHs XUV spectrum in the visible, and by a photomultiplier which acquires a HHs spectral interval selected with a slit. In this way, the single-shot intensity of a single harmonic, or group of harmonics, is used as feedback by the algorithm. A separate optical channel acquires in parallel the image of all at the MCP, from which the HHs spectrum is obtained.

The genetic algorithm used a population of 80 individuals, with a deterministic selection rule that saved the 13 best ones. Both mutations and crossover were used. The results showed an increase of the XUV photons by a factor of 5 when the algorithm was applied. Moreover, the cutoff region moved to shorter wavelengths as reported in Fig 3. The optimization process took about 20 iterations to converge.

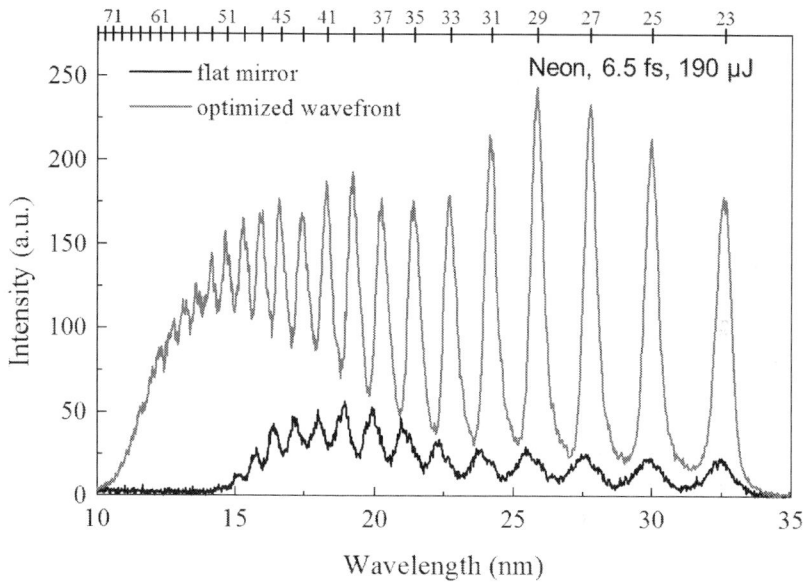

Figure 3. Result of the experimental optimization of the high order harmonics generation spectra in the case of the flat AO mirror (black line) and in the case of the optimized wavefront (red line).

2.2. Ant colonies

Ant colonies, in natural world, search the food by walking randomly. After having found it, they return to their colony leaving down a pheromone trail. If other ants cross the same trail they will not walk randomly but they will likely follow it and will reinforce the pheromone trail. The more ants will find food at the end of the trail, the more pheromone will mark it. However, since the pheromone evaporates reducing its strength, the described process will make the shortest path which will be the one with the highest density of pheromone, so providing a selection among all the possible paths, as illustrated in Fig. 4.

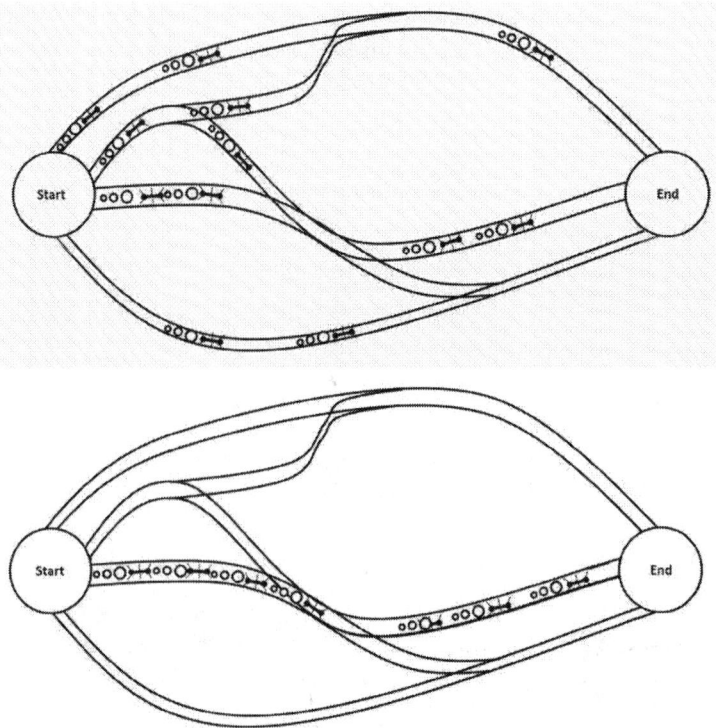

Figure 4. Ants start randomly their search for food, then the shortest path gets the higher content of pheromone. Finally, the ants will follow with larger probability the path having the highest content of pheromone.

The main essence of the Ant Colonies optimization algorithm [27] is to simulate the ant behavior for the optimization of a given problem. The algorithm steps necessary for running the optimization are listed in Table 2.

Table 2. Steps of an ant colony algorithm.

1_Set the initial ants position on the trail
2_Compute the paths length
3_Update pheromone
4_Move the ants
5_Go to step 2

As an example we show in Fig. 5 the simulation of the application of the ant colony strategy to a deformable mirror with 32 actuators and 8 bits control. In this example the actuators and their control values are the domain in which the ants can move. In the simulation the shortest path is a parabolic function, which is represented by the red line. Fig. 5 (top) shows the initial random pheromone distribution, while Fig. 5 (bottom) shows the pheromone distribution at the end of the optimization process.

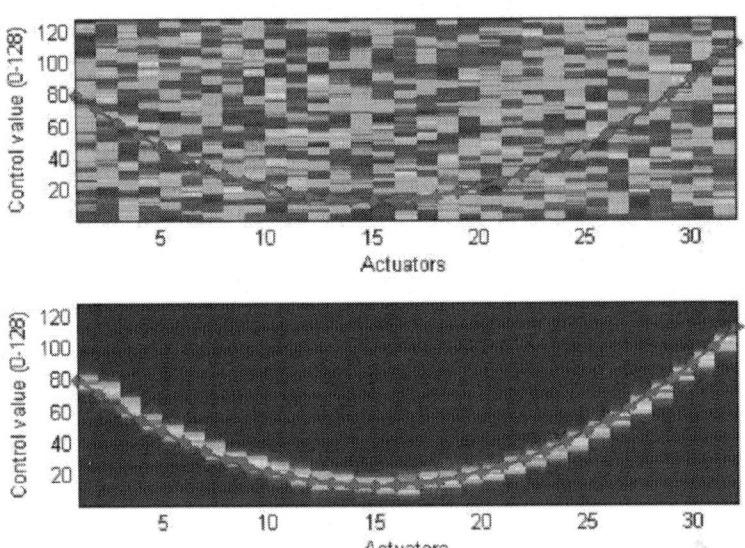

Figure 5. Implementation of an ant colony strategy for the optimization of a deformable mirror with 32 actuators and 8 bit control. The red curve represents the shortest (optimized) path. The top panel shows the initial random pheromone distribution while the bottom panel shows the pheromone at the end of the selection process.

2.2.1. Application example: Quantum optics

The quality of an optical wavefront plays an important role in Spontaneous Down Conversion (SPDC) process. As demonstrated by [31] the use of a deformable mirror can enhance the generation of photon pairs acting on the wavefront before the generation takes place in the nonlinear crystal. In that system the optimization was carried out by the use of an electrostatic DM (PAN, Adaptica srl) and the application of the ant colonies algorithm.

In the experiment, the pump beam is reflected by the DM to a BBO type-I nonlinear crystal. Then, the degenerate SPDC photons at 808 nm are selected and measured by a high efficiency SPADs (Single Photon Avalanche Diode).

Since the wavefront has a strong effect on the downconverted light, it can strongly affect the coupling in the fibers of the SPAD detectors. The feedback for the algorithm imposed the condition of photon coincidences. It was demonstrated in the experiment that the coincidences rate was increased by about 20% when the optimization algorithm was applied. The algorithm used about 80 ants and the convergence took place in about 800 iterations.

3. IMAGE BASED ALGORITHMS

Although the stochastic optimization algorithms have been demonstrated to represent important tools for optical experiments, new techniques, which demonstrated to be more effective, have recently been introduced. The use of a modal approach, based on the application of bias aberrations and of a suitable metrics, sorted out some of the limitations of the search algorithms, such as the long convergence time and the need of a training for the determination of the algorithm parameters. This new approach demonstrated to be effective both in visual optics and in laser optimization, as described later in this section. The arbitrary generation of aberrations can be achieved through the use of deformable mirrors, either thanks to a preliminary calibration of them or through the design of a suitable new class of wavefront correctors [32].

3.1. Devices for sensorless modal correction

Electrostatic membrane deformable mirrors rely on the electrostatic pressure between an actuator pad array and a thin metalized membrane [33]. Thus, the more the actuators the better the wavefront resolution that the mirror can control. The use of these deformable mirrors is, then, subjected to the acquisition of the deformation generated by each electrode. On the other hand, this kind of DMs can also be used with the optimization algorithms. The drawback, in this case, is that the higher the number of actuators the longer will take to the algorithm to converge.

Recently, a new type of deformable mirrors suitable for the direct generation of aberrated wavefronts was designed. The modal membrane deformable mirror, MDM, relies on the use of a graphite layer electrode arrangement (see Fig. 6) for the generation of a continuous distribution of the electric field which allows the generation of the low order aberrations (defocus, astigmatism, coma) and of the spherical aberration.

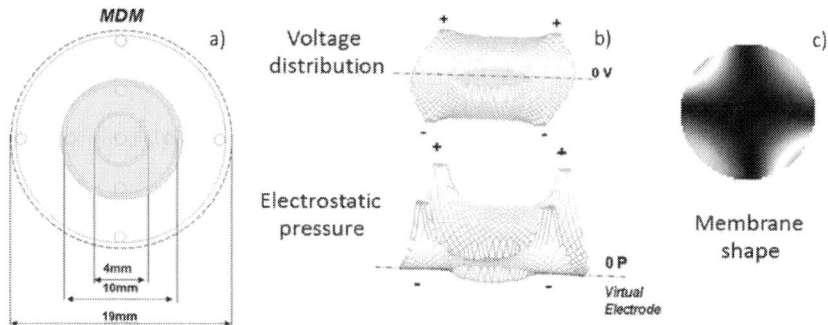

Figure 6. Electrostatic modal membrane deformable mirror, MDM. (a) Layout of the electrodes of the MDM; (b) voltage and electrostatic pressure distribution which generates the astigmatism shape illustrated in the interferogram shown in (c).

The MDM has already been demonstrated to be effective in several fields, as laser focalization [32], image sharpening and Optical Coherence Tomography (OCT), as it will be discussed later.

Another device for the generation of aberrations is the PhotoControlled Deformable Mirror (PCDM), which is schematically represented in Fig. 7. This deformable mirror [35, 36] is composed of an electrostatic membrane while the actuator pad array is replaced by a photoconductive material. Thus, the membrane shape depends on the light pattern projected on the photoconductor. Arbitrary actuator pads can be conveniently achieved by illuminating the photoconductive side of the mirror with a commercially available Digital Light Processing (DLP) hand-held projector.

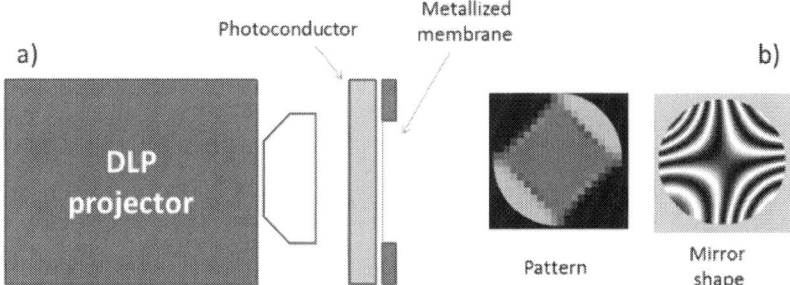

Figure 7. Photo-controlled deformable membrane mirror, PCDM. a) Schematic representation of the PCDM and the projection system allowing to achieve arbitrary actuator pads. b) Left: layout of the electrode pattern; right: correspondingly generated mirror shape; as an example the electrode pattern was chosen to generate astigmatism.

The calculation of the electrode pattern that generates a determined aberration is composed of the following steps:

a. division of the projector area into small subsets (i.e. 40 × 40);

b. calculation of the membrane shape for each of the 40 × 40 pixels, solving the Poisson equation by the iterative methods;

c. determination of the pattern by pseudoinversion of the matrix determined at point b.

A few examples of the realized electrode patterns are shown in Fig. 8, together with the corresponding measurements of the aberrated wavefronts.

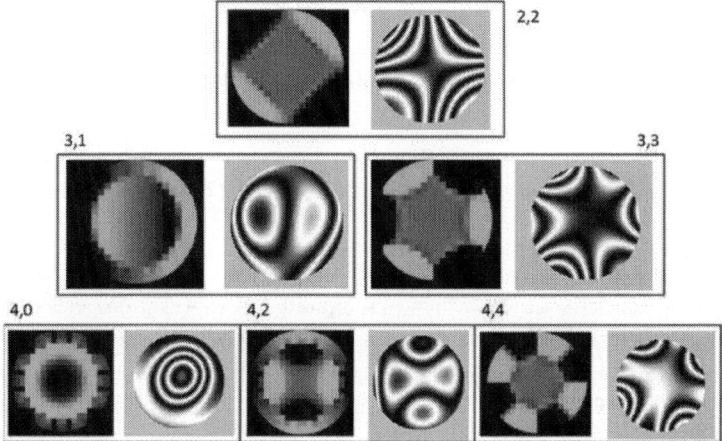

Figure 8. Generation of the first four Zernike orders with the photocontrolled deformable mirror; the light patterns necessary for their generation are on the left; the obtained corresponding interferograms are on the right.

We proved that the image quality can be considerably improved by using these adaptive devices in an image sharpening setup. For example, the MDM allowed achieving a significant image sharpening with just about 35 measurements, as illustrated in Fig. 9.

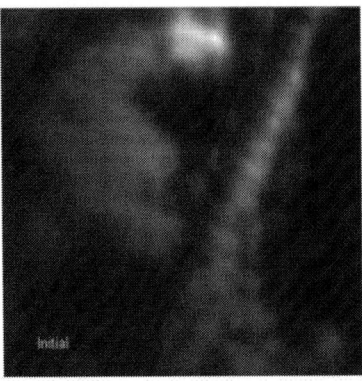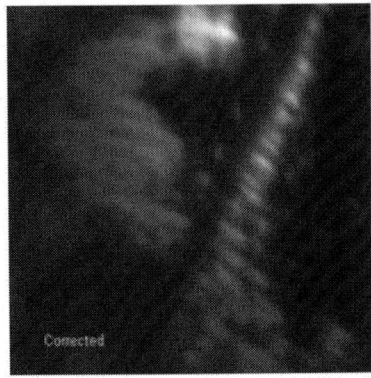

Figure 9. Optimization of an image deteriorated by aberrations; left: initial image; right: image corrected by the MDM (after 35 measurements).

3.2. Optimization of low spatial frequencies

The sharpness of an optical imaging system depends strongly on the wavefront quality. Recently [28] demonstrated that the low spatial frequency in an image can be used as a metric to perform the optimization. The process takes place by the acquisition of a series of images with the application of a predetermined aberration. The images, then, contain the information about the corrections which have to be applied to cancel the aberrations. This technique is very powerful, especially if coupled to the Lukosz modes aberration expansion. The Lukosz modes are similar to the Zernike polynomials: the difference is that the Zernike polynomials are normalized such that a coefficient of value 1 generates a wavefront with a variance of 1 rad^2, while the Lukosz functions are normalized such that a value 1 coefficient corresponds to a rms spot radius of $\lambda/(2\pi NA)$, where λ is the wavelength and NA is the numerical aperture of the focusing lens.

The peculiarity of this expansion is that the effect of the Lukosz polynomials coefficients $\{a_i\}$, on the image sharpness $I(a_i)$ is quadratic:

$$I(a_i) \approx \sum_i a_i^2.$$

This implies that the optimization of each mode can be performed independently and requires just the acquisition of three images. Then, the best point for each aberration can be found by interpolating the result with a quadratic function.

3.3. Point Spread Function (PSF) optimization

Another example of application of wavefront sensorless AO [29] consists in projecting a known point-like source through the optical system under test and then analyzing its image by means of a suitable software [36]. With the information obtained by the analysis of the point source image, the shape of a deformable mirror inserted along the optical path is modified. This process is iteratively repeated through a defined hierarchy, to gradually remove the optical aberrations.

With this technique, the point source image to be analyzed is not directly available and has to be somehow created. As an example, in the case of a fundus camera dedicated to the observation of the human retina, an illuminated pinhole can be projected on the retina itself through a dedicated optical path; this is a standard technique for this type of applications and is not going to introduce a significant complexity in the system. The light that is back-diffused by the retinal fundus acts as a point source, and its wavefront can be analyzed to estimate the aberrations present along the optical path from the retina to the detector.

3.3.1. Application of the PSF optimization in a visual optics setup

The closed loop method of correcting the aberrations of an optical system has been verified to be very stable, at least with respect to possible misalignments of the deformable mirror or aging of the mirror membrane that has been used. This stability is inherent in the adopted approach to the problem, which is less ambitious than correcting the wavefront aberration.

The described technique has been verified by means of the rather simple optical setup shown in Fig. 10. The radiation emitted by a LED diode source (SOU) is condensed by a microscope objective lens (L_{cond}) on a pinhole (PH). The radiation emerging from the pinhole is collected by a zoom collimating lens (L_{coll}). The collimated beam passes through a diaphragm (DIA) and a beam splitter (BS) and impinges normally onto a deformable mirror (M_{def}). After reflections from M_{def} and BS, the beam is compressed by an a-focal Newtonian system (L_{comp}^1 and L_{comp}^2) and can, then, follow two different paths: either a) a focusing two-lens system L_{foc} that makes the image of the pinhole on a CMOS digital camera (DET), or b) a flip mirror (M_{flip}) which deviates the beam on a wavefront analyzer (WFA). The latter has been used to measure the wavefront aberrations before and after the correction performed by the DM. With this system, both by varying the focal length of L_{coll} and tilting L_{comp}^1, it was possible to introduce controlled amounts of aberrations on the nominal pinhole image. Then, by the suitable image analysis and consequent estimate of the aberrations, the parameters needed to drive the deformable mirror to improve the image quality have been derived.

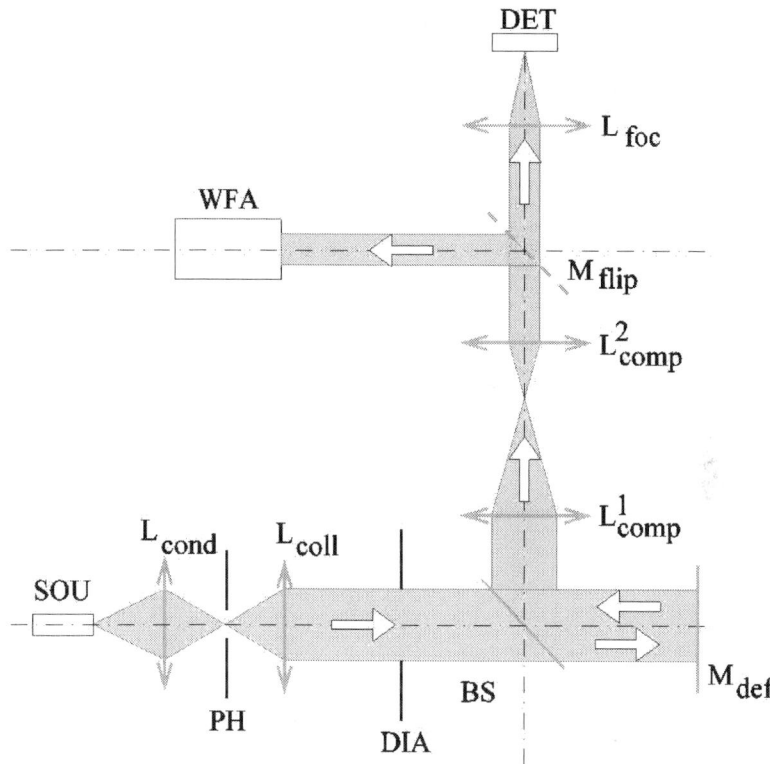

Figure 10. Schematic representation of the optical setup used for testing the capability of correcting the system aberrations with a sensorless technique. SOU: source LED diode; M$_{def}$: deformable mirror; DET: CMOS camera for detection; WFA: wavefront analyzer. See text for a complete description.

Even if the apparatus performance was constrained by the limited unidirectional sag of the deformable mirror, the obtained results proof the principle of the adopted methodology. This is clearly demonstrated in Fig. 11, which shows the wavefront error measured with the WFA before and after the deformable mirror correction for three different cases. From these graphs, and more quantitatively from the detailed analysis described in [29], it can be seen that a RMS wavefront error as low as $\lambda/10$ (@527.5 nm) can be obtained, which is a significant result for a sensorless AO system. The correction was not particularly effective only in those cases in which the unidirectionality of the mirror deformation did not allow aberration compensation, as in case of astigmatism. However, with a different choice of AO system, the system is very effective in identifying and correcting aberrations.

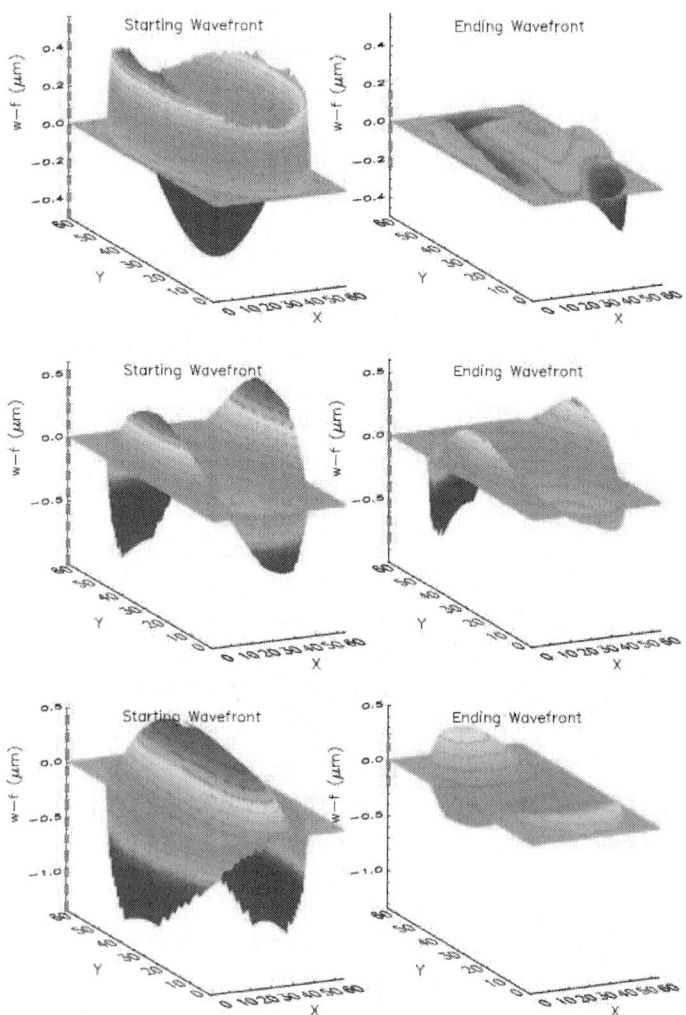

Figure 11. Wavefront plots (obtained with the wavefront analysis, WFA) before and after the correction applied by the deformable mirror for three considered cases. Top: the main aberration is defocus; Middle: the main aberrations are astigmatism and coma; Bottom: the main aberrations are defocus and astigmatism. The blue dashed lines over plotted to the Z axis represent the total wavefront excursion.

It has to be mentioned that these tests verified that the image analysis algorithm takes less than 100 cycles to reach the optimal condition; since one cycles takes approximately 1/20 - 1/25 s on a standard computer, the whole optimization takes just 4-5 s. Comparing this time with the typical times necessary to optimize other sensorless AO systems, it is evident the significant advantage of this technique, once implemented. The only limitation of this technique is that starting PSF image should give enough signal. In fact, if the point source image is too spread out, the signal to noise ratio can be very poor, substantially inhibiting the system to make a correct image analysis.

3.4. Optical Coherence Tomography (OCT)

Optical coherence tomography, OCT, is an imaging modality allowing acquisition of micrometer-resolution three-dimensional images from the inside of optical scattering media (e.g. biological tissue). OCT is analogous to ultrasound imaging, except that it makes use of light instead of sound. It relies on detecting interferometric signal created by the light back scattered from the sample and from a reference arm in a Michelson or Mach-Zehnder interferometer. OCT has many applications in biology and medicine and can be treated as a sort of optical biopsy without requirement of tissue processing for microscopic examination.

One of the interesting features of OCT is that, unlike in most optical imaging techniques, the axial and lateral resolutions are decoupled, thus allowing for an improved axial resolution, which is independent of transverse resolution. The axial resolution Δz is determined by the roundtrip coherence length of the light source and can be calculated from the central wavelength (λ_0) and the bandwidth ($\Delta\lambda$) of the light source as [37]:

$$\Delta z = \frac{2\ln2}{\pi} \frac{\lambda_0^2}{\Delta\lambda}.$$

The lateral resolution (Δx) in OCT is defined similarly to the confocal scanning laser ophthalmoscopy (cSLO), since OCT is based on a confocal imaging scheme. In many imaging systems, however, the confocal aperture exceeds the size of the Airy disc, which degrades the resolution to the value known from microscopy, i.e. [38]:

$$\Delta x = 1.22\lambda \frac{f}{D}.$$

Therefore, as for standard microscopy, AO enhanced devices might be necessary to achieve diffraction limited transverse resolution. As a result, only a combination of OCT with AO has the potential to achieve high and isotropic volumetric resolution. The use of broadband light sources that are necessary for OCT and the complexity of both the AO and the OCT technique, make the combination very challenging [39]. In general, any AO-OCT instrument can be divided into two subsystems: an adaptive optics subsystem, with wavefront sensing and wavefront correction, and an interferometric OCT subsystem. In every implementation of AO-OCT all the elements of the AO subsystem are located in the sample arm of the OCT interferometer. Indeed, there is no need to have AO correction in the reference arm because aberrations introduced within this part of the system will not influence the transverse resolution of the image. In most of the AO-OCT systems, a Shack–Hartmann wavefront sensor is used to measure aberrations and, then, to control adaptive optics correction.

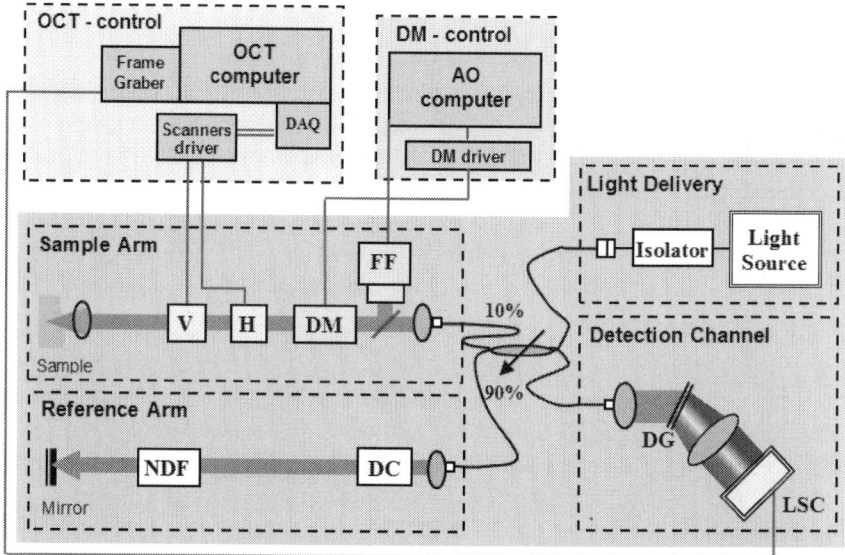

Figure 12. Schematic representation of the system for sensorless adaptive optics - optical coherence tomography. Note that there is no wavefront sensor in the sample arm. The far-field camera (FF) is used to check if the AO correction generates improved focal spots. DM : deformable mirror; V: vertical mirror galvanometer; H: horizontal mirror galvanometer. In the reference arm: NDF is a neutral density filter. The detection channel comprises a grating (DG) and a linear CCD detector (LSC). The quality of the image acquired with the OCT detection channel is used to search for DM shapes that correct aberrations in the imaged sample. The imaging system used to acquire the data was developed in the Vision Science and Advanced Retinal Imaging Laboratory (VSRI). Details of the OCT system components can be found in [40]. Here, we briefly describe the main characteristics of the system. In the current configuration, the light source for OCT was a superluminescent diode (Broadlighter) operating at 836 nm and with a 112 nm spectral bandwidth (Superlum LTD), allowing to achieve a 3.5 μm axial resolution. The beam diameter at the last imaging objective was 6.7 mm, allowing for up to 10 μm lateral resolution when a 50 mm focal length imaging objective was used. The AO correction was optimized by using the intensity of the AO-OCT en-face projection views during the volumetric data acquisition. In the current system configuration, we have used about 9 mm diameter of the modal deformable mirror. The light reflected from the sample is combined with the light from the reference mirror, and then sent to a spectrometer. There, a CCD line detector acquires the OCT spectrum.

Bonora and Zawadzki recently demonstrated that sensorless correction can be implemented in optical coherence tomography by using a specially developed resistive deformable mirror. This novel modal deformable mirror, MDM, was successfully employed in the UC Davis AO-OCT system to image static samples, test targets and tissue phantoms. Fig. 12 shows a schematic representation of the sensorless AO-OCT system used in the experiments.

To test the performance of our sensorless AO-OCT system, we evaluated the image quality of a sample, consisting of a USAF resolution test chart with an adhesive tape glued to its front side, after insertion of a trial lens with 0.5 Diopter astigmatism in front of the imaging objective. We were able to achieve improved resolution by using the following merit function S [41] on the OCT en-face projection images

$$S = \int I^2(x, y)\,dxdy,$$

where I(x,y) is the intensity in the OCT en-face image plane. This approach is simillar to PSF optimization. In fiber based OCT systems single mode fiber introduces OCT beam to the sample and also act as detector for back scattered light. Therefore we have a point source that is imaged by the optical system and the confocal pinhole that allows direct mesurment of light intensity trougput by the system. As expected, the algorithm performed the optimization by adjusting only defocus and astigmatism (see Fig. 13).

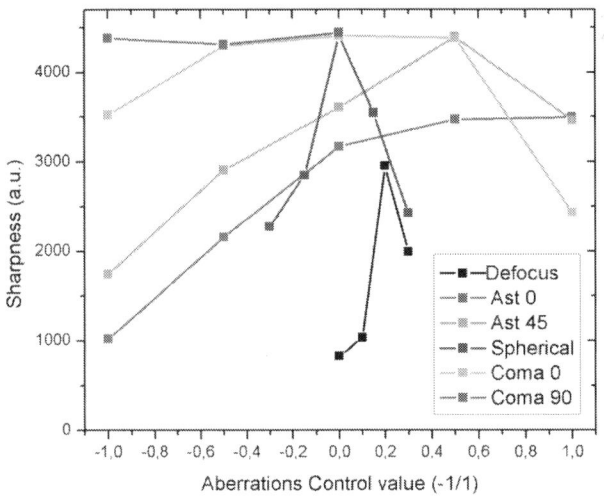

Figure 13. Graph of the Merit Function of AO-OCT images for different values of aberrations generated by the modal deformable mirror. Note that higher values correspond to better AO-corrections.

Fig. 14 shows some examples of the en-face projection views extracted from OCT volumes: there are the initial view acquired from the sample, and three improved views after correction of additional aberrations, namely, defocus and two astigmatisms. Clearly, at each correction step the images of the test target get sharper. Additionally, the features of the adhesive tape attached to the back of the Air Force test target become more visible as well.

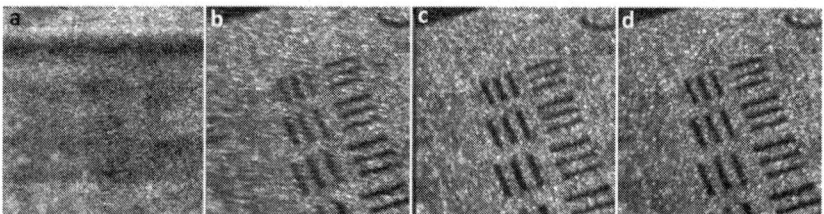

Figure 14. En-face projection views of the AO-OCT images of the test target for the best corrected values of the Zernike coefficients; (a) before correction, (b) after defocus correction, (c) after defocus and Ast 0° correction, (d) after defocus, Ast 0° and Ast 45° correction.

These recent results demonstrate that wavefront sensorless control is a viable option for imaging biological structures for which AO cannot establish a reliable wavefront that could be corrected by a wavefront corrector. Future refinements of this technique, beyond the simple implementation presented in this chapter, should allow its extension to in-vivo applications. An example of sensorless adaptive optics scanning laser ophthalmoscopy (AO-SLO) for imaging in-vivo human retina has been recently presented [42].

3.5. Laser process optimization

Similarly to the optimization process presented in section 2.1.1 [24], we report here about the optimization of a laser process by the use of a sensorless AO [43]. In the former case, the generation of harmonics from an ultrafast laser was improved by the use of a genetic algorithm. In the latter case, an algorithm derived from the image-based procedure was employed in conjunction with the use of a MDM deformable mirror similar to the one described in section 3.1. The advantages in terms of experimental complexity and convergence time are discussed in the given reference.

In the sensorless case, the laser source was a tunable high energy mid-IR (1.2μm-1.6μm) optical parametric amplifier with 10 Hz repetition rate [44]. The harmonics of the laser were generated by the interaction of the laser pulses with a krypton gas jet. In this system, the infrared pulses and the slow repetition rate made inconvenient, respectively, the use of a wavefront sensor and of an optimization algorithm needing hundreds of iterations.

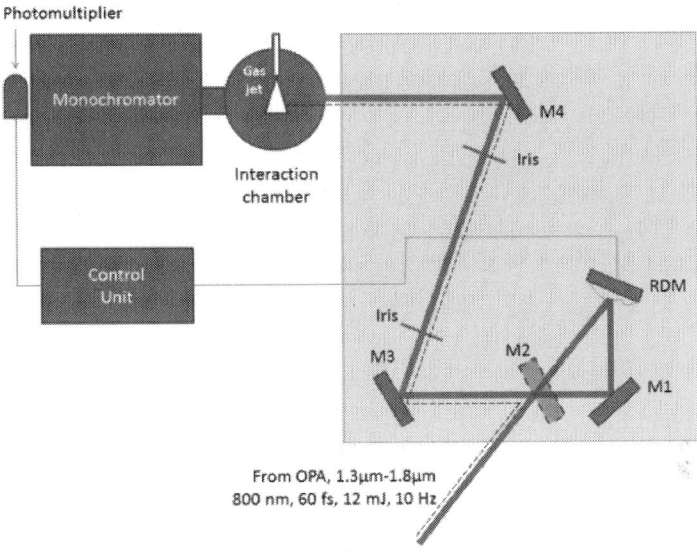

Figure 15. Experimental setup for the generation of harmonics from a femtosecond tunable high-energy mid-IR optical parametric amplifier, OPA. Dotted line: optical path before the insertion of the MDM. Red line: optical path realized for the experiment with the deformable mirror.

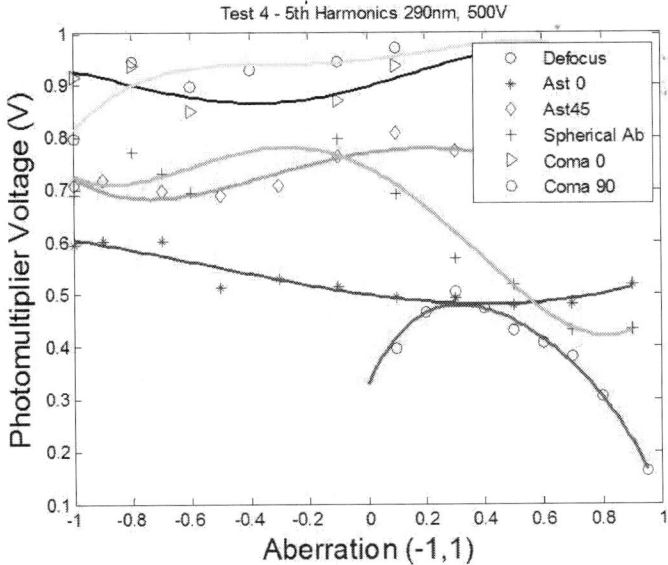

Figure 16. Optimization of the voltage generated by the photomultiplier over a 50 Ω load for the 5th harmonic at 290 nm, obtained by the use of krypton gas.

The experimental setup used for this application is illustrated in Fig. 15. To demonstrate the easiness of integrating the sensorless AO device within the experiment, the optical path before the DM is shown with a dotted line. The additional elements are simply a plane mirror and a resistive MDM, which have been introduced without any complex operations. The system optimization consisted in the increase of the harmonic signal detected by the photomultiplier at the output of the monochromator. The obtained result is illustrated in Fig. 16, where it is possible to see that the photon flux on the photomultiplier is doubled with respect to the one obtained after the correction of the defocus.

4. CONCLUSIONS

In adaptive optics the choice of the optimal correction strategy depends on the required application, desired image quality, and affordable complexity/cost of the final system. In this context, sensorless adaptive optics provides several solutions, most of them implementable at a simplified and relatively low-cost level, that can be exploited for a wide range of applications.

We have presented here both a review of the most diffused systems used in sensorless adaptive optics and some recently developed algorithms and devices. Essentially, two different approaches are employed: those based on random search and the subsequent application of evolutionary strategies, and those based on the application of some bias aberration. In general, the second class of algorithms present a faster convergence.

We have shown several application examples in different fields, such as the optimization of ultrafast nonlinear optical systems for the generation of high order harmonics, the image sharpening in microscopy applications and the enhancement of optical coherent tomography.

Sensorless adaptive optics appears, therefore, as having a great potential for finding new applications in current and future technologies. The continuous improvement of the optimization algorithms and development of novel deformable mirror devices, make the integration of AO into various optical systems increasingly easier. Particularly, the conjunction of sensorless AO with OCT might open the way to a new generation of diagnostic imaging.

REFERENCES

1.   Babcock H.W., The Possibility of Compensating Astronomical Seeing, Publication of the Astronomical Society of the Pacific Vol.65, No. 386, pp. 229, (1953).

2. Tyson R., Tharp J., Canning D., Measurement of the bit-error rate of an adaptive optics

3. Hardy J.W., Adaptive Optics

4. Liang J., Williams D., Miller D., Supernormal vision and high resolution retinal imaging through adaptive optics

5. Irwan R., Lane R., Analysis of optimal centroid estimation applied to Shack Hartmann sensing, Applied Optics Vol. 38, No. 32, pp. 6737-6743, (1999).

6. Porter J. , Queener H. , Lin J., Thorn K. E., Awwal A., Adaptive Optics

7. Roorda A., Adaptive optics for studying visual function: A comprehensive review, Journal of Vision, Vol. 11, No. 5, pp. 1-21 (2011).

8. Liang J., Williams D., Aberrations and retinal image quality of the normal human eye, Journal of the Optical Society of America A, Vol. 14, No. 11, pp. 2873-2883, (1997).

9. Zhu L., Sun P., Bartsch D., Freeman W., Fainman Y., Adaptive control of a micro-machined continuous-membrane deformable mirror for aberration compensation, Applied Optics, Vol. 38, No. 1, pp. 168-176, (1999).

10. Le Gargasson J.-F., Glanc M., Léna P., Retinal imaging with adaptive optics

11. Roorda A., Romero-Borja F., Donnelly W., Queener H., Hebert T., Campbell M., Adaptive optics scanning laser ophthalmoscopy, Optics Express, Vol. 10, No. 9, pp. 405-412, (2002).

12. Zawadzki R., Jones S., Olivier S., Zhao M., Bower B., Izatt J., Choi S., Laut S., Werner J, Adaptive-optics optical coherence tomography

13. Zommer S., Ribak E., Lipson S., Adler J., Simulated annealing in ocular adaptive optics

14. Gray D., Merigan W., Wolfing J., Gee B., Porter J., Dubra A., Twietmeyer T., Ahamd K., Tumbar R., Reinholz F., Williams D., In vivo fluorescence imaging of primate retinal ganglion cells and retinal pigment epithelial cells, Optics Express, Vol. 14, No. 16, pp. 7144-7158, (2006).

15. Fernandez E., Vabre L., Adaptive optics with a magnetic deformable mirror: application in the human eye, Optics Express, Vol. 14, .No. 20, pp. 8900-8917, (2006).

16. Tyson R., 1999, Adaptive Optics

17. Tyson R., Tharp J., Canning D., Measurement of the bit-error rate of an adaptive optics

18. Albert O., Sherman L., Mourou G., and Norris T., Smart microscope: an adaptive optics

19. Neil MAA., Juskaitis R., Booth M.J., Wilson T., Tanaka T., Kawata S., Adaptive Aberation correction in two-photon microscope, Journal of microscopy, Vol 200, pp 1-5-108 (2000).

20. Booth M.J., Neil M.A.A., Juskaitis R., Wilson T., Adaptive aberration correction in a confocal microscope, PNAS, Vol. 99, No. 9. Pp. 5788-5792 (2002).

21. Brida D., Manzoni C., Cirmi G., Marangoni M.,, Bonora S., Villoresi P., De Silvestri S., Cerullo G., (2010), Few-optical-cycle pulses tunable from the visible to the mid-infrared by optical parametric amplifiers, Journal of Optics, Vol. 12, No. 1, (January 2010), 2040-8978

22. Okada T., Ebata K., Shiozaki M., Kyotani T., Tsuboi A., Sawada M., Fukushima M., Development of adaptive mirror for CO_2 laser, in High-Power Lasers in Manufacturing, X. Chen, T. Fujioka, and A. Matsunawa, eds., Vol. 3888 of SPIE Proc., pp. 509-520, (2000).

23. Jackel S., Moshe I., Adaptive compensation of lower order thermal aberrations in concave-convex power oscillators under variable pump conditions, Optical Engineering Vol. 39, No. 09, pp. 2330-2337, (2000).

24. Villoresi P., Bonora S., Pascolini M., Poletto L., Tondello G., Vozzi C., Nisoli M., Sansone G., Stagira S., De Silvestri S., Optimization of high-order-harmonic generation by adaptive control of sub-10 fs pulse wavefront, Optics Letters, Vol. 29, No.2, pp. 0146-9592, (2004).

25. Zacharias R., Beer N., Bliss E., Burkhart S., Cohen S., Sutton S., Atta R.V., Winters S., Salmon J.T., Stolz M. L. C., Pigg D., Arnold T., Alignment and wavefront control systems of the National Ignition Facility, Optical Engineering, Vol. 43, No. 12, pp. 2873-2884, (2004).

26. Gonté F., Courteville A., Dandliker R., Optimization of single-mode fiber coupling efficiency with an adaptive membrane mirror, Optical Engineering, Vol. 41, No. 5, pp. 1073-1076, (2002).

27. Bonabeau E., Dorigo M. and Theraulaz G., Inspiration for optimization from social insect behaviour, Nature Vol. 46, pp. 39-42, (2000).

28. Debarre D., Booth M.J. and Wilson T., Image based adaptive optics

29. Naletto G., Frassetto F., Codogno N., Grisan E., Bonora S., Da Deppo V., Ruggeri A. (2007), No wavefront sensor adaptive optics

30. Judson R.S., Rabitz H., Teching lasers to control molecules, Phys. Rev. Lett., Vol. 68, No. 10, pp. 1079-7114 (1992).

31. Minozzi M., Bonora S., Vallone G., Segienko A., Villoresi P., Bi-photon generation with optimized wavefront by means of Adaptive Optics

32. Bonora S., Distributed actuators deformable mirror for adaptive optics

33. Bonora S., Capraro I., Poletto L., Romanin M., Trestino C., Villoresi P., Fast wavefront active control by a simple DSP-Driven deformable mirror, Review of Scientific Instruments, Vol. 77, No. 9, pp. 0034-6748 ,(2006).

34. Bortolozzo U., Bonora S., Huignard J.P., Residori S., Continuous photocontrolled deformable membrane mirror, Applied Physics Letters, Vol. 96, No.25, pp. 0003-6951, (2010).

35. Bonora S., Coburn D., Bortolozzo U., Dainty C., Residori S., High resolution wavefront correction

36. Grisan E., Frassetto F., Da Deppo V., Naletto G., Ruggeri A., No wavefront sensor adaptive optics

37. Fercher AF, Hitzenberger CK. Optical coherence tomography

38. Zhang Y, Roorda A. Evaluating the lateral resolution of the adaptive optics

39. Pircher M., Zawadzki R.J. , Combining adaptive optics

40. Zawadzki R.J., Jones S.M., Pilli S., Balderas-Mata S., Kim D., Olivier S.S., Werner J.S., Integrated adaptive optics

41. Muller R. A., Buffington A., Real-time correction of atmospherically degraded telescope images through image sharpening, J. Opt. Soc. Am., Vol. 64, No. 9, pp. 1200–1210, (1974).

42. Hofer H., Sredar N., Queener H., Li C., Porter J., Wavefront sensorless adaptive optics

43. Bonora S., Frassetto F., Coraggia S., Spezzani C., Coreno M., Negro M., Devetta M, Vozzi, C., Stagira, S. Poletto L., Optimization of low-order harmonic generation by exploitation of a deformable mirror, Applied Physics B, Vol. 106, No. 4, pp.905-909, (2011).

44. Vozzi C., Calegari F., Benedetti E., Gasilov S., Sansone G., Cerullo G., Nisoli M., De Silvestri S., Stagira S., Millijoule-level phase-stabilized few-optical-cycle infrared parametric source, Opt. Lett., Vol. 32, No. 20, pp. 2957-2959 (2007).

Chapter 5

A Solar Adaptive Optics System

Ren Deqing[1, 2, 3] and Zhu Yongtian[2, 3]

[1] Physics & Astronomy Department, California State University Northridge, USA
[2] National Astronomical Observatories/Nanjing Institute of Astronomical Optics & Technology, Chinese Academy of Sciences, China
[3] Key Laboratory of Astronomical Optics & Technology, National Astronomical Observatories, Chinese Academy of Sciences, China

1. INTRODUCTION

Solar activities are dominated by magnetic fields, which are arranged in small structure. The structure and evolution of small-size magnetic fields are the key component in a unified understanding of solar activities [1]. As such, a major application of a large solar telescope is for high-sensitivity observations of solar magnetic fields. The observation of solar dynamics of small-scale magnetic fields requires un-compromised high resolution, high magnetic field sensitivity, and high temporal resolution [2, 3]. The two important scales that determine the structuring of the solar atmosphere are the pressure scale height and the photon mean free path, which are of on the order 70 km or 0.1". Recently, structures as small as a few tens of kilometers on the solar surface corresponding to a few tens of milli-arcseconds on the sky have been predicted by sophisticated MHD models of the solar atmosphere [4-7]. For a ground-based telescope, however, the atmospheric turbulence will seriously degrade the actual performance for high-resolution imaging, and an adaptive optics (AO) system is needed to recover the theoretical diffraction-limited angular resolution in real-time scale [8].

Current major solar telescopes have been equipped with dedicated AO systems that adopt different techniques for real-time wave-front sensing and image signal processing [9]:

1. The AO system with the 0.76-meter Dunn solar telescope uses Digital Signal Processors (DSPs) for the real-time signal processing [10]. DSPs are superb for fast calculation for digital image processing. However it is time-consuming for the DSP programming, and it lacks flexibility.

2. The 0.7-m Vacuum Tower Telescope at Teide Observatory uses multiple Processors (CPUs) on workstation computers and low-level programming language such as C++ for AO programming [11, 12].

The performance of the AO systems with multi-CPUs is close to those with DSPs. However, the low-level C++ programming is also time-consuming. Recent CPU developments indicate that multi-core technique is superior over that of the multi-CPU in view of the calculation speed and power consuming. A detailed review of solar adaptive optics was discussed by Rimmele and Marino [13].

Due to the rapid development of multi-core personal computers and the powerful parallelism of the LabVIEW software, we proposed a novel solar AO system that is based on today's multi-core CPUs and "high-level" LabVIEW programing [14]. The Portable Solar Adaptive Optics (PSAO) system at California State University Northridge (CSUN) is designed to deliver diffraction-limited imaging with 1~2-m class telescopes which will cover the largest solar telescope currently operational. This AO is optimized for a small physical size, so that we can carry it to any available solar telescope as a visiting instrument for scientific observations. We use personal computers with Intel i7 multi-core CPUs for the AO real-time control, and use LabVIEW software for AO programming. LabVIEW, developed by the National Instruments (NI), is based on block diagram programming, which makes it inherently supporting multi-core or multi-thread calculation in parallel. LabVIEW also includes a large number of high-quality existing functions for mathematical operations and image processing, which makes the AO programing extremely efficient and is suitable for the real-time AO programming.

Since 2009, we have built and continually updated our PSAO system in our laboratory [15]. We have initially tested the PSAO with the 0.6-m solar telescope at San Fernando Observatory (SFO) as well as the 1.6-m McMath-Pierce telescope (McMP). In this paper, we will present recent results in the development of the PSAO in the laboratory and the on-site trial observations.

2. DESIGN PHILOSOPHY

2.1. Optical Design

The PSAO must be able to work with any solar telescopes with different aperture size and focal ratios, although it was initially developed for testing with the 0.6-meter vacuum solar telescope located at the San Fernando Observatory, CSUN. For such an application, the PSAO optics consists of two individual parts. The first part is the fore-optics, while the second part is the main AO optics. The PSAO optics layout is shown in Figure 1. The fore-optics consists of L1, M1, L2 and M2, while the main AO optic consists of the remaining optical components. Where, L1 and L2 are two lenses and M1 and M2 are two fold mirrors. All the optical components are off-the-self parts. The function of the fore-optics is to convert a telescope's focal ratio to f/54, and create an exit pupil at infinite distance. i.e. create a telecentric image at f/54: in Figure 1, the telescope focal plane image IM0 is first collimated by lens L1, which forms a pupil image on the fold mirror M1. The pupil image is located one focal length distance from lens L2. In such a way, lens L2 forms a solar image at IM1 with the exit pupil at infinite. By adjust the focal length ratio between L1 and L2, one can convert the telescope image IM0 to a telecentric image of f/54 at the IM1; the main AO optics is fixed even with different telescopes, except that the wave-front sensor lenslet array (L6 in Figure. 1) can be chosen from a set of lenslet arrays for different telescopes and seeing conditions. In this way, we only need to adjust the fore-optics without any change for the main AO optics, which makes the PSAO suitable with any solar telescope. For example, the 1.5-m McMP telescope, located at the Kitt Peak National Solar Observatory (NSO), has a focal ratio of f/54 at the focal plane. When working with the McMP, both lenses L1 and L2 are identical and have a focal length of 250mm. As shown in the Figure 1, we use two lenses L1 and L2 as the fore-optics which is of a typical telecentric optics design. The whole AO optics uses several flat fold mirrors (M1, M2, M3, M4) to fold the optical path and reduce the overall physical size. The fold mirror M1 in the fore-optics also serves as a tip-tilt mirror (TTM). The output focal plane image after the fore-optics (L2) is collimated by the lens L3, which creates a pupil image with a size of ~ 4.4-mm on the deformable mirror (DM). Please note that the fold mirror M4 also serves as the DM. After the DM, the beam is split as several parts by two beam splitters B1 and B2, which are used for DM wave-front sensing, tip-tilt sensing and focal plane imaging, respectively. Currently, our AO system has its individual optical channels for DM wave-front as well as tip-tilt sensing, respectively. The DM wave-front sensor (WFS) consists of lenses L4, L5, a lenslet array L6 (for clarity, only one lenslet is shown) and a WFS camera, while the tip-tilt sensor (TTS) consists of the lens L8 and the tip-tilt camera only.

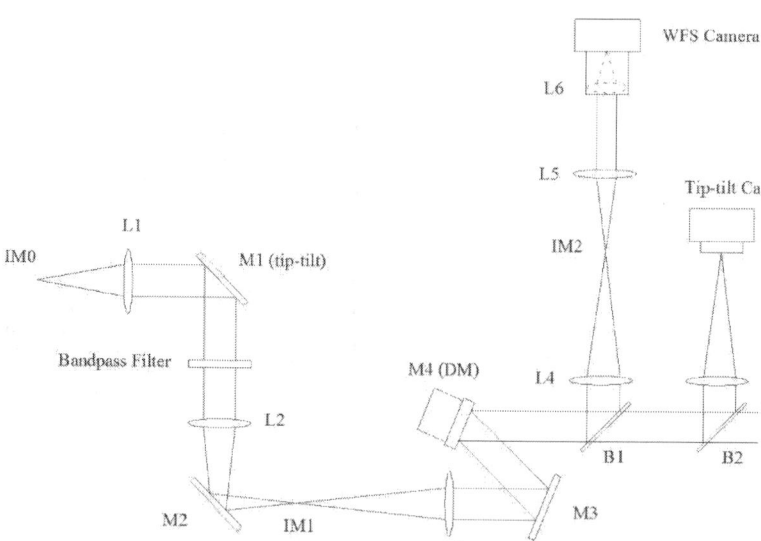

Figure 1. The optical layout of the PSAO.

For the DM WFS channel, lens L4 forms a telecentric solar image IM2, which is collimated subsequently by lens L5. A pupil image is formed one focal length distance behind lens L5, where the lenslet array L6 is located to sample the pupil image for proper wave-front sensing. This is a typical configuration of a Shack-Hartmann wave-front sensor, except that the field of view (FOV) formed by each lenslet must have a suitable size for wave-front sensing with a two-dimensional solar structure. A typical field size for solar wave-front sensing is in a range of 15″ ~ 20″, which is a compromise between the wave-front sensing speed and the sensing accuracy. The TTS is very simple. A lens L8, which is a zoom lens, is used to form a solar image directly on the tip-tilt camera for the calculation of the overall image movement. The corrected image is fed directly to the science camera via the lens L9, which is changeable for different image scales with different telescopes.

The WFS - DM and TTS - TTM channels are controlled by two high performance personal computers to form two individual correction loops, respectively; one is for the DM wave-front correction and the other is for the tip-tilt correction. The DM and TTM are both conjugated on the telescope pupil, which eliminates the pupil wander problem and ensures the AO correction extremely stable. A field stop is placed on the telescope focal plane IM0 or on the solar image plane IM1. An adjustable field stop is also located on the solar image plane IM2 to limit the field size for wave-front sensing. The lenslet array is integrated with the WFS camera directly via the camera's C mount, and can be replaced for different focal lengths, which makes the PSAO suitable for different telescopes and seeing conditions.

A band-pass filter is located just before lens L2. The filter has a band-pass width of ~ 100 nm, which will limit the light energy on the small DM. This is not a problem for solar scientific observations, which only need to work on narrow band in most situations. In fact, a filter wheel can be used for the observations at any band. The size of the AO field of view is controlled by the aperture size of a field stop located on the IM0 (or IM1), which limits the AO field of view as 60"x60". The field of view for the TTS is also set the same as that of the AO field of view, and is sampled by 60x60 pixels of a "region of interest" of the tip-tilt camera, which results in a sampling scale of 1"/pixel. The primary optical specifications are listed in Table 1.

Table 1.PSAO Optical Specifications.

AO FOV	TTS FOV	WFS FOV	Wavelength range
60"x60"	60"x60"	8"x8" ~ 30"x30"	0.6 ~ 1.5μm

The use of an individual DM wave-front sensor as well as a tip-tilt sensor has some benefits. In addition to avoid the pupil wander for the wave-front sensing, which will deliver a super stable AO system at different seeing conditions, it will allow the use of small field of view for wave-front sensing at good seeing condition, which can further improve the wave-front sensing sensitivity or accuracy. Since the PSAO has its individual DM wave-front and tip-tilt sensors, the TTS can be used to measure the overall wave-front movement in the large 60"x60" FOV. As the wave-front tip-tilt component is corrected by the tip-tilt mirror, the DM WFS can use a small FOV, such as 8"x8", for accurate wave-front sensing. Since each WFS lenslet sub-aperture is sampled by 30x30 pixels in our WFS, a 8"x8" FOV corresponds to a WFS sampling scale of 0.27"/pixel, compared to the 1.0" /pixel sampling scale for a 30"x30" WFS FOV that may be used in poor seeing conditions. This can significantly improve the wave-front sensing accuracy and thus deliver a better AO performance.

2.2. Wave-Front Sensing

Solar wave-front sensing uses a Shark-Hartmann wave-front sensor. The wave-front gradient or slope vector at each sub-aperture of the lenslet array is solved by the cross-correlation calculation of a two-dimensional pattern over a field of view. The correlation function C(x,y) of a sub-aperture S(x,y) and a reference pattern R(x,y) can be calculated over the two-dimensional sub-aperture as

$$C(x, y) = FFT^{-1}[FFT(S) \cdot (FFT(R))^*]$$

(1)

the asterisk denotes the complex conjugate. FFT and FFT^{-1} represent Fourier and inverse Fourier transfers, respectively. The resulting correlation function is "star-like" and can be treated like a star in the stellar WFS. The position of the maximum value of the correlation function corresponds to a slope vector where the reference pattern of R(x,y) best matches the sub-aperture pattern of S(x,y). The calculations of the Fourier and inverse Fourier transfers for the so-called pattern match are time consumed, which makes the high-speed AO correction challenging.

If wave-front phase ϕ is described by the Zernike polynomial expansion as

$$\varphi = \sum_{k=1}^{K} a_k Z_k (x, y)$$

(2)

From equation 2, the slope vector s of the WFS and mode coefficient vector a are associated as

$$s = [B] a$$

(3)

where,

$$S = (\frac{\partial \varphi}{\partial x}\Big|_1 \dots \dots \frac{\partial \varphi}{\partial x}\Big|_M \frac{\partial \varphi}{\partial y}\Big|_1 \dots \dots \frac{\partial \varphi}{\partial y}\Big|_M)^T$$

(4)

$$a = (a_1 \quad a_2 \dots \dots a_k)^T$$

(5)

while the matrix [B] is

$$[B] = \begin{pmatrix} \frac{\partial Z_1(x,y)}{\partial x}\Big|_1 & \frac{\partial Z_2(x,y)}{\partial x}\Big|_1 & \dots & \frac{\partial Z_K(x,y)}{\partial x}\Big|_1 \\ \vdots & & & \\ \frac{\partial Z_1(x,y)}{\partial x}\Big|_M & \frac{\partial Z_2(x,y)}{\partial x}\Big|_M & \dots & \frac{\partial Z_1(x,y)}{\partial x}\Big|_M \\ \frac{\partial Z_1(x,y)}{\partial y}\Big|_1 & \frac{\partial Z_2(x,y)}{\partial y}\Big|_1 & \dots & \frac{\partial Z_K(x,y)}{\partial y}\Big|_1 \\ \vdots & & & \\ \frac{\partial Z_1(x,y)}{\partial y}\Big|_M & \frac{\partial Z_2(x,y)}{\partial y}\Big|_M & \dots & \frac{\partial Z_1(x,y)}{\partial y}\Big|_M \end{pmatrix}$$

$$[B] = \begin{pmatrix} \dfrac{\partial Z_1(x,y)}{\partial x}\bigg|_1 & \dfrac{\partial Z_2(x,y)}{\partial x}\bigg|_1 & \cdots & \dfrac{\partial Z_K(x,y)}{\partial x}\bigg|_1 \\ \vdots & & & \\ \dfrac{\partial Z_1(x,y)}{\partial x}\bigg|_M & \dfrac{\partial Z_2(x,y)}{\partial x}\bigg|_M & \cdots & \dfrac{\partial Z_1(x,y)}{\partial x}\bigg|_M \\ \dfrac{\partial Z_1(x,y)}{\partial y}\bigg|_1 & \dfrac{\partial Z_2(x,y)}{\partial y}\bigg|_1 & \cdots & \dfrac{\partial Z_K(x,y)}{\partial y}\bigg|_1 \\ \vdots & & & \\ \dfrac{\partial Z_1(x,y)}{\partial y}\bigg|_M & \dfrac{\partial Z_2(x,y)}{\partial y}\bigg|_M & \cdots & \dfrac{\partial Z_1(x,y)}{\partial y}\bigg|_M \end{pmatrix}$$

$$[B] = \begin{pmatrix} \dfrac{\partial Z_1(x,y)}{\partial x}\bigg|_1 & \dfrac{\partial Z_2(x,y)}{\partial x}\bigg|_1 & \cdots & \dfrac{\partial Z_K(x,y)}{\partial x}\bigg|_1 \\ \vdots & & & \\ \dfrac{\partial Z_1(x,y)}{\partial x}\bigg|_M & \dfrac{\partial Z_2(x,y)}{\partial x}\bigg|_M & \cdots & \dfrac{\partial Z_1(x,y)}{\partial x}\bigg|_M \\ \dfrac{\partial Z_1(x,y)}{\partial y}\bigg|_1 & \dfrac{\partial Z_2(x,y)}{\partial y}\bigg|_1 & \cdots & \dfrac{\partial Z_K(x,y)}{\partial y}\bigg|_1 \\ \vdots & & & \\ \dfrac{\partial Z_1(x,y)}{\partial y}\bigg|_M & \dfrac{\partial Z_2(x,y)}{\partial y}\bigg|_M & \cdots & \dfrac{\partial Z_1(x,y)}{\partial y}\bigg|_M \end{pmatrix}$$

$$[B] = \begin{pmatrix} \dfrac{\partial Z_1(x,y)}{\partial x}\bigg|_1 & \dfrac{\partial Z_2(x,y)}{\partial x}\bigg|_1 & \cdots & \dfrac{\partial Z_K(x,y)}{\partial x}\bigg|_1 \\ \vdots & & & \\ \dfrac{\partial Z_1(x,y)}{\partial x}\bigg|_M & \dfrac{\partial Z_2(x,y)}{\partial x}\bigg|_M & \cdots & \dfrac{\partial Z_1(x,y)}{\partial x}\bigg|_M \\ \dfrac{\partial Z_1(x,y)}{\partial y}\bigg|_1 & \dfrac{\partial Z_2(x,y)}{\partial y}\bigg|_1 & \cdots & \dfrac{\partial Z_K(x,y)}{\partial y}\bigg|_1 \\ \vdots & & & \\ \dfrac{\partial Z_1(x,y)}{\partial y}\bigg|_M & \dfrac{\partial Z_2(x,y)}{\partial y}\bigg|_M & \cdots & \dfrac{\partial Z_1(x,y)}{\partial y}\bigg|_M \end{pmatrix}$$

(7)

Here, M is the number of WFS sub-apertures. K is the number of Zernike modes. The mode coefficient vector is found by finding the pseudo-inverse of [B], which is solved by using the singular value decomposition (SVD) as,

$$[B] = UDV^T$$

(7)

and,

$$[B]^{-1} = VD^{-1}U^T$$

(8)

Once the DM's influence function is known, the measured wave-front is used to find the DM' signals (i.e. voltages) that are required to correct the wave-front error. The Zernike polynomials act as a low-band pass filter, which is used to measure and control the actual wave-front error up to the mode number K. The choice of actual mode number that an AO system can correct should consider the system's stability, which can be determined by the condition number as discussed by Kasper et al. [16].

2.3. Electrics and Programming

All the PSAO's hardware is based on off-the-shelf commercial components, which makes a low cost system possible. The performance of our AO system can continue to improve once better components are available on the market. The current AO WFS loop uses a computer equipped with a first-generation Intel i7 -990X CPU, which has 6 cores and 12 threads for parallel computation. This computer can be updated to a second-generation Intel i7 CPU that should deliver a better performance, or even updated to a computer with two recent Xeon CPUs that will have 16 cores and 32 threads in total, which is expected to be two times faster than the current system. The specifications of current hardware components are listed in the Table 2.

Table 2.PSAO Hardware specifications.

Hardware	Specifications
DM	140 actuators, 3.5μm stroke, 14-bit resolution, 4.4-mm clear aperture, 8000 Hz frame rate.
TTM	PI S-330.4SL, 5mrad stroke range, 0.25μrad resolution, 1600 Hz frame rate.
WFS & Tip-tilt Camera	1024x1024 pixels, 10.6μm pixel size, 150Hz frame rate at full resolution.
Image Grabbers	NI PCIe-1429 Camera-Link image grabber.
WFS lenslet arrays	0.3mm pitch, 4.7mm, 8.7mm, 18.8mm focal lengths.
Computer 1 (for WFS loop)	Intel Core i7-990X @ 3.47GHz, 8GB RAM.
Computer 2 (for TTS loop)	Intel Core i7-980X @ 3.33GHz, 4GB RAM.

The DM and TTM are two critical components for the AO system. We use the high-speed Multi-DM from Boston Micronmachines Corporation (BMC), which is a micro-electro-mechanical-systems (MEMS) deformable mirror and has 140 actuators (in a 12x12 array without 4 corners) with 3.5μm stroke and can deliver a frame rate up to 8000 Hz. The DM has a clear aperture of 4.4 mm only. Although this allows for a small beam size and makes the whole AO system smaller in physical size, a DM with a clear aperture on the order of 8~10mm will be preferred for our AO system, which will not significantly increase the physical size and will be more robust to the alignment error between the DM and the WFS, such as the error introduced by the vibration from where the AO is located. If the DM had an 8-mm clear aperture, for example, the focal ratio at IM1 in Figure 1 could be f/27, instead of f/54, and in this case lens L3 will has the same focal length with that for the 4.4-mm DM, and thus will not increase the overall AO physical size. The TTM is a flat mirror mounted on a S-330.4SL piezo-tilt platform from Physik Instrumente (PI), which has 5-mrad stroke and can deliver an actual frame rate of 700 Hz only with the digital USB input port, although the datasheet claims that it has a unloaded resonant frequency of 3.3 kHz, and a resonant frequency of 1.6 kHz loaded with a 25 x 8 mm glass mirror. The strokes of our DM and TTM are both sufficient for the AO requirements. In theory, the WFS can simultaneously measure tip-tilt and high-order wave-front errors so that the DM and TTM can be controlled by one computer only. However the current TTM maximum frequency is too slower, comparing to that of the DM, which will reduce the overall AO correction speed, if both were controlled by a single closed-loop. In order to keeping the DM correction fast, we split the DM wave-front and tip-tilt corrections as two individual close-loops, and use two computers to control the DM and TTM, separately.

Both the DM wave-front sensor and the tip-tilt sensor adopt a high-speed MV-D1024E-CL160 camera made by Photonfocus, respectively. The camera transfers image data via the base camera-link interface at a speed of 255MB/s. The output data from each camera is acquired by a high-performance NI PCIe-1429 Camera-Link image grabber connected to the associated controlling computer via a PCIe slot. We chose the NI image grabbers for both WFS and TTS cameras, since many existing LabVIEW standard functions for image acquisition are supported by this grabber. The NI PCIe-1429 image grabber supports full, medium, and base camera-link interfaces. The camera can achieve a rate up to 150 frames/second at full-resolution with 1024x1024 pixels. Since we only use a small region of interest with ~ 300x300 pixels for wave-front sensing, the acquisition speed can achieve 1800 frames/s. For the TTS, we only use 60x60 pixels. In a future update, we schedule to use a full camera-link camera, which should be able to deliver an image data acquisition at a speed of ~800 MB/s.

The most time-consuming part in the PSAO is the wave-front sensing and the calculation of control signals for the DM wave-front correction, which must be executed with a high performance computer. The computer used for DM wave-front correction loop has an Intel Core i7-990X CPU with 6 cores, at a clock frequency of 3.47GHz. The computer used for the tip-tilt correction loop has an Intel Core i7-980X CPU with 6 cores, at a clock frequency of 3.33GHz. Here we choose the computer with a single CPU with multiple cores other than the multiple CPUs, since these multiple CPUs are mainly optimized for large data handling such as for internet servers, and are not optimized for real-time calculation and control. This problem will be solved by the latest Intel Xeon E5-2600 series processors which adopt the same architecture with the Core i7 processor, and have up to 8 cores per CPU. A computer equipped with 2 Intel

Xeon E5-2687W CPUs for the DM wave-front correction loop is expected to increase the closed-loop bandwidth of our AO system better than 100 Hz, which will be on the state-of-the-art of the current solar AO systems.

Our AO software is written in LabVIEW codes. LabVIEW's Graphical programs inherently contain information about which parts of the code should execute in parallel. Parallelism is important in AO programing because it can unlock performance gains relative to purely sequential programs, due to recent changes in computer processor designs, in which CPUs are moved to multi-cores. LabVIEW also has a large number of standard functions for image processing, mathematical operations and hardware control. These high-quality functions are optimized for high-speed calculation as well as real-time control. For example, the standard function of pattern match in LabVIEW is about 6 ~ 9 times faster than the conventional cross-correlation algorithm. To fully take advantage of the power of today's multi-core CPU and high-quality LabVIEW's graphic programming, we use LabVIEW's parallel programming, which makes rapid development of a high-performance AO system possible. LabVIEW has greater flexibility and capability for real-time hardware system control than other general-purpose programming languages. LabVIEW programming is performed by wiring together graphical icons on a diagram, in which each icon is a built-in function. This makes programming extremely easy and efficient. In addition, dataflow languages like LabVIEW allow for automatic parallelization. Graphical programs inherently contain information about which parts of the code should execute in parallel. In the future, our system can also be easily updated to a full field-programmable gate array (FPGA) system by using NI's PXI system that is fully supported by NI's LabVIEW parallel programming, which may further increase the AO correction speed. Historically, FPGA programming was the province of only a specially trained expert with a deep understanding of digital hardware design languages. LabVIEW's FPGA programming makes it possible for engineers without FPGA expertise to use it, and makes the software development efficient.

The AO software is composed of two parts. The first part is for the AO calibration, which automatically searches for all effective WFS sub-apertures, calculates DM influence function, and record all the calibration data; the second part is for AO real-time correction, which first reads the calibration data and then performs wave-front correction in real-time. Due to the intrinsic support for parallel processing, LabVIEW automatically assigns the calculation tasks as multiple threads for each core, so that the program can be run in parallel, which greatly increases the running speed for the AO wave-front sensing and correction.

2.4. Tip-Tilt and Deformable Mirror Requirements

Since the atmospheric turbulence is corrected by the tip-tilt and deformable mirrors, the strokes provided by the tip-tilt or deformable mirror must be sufficiently large, so that the wave-front errors can be effectively corrected. Here, we use the 1.6-meter McMP as an example to calculate the tip-tilt and DM requirements. The total minimum stroke required for the tip-tilt mirror is given by [17]

$$Stroke_{min} = 1.25 \frac{D}{D_{tilt}} \sqrt{0.184(\frac{D}{r_0})^{5/3}(\frac{\lambda}{D})^2}$$

(9)

where, D is the telescope aperture size and is equal to 1.6 meters for the McMP. D_{tilt} is the diameter of the telescope aperture de-magnified on the tilt mirror. Assume that the telescope aperture is de-magnified as ~5.0 mm on the tip-tilt mirror, D_{tilt} is equal to 5.0 mm. At 1.25-μm wavelength, r_0 is equal to 150 mm (see next subsection). This results in a minimum stroke of 0.001 radians. The tip-tilt mirror will be mounted on a S-330.4SL piezo-tilt platform that can provide a maximum tilt of 0.005 radians that is larger than the minimum stroke requirement.

Similarly, the required stroke for the deformable mirror is calculated as [17]

$$Stroke \; (waves) = 2.5 \sqrt{0.00357(\frac{D}{r_0})^{5/3}}$$

(10)

This results in a required stroke of 1.08 waves at 1.25-μm wavelength. The deformable mirror from BMC can provide a maximum stroke of 3.5 μm that is larger than the required stroke. BMC's deformable mirrors are being used for astronomic adaptive optics systems, where large amount of actuators or a compact design is required. For example, the "Extreme Adaptive Optics" will use a BMC deformable mirror for the 8-meter Gemini telescope, where a deformable mirror with 4096 actuators is required [18]. The high-speed MBC Multi-DM was also used for stellar adaptive optics with a small physical size [19]. The precision and stability of the BMC's deformable mirror have been proved by our past experiences.

2.5. Performance Estimation

As an example, the AO performance estimation will only focus on the Kitt Peak 1.6-meter McMP that has a large aperture size. The estimated performance should be better for other telescope with a smaller aperture size or a site with a better seeing condition. The Fried parameter r_0 is equal to ~ 5 cm at 0.5 μm for the daytime median seeing conditions at Kitt Peak, which is available almost every week for a clear sky. The seeing r_0 is scalable with wavelength as

$r_0 \propto \lambda^{\frac{6}{5}}$ for Kolmogorov turbulence, and it reaches 70 mm and 150 mm at the 0.65 μm and 1.25 μm wavelengths, respectively.

A DM surface with a finite actuator number cannot exactly match the wave-front patterns of the atmospheric turbulence. For an atmospheric wave-front with Kolmogorov spectrum, the fitting error variance of a DM with finite actuator number is derived by Hudgin [20] and is given as

$$\sigma^2_{fit} = \kappa (r_s / r_0)^{5/3}$$

(11)

where r_s and r_0 are the spaces between two actuators and the Fried parameter, respectively. κ is the fitting parameter. Since there are 12 actuators across the DM aperture that is conjugated onto the 1.0-meter effective telescope aperture (see Section 4) although McMP has a 1.6-meter aperture, r_s is equal to 83 mm. An extensive analysis of the fitting error showed that $\kappa=0.349$ was applicable for many influence functions that are not constrained at the edge [21]. Therefore, the fitting error variance

σ^2_{fit} is 0.46 radians^2 and 0.25 radians^2 at the 0.65 μm and 1.25 μm wavelengths, respectively.

The temporal error of the wave-front correction is determined by the Greenwood frequency and the bandwidth of the AO system. The Greenwood frequency can be calculated as $0.43 v/r_0$ [22], where v is the average wind speed. At McMP, when wind speed reaches 20 m/s, the telescope will be closed and observations will not take place. We assume that the AO system will operate with an average wind speed of 8.0 m/s. This results in a Greenwood frequency of 49 Hz and 23 Hz at the 0.65 μm and 1.25 μm wavelengths, respectively. The wave-front variance due to the temporal error of the correction can be calculated as

$\sigma^2_{tem}=(f_G / f_{BW})^{5/3}$, where f_G is the Greenwood frequency and f_{BW} is the bandwidth of the AO system. Since the AO closed-loop bandwidth is 80 Hz, the temporal wave-front variance $\sigma2_{tem}$ is 0.44 radian^2 and 0.13 radian^2 at the 0.65 μm and 1.25 μm wavelengths, respectively.

Read out noise is not a problem for solar wave-front sensing since plenty of photons are available for the wave-front sensing [10]. The corrected wave-front variance is the sum of all the error contributors. If only the fitting and temporal errors are considered, the wave-front variance can be calculated approximately as $\sigma^2 \approx \sigma^2_{fit} + \sigma^2_{tem}$. This results in a wave-front variance of 0.90 radian^2 and 0.38 radian^2 at the 0.65 μm and 1.25 μm wavelengths, respectively. The overall performance of an AO system can be evaluated in terms of the Strehl ratio, which defines the peak of the actual point spread function normalized to the peak of the diffraction-limited point spread function. The Strehl ratio is calculated as $S = e^{-\sigma^2}$. At 0.65 μm, the AO system is expected to achieve a Strehl ratio of ~ 0.41; at 1.25 μm, it will deliver a Strehl ratio of ~ 0.68, and should deliver a better correction at longer wavelengths.

3. RECENT LABORATORY TESTS

The PSAO was first built in CSUN laboratory in 2009, with an OKO 37-actuator DM for software development and test purpose. In 2010, the DM was upgraded with the current BMC 140-actuator model. A cross target illuminated by a white-light fiber bundle is used for the 2-dimensional object test. A 32-actuator DM from Edmund Optics is used to generate a real-time wave-front aberration at a desired frequency, which is subsequently fed into the AO system, so that the AO performance can be evaluated in real-time. The whole optics of PSAO is built on a 900mm x 600mm optical breadboard as shown in the Figure 5, and can be easily carried to any solar observatory.

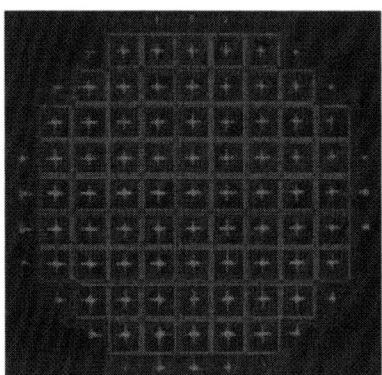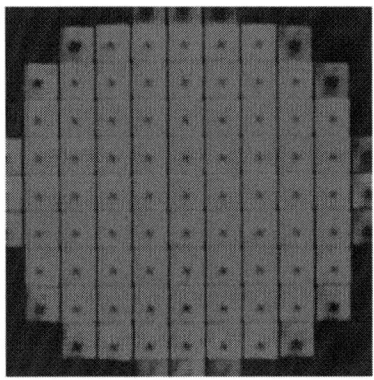

Figure 2. Graphic WFS interfaces for AO calibration with a point source (left) and for real-time correction with a cross target (right).

The AO software consists of two parts. The first part is used for the AO calibration, while the second part is the code for AO real-time correction. In principle, the calibration only needs to be done once a time, provided that the hardware has not been realigned. In the calibration, a single-mode fiber is served as a point source. Any possible incoming wave-front error will be filtered out by the single-mode fiber, and only the fundamental mode can propagate through the fiber. The output wave-front can be viewed as a perfect wave for the calibration. In the calibration, the fiber is switched into the optical path on the telescope focal plane just before the lens L1 (see Figure 1). In AO real-time correction, the fiber is switched out the optical path, and an extended target (a cross herein), including the wave-front error generated by the 32-element Edmund Optics DM, is allowed to input into the AO for testing. A lenslet array with 8.7-mm focal length is used for the WFS. Figure 2 shows the WFS interfaces in the AO calibration with a point source (Figure 2 left) and in real-time correction with a cross target (Figure 2 right): there are 69 effective WFS sub-apertures arranged in the 9x9-lenslet grid configuration and wave-front correction up to 65 modes of Zernike polynomials can be chosen. Please note that in this test, there is no central obstruction on the pupil, although our software can be used with a telescope with central obstruction.

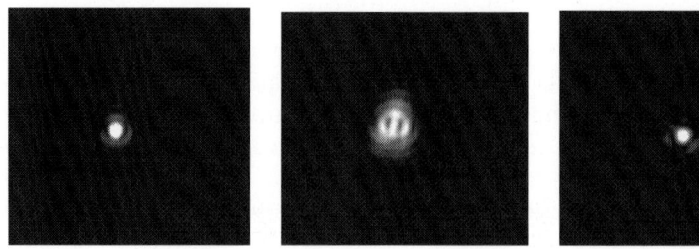

Figure 3. Point-source images: original image without aberration applied (left), with aberration applied and AO off (center), with aberration applied and AO on (right). 9x9 lenslets (without those on the four corners) are used for wave-front sensing, and 25 Zernike modes are corrected.

The PSAO had achieved excellent corrections in the test with a point source target. Figure 3 shows the original point-source image (left) without aberration applied, as well as the AO-off (center) and AO-on (right) images with the aberration applied, respectively. In this test, the wave-front error was introduced by the 32-actuator Edmund Optics DM in real-time, in which the wave-front is variable as a sinusoidal function with a frequency up to 80Hz. The central image of the distorted point-spread function shows that the wave-front error applied is significantly large, while the right image clearly demonstrates that the AO system can recover the diffraction-limited image. The AO system demonstrated good results in different situations:

1. the amplitude of the applied wave-front error is less or equal to the AO DM's maximum stroke (see Figure 3 and Figure 4). In this situation, the AO system can almost completely correct the wave-front and deliver the same image quality as that there is no wave-front error (i.e. the wave-front error is not been applied);

2. wave-front error with amplitude larger than the AO DM's maximum stroke is applied, which simulates the bad seeing condition on a site. The AO system can still compensate part of the wave-front error, which is consistent to the DM's maximum stroke. Therefore, the image quality can still be improved.

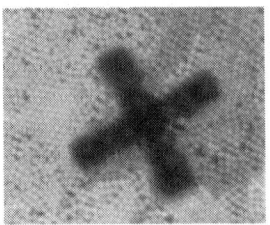 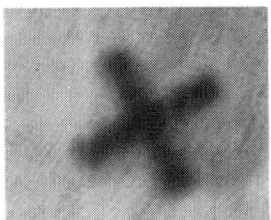 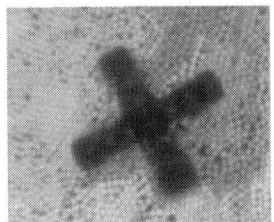

Figure 4. Cross-target images: original image without aberration applied (left), with aberration applied and AO off (center), with aberration applied and AO on (right). 9x9 lenslets (without those on the four corners) are used for wave-front sensing, and 25 Zernike modes are corrected.

The AO system was also tested with a 2-dimensional extended target with the same procedure for the point-source target. In this test, a cross was printed on a transmission film and was used as a 2-dimensional target. A fiber bundle light source located immediately behind the cross target was used to illuminate it, so that the image of the fibers was almost overlapped on the cross target, except for a slight defocus between them. Figure 4 shows the original cross-target image (left) without aberration applied, as well as the AO-off (center) and AO-on (right) images with the aberration applied, respectively. Please note that the image of the cross target and background small fibers is seriously blurred by the wave-front error applied (center image). Compared the left and right images, however, it is clear that the AO recovers the original image perfectly, indicating that the AO's performance is excellent. In fact, our AO system can recover the original wave-front that is associated with the original image with accuracy up to 1/1000 wave-length in the visible [23].

4. RECENT ON-SITE OBSERVATIONS

The PSAO's small size makes it can be easily brought to any observatory for science observations. Since 2010, we have carried out observations at two different sites. To demonstrate the AO's feasibility, an initial on-site observation was conducted by using the 0.61-m solar telescope at the San Fernando Observatory, California State University Northridge. This solar telescope is a three-aspheric mirror system with a central obstruction area of 14%, and a focal ratio of f/20. Because of the poor seeing conditions at the San Fernando Observatory, WFS with 9x9 sub-apertures (exception those in regions of the four-corners and the central obstruction) was used. The best observational results were acquired in October 2011. Using a sunspot as a target, the AO system was able to lock on the sunspot for wave-front sensing and provides high-resolution images at the wavelength of 0.75 μm [24], which indicates that our PSAO is able to provide wave-front correction at a site with a poor seeing condition.

After the successful observations on the San Fernando Observatory, we continued to test this system with the 1.6-m McMP. The McMP is located at the Kitt Peak, and is operated by the National Solar Observatory. It is one of the largest solar telescopes and is accessible to the solar community around the world. The medium seeing at the Kitt Peak is better than that at the SFO, but is still poor with a Fried parameter of ~ 5 cm at median seeing condition. The poor seeing condition at the Kitt Peak is a great challenge for an adaptive optics system. There are no AO available for routine observations with the McMP, although a prototype with 36-actuator DM was developed many years ago [25]. McMP is an off-axis telescope without central obstruction. The solar image is formed on a rotational station at f/54, where our PSAO can be conveniently placed for observations. Figure 5 shows the PSAO loaded on the McMP rotation station for an observation, in which the small size of the AO system is clearly referred. The non-common optical path error between the WFS and the science camera, which the WFS cannot measure, was calibrated by an approach we proposed recently [23].

During the latest observations in May 2012, the PSAO delivered excellent performance in the visible with the McMP. Solar images were captured at 0.6-μm visible wavelength. Figure 6 shows the typical images of Sunspot 1492 with the AO off and AO on respectively, on the May 28 run. For the AO off images, the AO still provides the tip-tilt correction, so that the overall image movement is corrected. Compared with the poor image quality when the AO is switched off, the AO system provides significant improvement for the image quality when the AO is switched on, which demonstrates the power of the AO correction: the granules around the sunspot can be clearly seen with the AO correction, while they are totally blurred and disappeared without the correction. The AO off image clearly shows how poor the seeing condition was during our observation run. In the observation, only 7x7 sub-apertures were used for the WFS, and only 1.0-meter of the McMP aperture was used for imaging, because a small area on the edge of the telescope primary mirror was damaged and the telescope heliostat was tilted at a large angle during the observation, which delivered an useful circle aperture on the order of 1.0-meter in diameter. Each sub-aperture was sampled by 30x30 pixels of the WFS camera. The AO delivered an open-loop bandwidth of 800 Hz, which corresponds to a closed-loop bandwidth of ~ 80 Hz. The improvement of image performance with the AO correction was significant. This was the first time demonstration that an AO system can be effectively used for high-resolution imaging in the visible with the McMP.

Figure 5. PSAO setup (the black breadboard) on the McMP rotation station. The two red cameras are used for WFS and TTS, while the grey one is the science camera.

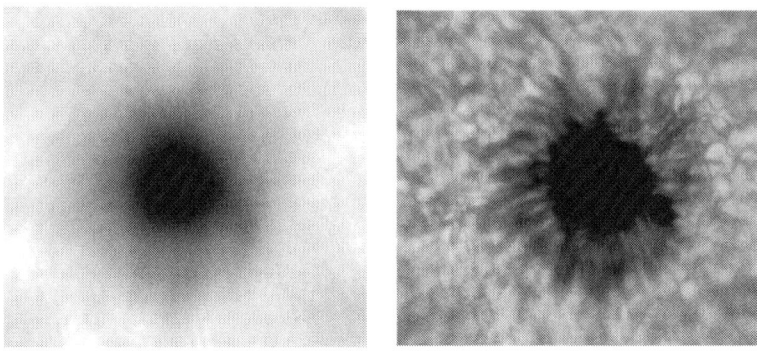

Figure 6. Sunspot 1492 image captured on the McMP with the AO off (left) and AO on (right).

Because of the poor seeing conditions at the Kitt Peak, only large sunspots can be used for wave-front sensing. Other small fine structures such as solar granules and pores are seriously distorted by the strong atmospheric turbulence and cannot be resolved by the WFS, which prevent accurate wave-front sensing by using a small WFS field of view. In this observation, lenslet array with 8.7-mm focal length was used for the WFS. The WFS field of view is 30″x30″ and is sampled by 30x30 pixels, which results in a sampling scale of 1.0″/pixels. Although the wave-front sensing of the AO software can execute with sub-pixel accuracy, such an improvement is very limited because of a number of reasons, such as the distortions of the sunspot images in each sub-aperture as well as the low contrast image resulted from the strong wave-front error. Better performance should be achievable with telescopes with good seeing conditions, where small fine solar structures can be used for accurate wave-front sensing.

4. CONCLUSIONS

We have fully demonstrated the feasibility of a portable AO system, both in the laboratory and on-site observations. The system is able to provide a wave-front correction with different telescopes with the aperture size up to 1.6 meters. Our AO system features low cost, high-performance, and is compact. Combining the multi-core computer and LabVIEW parallel programming, the AO system is particularly flexible and can achieve good performance. The open-loop correction speed can achieve 800Hz with sub-pixel accuracy for wave-front sensing, for the 7x7 sub-aperture WFS when 25 modes of Zernike polynomials of the wave-front are corrected. It can further achieve 1100 Hz, if sub-pixel wave-front sensing accuracy is not required. Higher wave-front correction speed should be able to achieve by using more CPU cores with a computer. The commercial CPU market for personal computers is being evolved rapidly, with efforts focusing on multi-core CPUs. For example, two Eight-Core Intel Xeon E5-2687W CPUs can be installed in a computer, which will deliver 16 cores in total and each core can run at 3.1GHz clock frequency. In another approach, we are also developing LabVIEW based FPGA technique, which may dramatically increase the running speed of the AO system. The 12x12-actuator DM is also being updated to a 24x24-actuator DM that will have a clear aperture of 9.0 mm, and should deliver better performance. The PSAO is being upgraded accordingly, and we will report our progresses in the near future.

5. ACKNOWLEDGEMENTS

This work is supported by the National Science Foundation under the grant ATM-0841440, the National Natural Science Foundation of China (NSFC) (Grant 10873024 and 11003031), the National Astronomical Observatories' Special Fund for Astronomy-2009, as well as the Strategic Priority Research Program of the Chinese Academy of Sciences (Grant No. XDA04070600). We thank Dr. Xi Zhang for his contribution to this AO project, and we gratefully acknowledge the assistances from the staff at Kitt Peak National Solar Observatory during our observations with the McMP.

REFERENCES

1. Jan. Olof Stenflo, 2004 Solar physics: Hidden magnetism Nature 430 6997 304 305
2. F. Paletou, G. Aulanier, 2003 On the Need of High-Resolution Spectropolarimetric Observations of Prominences in Current Theoretical Models and Future High Resolution Solar Observations: Preparing for ATST, ASP Conference Series286, held March 2002 at NSO, Sunspot, New Mexico, USA, Alexei A. Pevtsov and Han Uitenbroek (eds.), (NSO, Sunspot, New Mexico, USA) 11 15
3. H. Lin, 2003 ATST near-IR spectropolarimeter Proc. SPIE 4853 215 222
4. F. Cattaneo, T. Emonet, N. Weiss, 2003 On the Interaction between Convection and Magnetic Fields ApJ 588 1183 1198
5. A. Vögler, M. Schüssler, 2007 A solar surface dynamo Astron. Astrophys 465 L43 L46
6. Å. Nordlund, R. F. Stein, M. Asplund, 2009 Solar Surface Convection Living Rev.Solar Phys 6 2
7. Å. Nordlund, R. F. Stein, 2009 Accurate Radiation Hydrodynamics
8. F. Woger, O. von der Luhe, K. Reardon, 2008 Speckle interferometry with adaptive optics
9. T. R. Rimmele, 2004 Recent advances in solar adaptive optics
10. T. R. Rimmele, K. Richards, S. Hegwer, D. Ren, S. Fletcher, S. Gregory, 2003 Solar adaptive optics
11. O. van der Luhe, D. Soltau, T. Berkefeld, T. Schelenz, 2003 KAOS: Adaptive optics system for the Vacuum Tower Telescope at Teide Observatory Proc. SPIE 4853 187 193
12. T. Berkefeld, et al. 2010 Adaptive optics development at the German solar telescopes Applied Optics 49 31 G155 G166
13. T. R. Rimmele, J. Marino, 2011 "Solar Adaptive Optics
14. D. Ren, M. Penn, H. Wang, G. Chapman, C. Plymate, 2009 A Portable Solar Adaptive Optics
15. D. Ren, M. Penn, C. Plymate, H. Wang, X. Zhang, B. Dong, N. Brown, A. Denio, 2010 A portable solar adaptive optics
16. K. Kasper, et al. 2000 ALFA: adaptive optics
17. R. K. Tyson, 2000 Introduction to adaptive optics
18. B. Macintosh, et al. 2006 The Gemini Planet Imager Proc. SPIEL 6272 62720L

19. K. Morzinski, L. C. Johnson, D. T. Gavel, B. Grigsby, D. Dillon, M. Reinig, B. A. Macintosh, 2010 Performance of MEMS-based visible-light adaptive optics

20. R. H. Hudgin, 1977 Wave-front compensation error due to finite correction-element size, J. Opt. Am 67 393 395

21. R. K. Tyson, D. P. Crawford, R. J. Morgan, 1990 Adaptive optics system considerations for ground-to-space propagation Proc.SPIE 1221 146 156

22. R. K. Tyson, 1998 Principles of Adaptive Optics

23. D. Ren, B. Dong, Y. Zhu, J. C. Damian, 2012 Correction of non-common path error for extreme adaptive optics

24. D. Ren, B. Dong, 2012 Demonstration of portable solar adaptive optics

25. C. U. Keller, C. Plymate, S. M. Ammons, 2003 Low-cost solar adaptive optics

Chapter 6

Impact of Liquid Crystals in Active and Adaptive Optics

Justo Arines

Departamento de Física Aplicada (Área de Óptica), Facultad de Ciencias /
Universidad de Zaragoza, Zaragoza, Aragón, 50009, Spain

ABSTRACT

Active and dynamic modulation of light has been one of major contributions of liquid crystals to Optics. The spectrum of application range from signposting panels to high resolution imaging. The development of new materials is the key to continued progress in this field. To promote this we will present in this paper recent uses of liquid crystals as active or adaptive modulators of light. Besides, we will reflect on their current limitations. We expect with this to contribute to the progress in the field of liquid crystals and thus the development of new useful tools for Active and Adaptive Optics.

Keywords: liquid crystals; spatial light modulators; adaptive optics; wavefront correction

1. INTRODUCTION

Liquid Crystal (LC) materials have dramatically impacted our daily life in the last decades thanks to their optical properties. They can be found in a wide spectrum of applications: signpost panels, digital watches, calculators, cell phones, laptop displays... Nearly all the devices used in our daily life nowadays use a liquid crystal display. What is not so well known is the increasing use of liquid crystals in scientific optical applications: optical tweezers, beam shaping, tunable waveguides, digital holography, optical processors, wavefront sensing and correction, visual psychophysics, gray masks for photolithography, diffraction gratings... There are a huge variety of applications where the properties exhibited by the interaction of light with liquid crystals offer enormous design capabilities.

Which is the secret of Liquid Crystals? The answer lies the anisotropic organization of the molecules that provides a direction-dependant interaction with light. An electromagnetic field propagating through the material experiences different relative susceptibilities, depending on the relative orientation of liquid crystal directors and the direction of propagation and plane of vibration of the optical field. Additionally, due to its liquid like behavior, this relative orientation can be easily tuned by using electric, magnetic or optical fields [1,2,3,4,5,6], so anisotropy and tuning capability are the two main properties that make LCs so attractive.

In the preceding paragraph we spoke about relative susceptibility. This magnitude is related with two main properties: polarization and birefringence. The term polarization (P) refers to two different magnitudes. In one hand it describes the dipole moment per unit volume of a material that points in the same direction as the average molecular dipole moment. On the other hand it refers to the vibrational plane of an electromagnetic wave. The other term, birefringence, is an optical property of a material and refers to the anisotropic behavior of the refractive index [7]. This name comes from the splitting of the incident electromagnetic wave into two components, the ordinary and extraordinary rays. These rays experience different refractive indexes (n_o and n_e) and thus they propagate through the birefringent media with different velocities. At the end of the media one of the two electromagnetic components is delayed with respect to the other. If $n_e > n_o$ we have positive birefringence, otherwise it would be negative. Anisotropic materials can be classified in uniaxial or biaxial, depending on the eigenvalues of the refractive index (uniaxial presents two different values and biaxial three) [8,9,10].

Nearly all the applications of LCs in optics are related with their capability of inducing controllable amounts of retardation (or phase delay) and changes in the polarization state of the incident optical field [1,2]. In the next sections we will see the use of these principles in different optical applications.

Liquid Crystals are classified into two main families: thermotropic and lyotropic. Nearly all the LCs used in optical applications are thermotropic. Within this group the most outstanding phases are: nematic, smetic, and cholesteric (twisted nematic) [1,2,3,4,5,6].

Nematic liquid crystals present uniaxial positive birefringence. Smetic liquid crystals are normally uniaxial positive crystals, but they can also act as uniaxial negative or biaxial positive/negative crystals. Cholesteric liquid crystals (spontaneously twisted nematic) behave like uniaxial negative crystals. The main propertiy of this phase is its ability to rotate linearly polarized light. They present dichroism too. In these media one of the polarized components of light is much more absorbed than the other. This property is responsible for the typical iridescent color of cholesteric LCs [1,2,3,4].

Nematic, smetic and cholesteric LCs can be considered as waveplates with controllable continuous retardance. Within the smetic group there is an important phase that presents important optical characteristics, the chiral Smetic C. This subgroup is called Ferroelectric LC due to its permanent polarization without the need of an external electric field. They may be thought of as a waveplate with fixed retardance, but with an electrically controllable optical axis. They are faster than nematic ones and thus they are of interest for optical applications like optical switches or binary optics [11,12].

Lyotropic LCs also exhibit interesting optical properties [6]. As part of the the thermotropic group they can be uniaxial or biaxial negative and positive. Their main uses are restricted to the cosmetic and soap industry, but they have recently experienced a growing interest for optical applications [13,14,15,16,17,18,19,20].

Going back to birefringence, the existence of biaxial LCs has been theoretically proposed since the seventies [21]. Claims of achieving this in lyotropic phases date back as far from the eighties [15], but it took more than twenty years to achieve it in thermotropic ones [15,16,17,18,19,20,21,22,23,24,25,26,27]. The first such claims were demonstrated to be erroneous by the use of deuterium NMR spectroscopy [22]. It was only in 2004 that Madsen et al. and Acharya et al. provided firm evidence of the first biaxial nematic phase in thermotropic LCs [22,26]. Since then research on this topic has increased significantly. The potential of this kind of LC are enormous, considering the additional degree of freedom that it offers with respect to uniaxial LCs. Current proposals for the use of biaxial crystals concern their use as display compensators for improving the field of view or the color appearance [28]. Thus, the extension to the use of biaxial LC to these applications is only a question of time. Additionally, recent studies have shown the possibility of increasing the switching speed of LC valves by at least one order of magnitude with the use of biaxial LCs. This increase is obtained by rotating the LC along the molecules' small axes instead of the longest ones [24,27].

Nonlinear optics has also benefited from liquid crystals. High second and third relative susceptibility values allow for nonlinear effects like second harmonic generation and intensity dependant refractive index change (Pockels and Kerr effects), photorefractive effect, molecular reorientation and thermal and density effects [1,2,12,29,30,31].

Thus, liquid crystals offer great flexibility in their optical behavior. Index of refraction, birefringent magnitude (nearly one order of magnitude higher than that of solids) and polarization properties can be selected by mixing different components. In the next sections we will present some examples related with the use of liquid crystals in different fields of optics. Then we will focus on their use in active and adaptive optics systems. Finally we present the conclusions.

2. LIQUID CRYSTALS IN OPTICS

Optical properties of liquid crystals make them suitable for many applications [1,2,5,32]. Additional properties that have contributed to their expansion are: low cost, low power consumption, low weight, the fact they are non-mechanical devices, and easy addressability.

One of the first uses of liquid crystals in optical devices was as variable retarders and polarization rotators [33,34]. Nematic and smetic liquid crystals allow for a continuous variation of the retardance and polarization angle so they are the first options for these tasks. However these devices are too slow for certain applications (their switching speed is above 10 ms), so a lot of work has been devoted to increasing their time response by modifying the mechanical properties of the LC or by taking advantage of their nonlinear optical properties (for example by exploiting the Kerr effect in cyanobiphenyl molecules around their long axis in the smectic-A phase, which provides relaxation times of subpicosends [35]). A recent commercial result of the carrier for obtaining faster LC devices is a variable retarder for optical assemblies based on bulk-stabilized polymer liquid crystals, which have achieved switching speeds of less than 150 microseconds [36,12,27].

It is well known that ferroelectric LCs are faster (speeds over nanoseconds), but they present a big drawback, their binary behavior, so they provide bistable polarization orientation [11,12]. Considering switching applications they are the first option. However, there is a new trend in the search of faster LCs, the biaxial nematic phase. As said in previous paragraphs, their time response is one order of magnitude faster that uniaxial ones, while maintaining continuous tuning of optical retardance [26,27,28].

Most recent uses of liquid crystals are as tunable waveguides or active elements in integrated optical devices [37,38,39,40,41,42]. The guidance properties joined with their nonlinear response make LC suitable for integrated optical switchers, or logic gates [41,42]. The tunability of LC waveguides makes them extraordinarily useful for the study of optical propagation in complex structures. Recent studies have shown the possibility of using an array of channel waveguides in undoped nematic LCs to drive the system from one-dimensional bulk diffraction to discrete propagation and, even to nematicons (discrete spatial solitons in bulk nematic LC waveguides) [43]. A similar arrangement can be used to achieve multiband optical breathers [44]. Nonlinear optical responses of LCs can also be used for make optical routers by using the mode mixing approach, which consists of tuning the relative phase shift between the two propagation modes allowed by the waveguide in order to modulate their interference (see [45,46] for deeper analysis). Another example is the use of LCs in a tunable photonic-crystal waveguide Match-Zehnder interferometer where the LC changes the phase difference between the two arms of the interferometer (this device can be used as optical switcher or amplitude modulator in optical circuits [37]). Three-dimensional data storage, or holographic storage is another field where LCs play an important role. Data recording with density of 204.8 Gbits/cm^3 was recently achieved using the two-photon excitation technique [47].

3. SPATIAL LIGHT MODULATORS, ACTIVE AND ADAPTIVE OPTICS

There is one application of liquid crystals that stands out above the others, their use in Spatial Light Modulators (SLMs). A SLM is a device that modulates the spatial characteristics of light (for example LC displays). Depending on its configuration it can modulate the amplitude, the phase or both magnitudes of an impinging optical field. Liquid Crystal Spatial Light Modulators (LCSLMs) are classified into two main categories, depending on how they are addressed. If the electric field is electrically controlled by a matrix of pixels, then it is an electrically-addressed LCSLM. If an optical field is used to control the SLM, then it is called an analogue-addressed or optically-addressed LCSLM [48].

Liquid crystal display technology employs electrically driven LCSLMs. The commercial applications of this technology have increased the research efforts and the investment in new materials. Contrast ratio, viewing angle, polarization sensitivity, electronic consumption, temperature stability, operating speed and computer configuration are some of the characteristics of the new devices. The main disadvantages of these LCSLMs are: polarization sensitivity, time response, and above all pixelization, which reduces the fill factor and induces light diffraction. In contrast, optically-addressed LCSLMs were developed for scientific applications in order to create an SLM free of the diffraction effect induced by the pixel matrix used in the electrically driven LCSLMs. However these are less flexible in their configuration and use.

Scientific research has benefited from the developments in LC displays. Application of LCSLMs to photolithography, optical tweezers, optical processing, beam shaping and active and adaptive optics are some of the fields that have experienced a big increase due to the use of SLMs in the last decade [49,50,51,52].

Photolithography is a very important technique for manufacturing microelectronic and microoptical devices. It consists of exposing a photoreactive material to light through an amplitude mask. This mask can be binary or continuous. The SLM offers the flexibility of varying the mask type and its spatial distribution just by sending different images to them [49].

An optical tweezer is a device used to trap and guide particles or cells [50]. As photolithography the SLM is used for modulating the spatial distribution of the irradiance of the focal distribution of a light beam by changing the spatial distribution of the amplitude and or phase of the optical field at the exit pupil of the system [51]. This procedure is also used for beam shaping [52].

3.1. Active and Adaptive Optics

Active and Adaptive Optics are two technologies characterized by the use of a configurable electro-optical device that can be adapted for the correction of the wavefront degradation [53,54,55,56]. These techniques were first used for non military purposes for the correction of the atmospheric turbulence degradation in order to obtain high resolution astronomic images. The main difference between active and adaptive optics is their time response. While the term "active optics" is used when the correction is adjusted at low frequency and the changes in the degradation function are slow, the adaptive term is used in cases for which degradation changes continuously and the correction must be quickly adapted. Thus, in adaptive optics systems (AOs) we need not only the electro-optical device for compensation, but also a wavefront sensor.

Traditionally AO used deformable mirrors for accomplishing the wavefront compensation. These devices work by introducing optical path differences between different parts of the wavefront. Two kind of approaches has been used: segmented and membrane deformable mirrors. Segmented deformable mirrors can be of bigger aperture than membrane ones, and are easy to control. However they present diffraction effects for the individual segmented edges and difficulty for achieving inter-segment alignment. On the other hand membrane deformable mirrors are difficult to control, and present smaller aperture, but they avoid the diffraction effect [53,55].

Liquid crystal spatial light modulators operating in pure phase modulation mode have been used in the last twenty years as an alternative to deformable mirrors [54,57,58,59,60,61,62,63,64,65,66,67,68,69]. These devices introduce controlled amounts of retardance at different parts of the wavefront. Their initial limitation concerning speed, polarization sensitivity, retardance range, and diffraction effect reduced their expansion. However they have been overcome nowadays by using different approaches. For example, polarization sensitivity can be avoided by using a quarter wave plate and a mirror as proposed by Love [69] or by using a device with two active LC layers that are orthogonal to each other [70]. Speed can be increased by using the double frequency scheme [71]. The diffraction effect is not present in optical-addressed LCSLMs [48,72]. The use of LCSLMs for astronomical proposes is still a challenging question. Management of small amounts of polarized light of different wavelengths is still better accomplished with deformable mirrors, although the response of LCSLMs has been greatly increased.

The future of LCSLMs is not restricted to astronomical applications. In fact, their main uses are out of that field. The active and adaptive optics philosophy has been applied in different fields of research. Clinical applications for early diagnosis and patient monitoring as well as basic studies on physiological optics and vision science have also benefited in the last few years [73,74,75,76,77,78,79,80,81]. Low price, easy handling and easy control of LCSLMs are the main reasons that have encouraged different research groups to initiate lines of research based on SLMs that were initially restricted to well funded groups capable of acquiring the expensive deformable mirrors. Additionally, the flexibility in the design of the coded phase allowed to uses the same LCSLMs to perform different tasks. The same device can be used to code wavefront sensors, amplitude masks, or wavefront modulators, different optical elements that are important parts of complex optical assemblies.

3.2. Low Cost Active and Adaptive Optics

Within the LCSLM market there are some devices especially manufactured for wavefront sensing and control, like those manufactured by Hamamatsu Photonics, Meadowlark Optics or Holoeye. These devices are continuously evolving, increasing the sampling rate, wavelength range, time response, and amount of retardance, among other features. They are not as expensive as deformable mirrors, but are not especially cheap. Another important user of LCSLMs is the video projector industry. Although the LCSLMs used in that devices are specially designed for providing high quality images, they can be also used for scientific applications. Their specifications are not as good as those built for wavefront applications, but they satisfy the minimum necessary requirements. In addition to this, their price is continuously decreasing following the trends of the audiovisual market. Besides, these VPTNLCs are robust and use the same interconnection hardware and software as video projectors, which increases their flexibility and simplicity of control.

One of the main limitations of LCSLMs, and off course VPTNLCs, is the retardation range, normally ranging under one wavelength. Optical researchers had overcome this limitation by projecting a wrapped image instead of the continuous one in the same way as the diffractive optics elements (see [7] and [80] for more details). In a recent study our group, in collaboration with researchers from the University Jaume I in Castellón (Spain), used a VPTNLC with a four level encoding scheme in an adaptive optics system used to compensate aberrations like those presented by human eyes [80], which in general are above one wavelength of magnitude. The authors used in that study an optical arrangement for operating with the VPTNLC in a pure phase modulation mode. See Figure 1 for details.

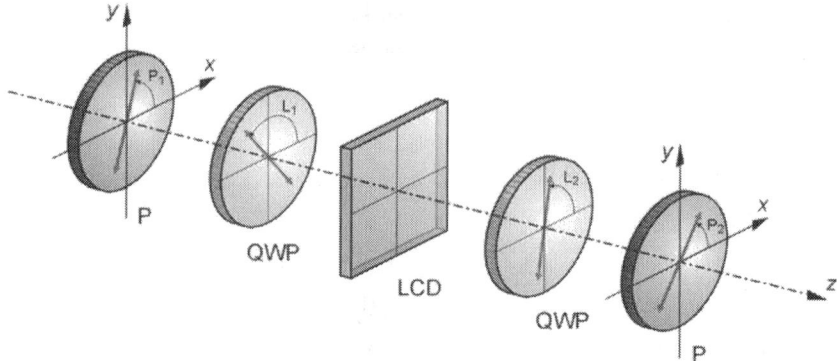

Figure 1. Phase-only modulation experimental setup with a VPTNLC. P: polarizer; QWP: a quarter-wave plate. In the diagram, the x-axis coincides with the input molecular director of the liquid crystal cell. P1 and P2 are, respectively, the orientation of the polarizer and the analyzer. L1 and L2 are the angles of the slow axis of the quarter-wave plates with respect to x-axis.

That system presented an additional characteristic, the same VPTNLC was used for the measurement of the wavefront aberration and its compensation. This means that we encoded in the VPTNLC a microlens array (sampling element used in Hartmann-Shack wavefront sensors [53,54,55]) for measuring the distorted wavefront and then the aberration compensation. These two steps were repeated in a close-loop in order to correct the wavefront aberration. In Figure 2 we show the two images sent to the VPTNLC.

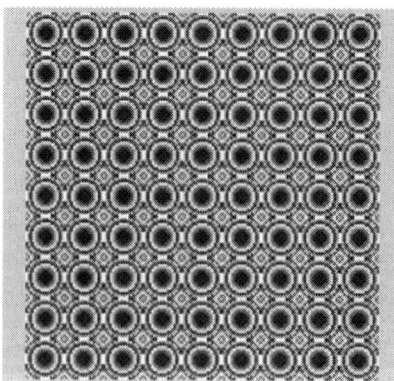 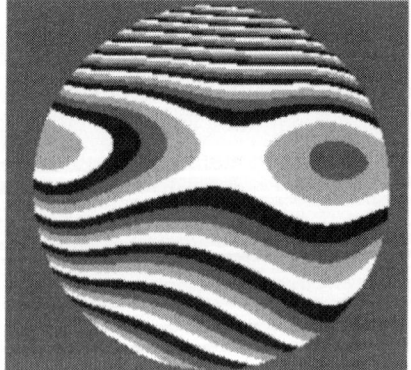

Figure 2. Grayscale representation of the four-level VPTNLC patterns: (left) for generating a 9 x 9 diffractive microlens array; (right) for compensating the aberration produced by an artificial eye.

In the next figure we show the system operating in close-loop. We can see the focal spots of the microlens array, the four level encoded compensation, and degraded and restored image.

Figure 3. A real time video of the performance of the system when compensating continuously varying amounts of defocus. See the supplementary video for observing the system operating in close-loop.

At this point we want to remark one of the characteristic of VPTNLCs, flexibility. We can use the same VPTNLC cell for measuring and compensating the wavefront aberration. Additionally we can align by software the sampling array and the compensation element with the exit pupil of the optical system easily by displacing the projection coordinates of the images, alignment with a precision of one pixel.

4. FINAL REMARKS

Liquid crystal materials offer a variety of properties that make them attractive for a lot of applications. The secret of their success is their anisotropic molecule organization and the capability of change their orientation by the application of controlled electric, magnetic or optic fields.

Their anisotropy is responsible for their most important optical properties. Birefringence, polarization rotation and significant values of 2nd and 3th order relative susceptibilities make liquid crystals one of the most interesting materials for the future optics.

Thermotropic liquid crystals are the most used for optical applications and among them the nematic and twisted nematic phases are the preferred ones. The main advantage of these phases over smetic ones is their capability of continuous variation of the induced retardation. However in applications like optical switching were to states (on/off) are desired, ferroelectric liquid crystals are the first option due to their fast time response.

In this paper we have shown the variety of applications that liquid crystals had found in optics. Variable retarders and polarizers, optical switchers and spatial light modulators are some of the most outstanding ones.

Special emphasis has to be made with respect to spatial light modulators. Their use for phase and/or amplitude modulation has experienced an enormous increase in the last decade, mainly due to their low cost, easy addressability and flexibility. Different applications have been found for SLM, among them Active and Adaptive Optics for astronomical and non astronomical tasks must be remarked. The high potential of LC in this field has not been reached yet. The application of this technique for the manipulation of light offers a lot of applications that still remains to be proposed. Light traps, pholitography, digital holography, optical tweezers and tunable aberrated lenses, are part of the present. Vision sciences has also benefited from VPTNLCs. Visual tests devoted to determine the influence of controlled amounts of aberration on vision [73], the use of adaptive wireless intraocular LC lens that provides an accommodation amplitude of 2.5 diopters [75], and high resolution imaging of the retina are some of the applications in this field, [82,83].

The optical demands that fall on liquid crystal science encourage the research of new materials. Polymer stabilized liquid crystals [36] or LC doping with dichroic dye dopants [84] or nanoparticles [85], and biaxial phases [25,26] are some of the lines of research that are now being exploited. These studies are mainly devoted to increase the projection devices. However as has been shown in this paper, there are other specific applications that can take advantage of those efforts.

Liquid crystal science is one of the most interdisciplinary fields that exist today. The variety of applications found for liquid crystals, urge the collaboration between scientists of different disciplines. In this way, we hope that the presentation of some of the most relevant applications of liquid crystals in optics had contributed to the development of new liquid crystals useful for optical applications.

ACKNOWLEDGEMENTS

This work has been supported by the Spanish Ministerio de Ciencia e Innovación (MICINN), grant FIS2008-03884.

REFERENCES AND NOTES

1.   Collins, P.J.; Hird, M. Introduction to liquid crystals: Chemistry and Physics, 3rd Ed. ed; Taylor & Francis: London, UK, 2004.

2. Dunmur, D.; Fukuda, A.; Luckhurst, G. Physical properties of liquid crystals: nematics; INSPEC: London, UK, 2001.

3. Gennes, P.G.; Prost, J. The physics of liquid crystals, 2nd Ed. ed; Oxford University Press: Oxford, UK, 1993.

4. Chandrasekhar, S. Liquid Crystals, 2nd Ed. ed; Cambridge University Press: Cambridge, UK, 2008.

5. Khoo, I.C. Liquid Crystals: Physical properties and Nonlinear Optical Phenomena; Wiley-VCH: Hoboken, NJ, USA, 1994; ISBN 0471303623.

6. Figueiredo, M.; Azevedo, S.R. The physics of lyotropic liquid crystals: phase transitions and structural properties; Oxford University Press: Oxford, UK, 2005; ISBN 0198525508.

7. Born, M.; Wolf, E. Principles of Optics, 6th Ed. ed; Pergamon: Oxford, UK, 1992.

8. Brown, G.H. Structure, properties, and some applications of liquid crystals. J. Opt. Soc. Am.**1973**, 63, 1505–1514.

9. Clark, M.; Harrison, K.; Raynes, E. Liquid crystal materials and devices. Phys. Technol.**1980**, 11, 232–240.

10. Madhusudana, N. Recent advances in thermotropic liquid crystals. Curr. Sci.**2001**, 80, 1018–1025.

11. Love, G.; Bhandarib, R. Optical properties of a QHQ ferroelectric liquid crystal phase modulator. Opt. Commun.**1994**, 110, 475–478.

12. Mishina, E.D.; Sherstyuk, N.E.; Stadnichuk, V.I.; Sigov, A.S.; Mukhorotov, M.; Golovko, Y.I.; van Etteger, A.; Rasing, T. Nonlinear-optical probing of nanosecond ferroelectric switching. Appl. Phys. Lett.**2003**, 83, 2402.

13. Bertolotti, M.; Ferrari, A.; Sibilia, C.; Alippi, A.; Nesrullajev, A.N. Optical nonlinearities in a lyotropic liquid crystal. Opt. Lett.**1987**, 12, 419–421.

14. Alves, S.; Cuppo, F.L.; Figueiredo, A.M. Determination of the nonlinear refractive index of lyotropic mixtures with and without ferrofluid doping: a time-resolved Z-scan experiment in millisecond time scales. J. Opt. Soc. Am. B**2006**, 23, 67–74.

15. Yu, L.J.; Saupe, A. Observation of a Biaxial Nematic Phase in Potassium Laurate-1-Decanol-Water Mixtures. Phys. Rev. Lett.**1980**, 45, 1000–1003.

16. Bartolino, R.; Chiaranza, T.; Meuti, M. Uniaxial and biaxial lyotropic nematic liquid crystals. Phys. Rev. A**1982**, 26, 1116–1119.

17. Galerne, Y.; Figueiredo, A.M.; Liebert, L. Optical birefringence and temperature range of the biaxial nematic phase in a lyotropic liquid cristal. J. Chem. Phys.**1986**, 84, 1732–1734.

18. Nastishin, Y.A.; Liu, H.; Schneider, T.; Nazarenko, V.; Vasyuta, R.; Shiyanovskii, S.V.; Lavrentovich, O.D. Optical characterization of the nematic lyotropic chromonic liquid crystals: Light absorption, birefringence, and scalar order parameter. Phys. Rev. E**2005**, 72, 041711.

19. Boiko, O.P.; Vasyuta, R.M.; Nazarenko, V.G.; Pergamenshchik, V.M.; Nastishin, Y.A.; Lavrentovich, O.D. Polarizing properties of functional optical films based on lyotropic chromonic liquid crystals. Mol. Cryst. Liq. Cryst.**2007**, 467, 181–194.

20. Park, H.S.; Tortora, L.; Vasyuta, R.M.; Golovin, A.B.; Augustin, E.; Finotello, D.; Lavrentovich, O.D. Lyotropic Chromonic Liquid Crystals: Effects of Additives and Optical Applications. IMID '07 Techical Digest**2007**, 307–310.

21. Taylor, T.R.; Fergason, J.L.; Arora, S.L. Biaxial liquid crystals. Phys. Rev. Lett.**1970**, 24, 359–361.

22. Luckhurst, G.R. Biaxial nematic liquid crystals: fact or fiction? Thin Solid Films**2001**, 393, 40–52.

23. Madsen, L.A.; Dingemans, T.J.; Nakata, M.; Samulski, E.T. Thermotropic biaxial nematic liquid crystals. Phys. Rev. Lett.**2004**, 92, 145505.

24. Dingemans, T.J.; Madsen, L.A.; Zafiropoulos, N.A.; Lin, W.; Samulski, E.T. Uniaxial and biaxial nematic liquid crystals. Phil. Trans. R. Soc. A**2006**, 364, 2681–2696.

25. Severing, K.; Saalwachter, K. Biaxial nematic phase in a thermotropic liquid-crystalline side-chain polymer. Phys. Rev. Lett.**2004**, 92, 125501.

26. Acharya, B.R.; Primak, A.; Kumar, S. Biaxial nematic phase in bent-core thermotropic mesogens. Phys. Rev. Lett.**2004**, 92, 145506.

27. Berardi, R.; Muccioli, L.; Zannoni, C. Field response and switching times in biaxial nematics. J. Chem. Phys.**2008**, 128, 024905.

28. Gu, C.; Yeh, P. Extended Jones matrix method and its application in the analysis of compensators for liquid crystal displays. Displays**1999**, 20, 237–257.

29. Khoo, I.C. Nonlinear optics of liquid crystalline materials. Phys. Rep.**2009**, 471, 221–267.

30. Pishnyak, O.P.; Lavrentovich, O.D. Electrically controlled negative refraction in a nematic liquid cristal. Appl. Phys. Lett.**2006**, 89, 251103.

31. Lee, J.H.; Lim, T.K.; Kim, W.T.; Jin, J.I. Dynamics of electro-optical switching processes in surface stabilized biaxial nematic phase found in bent-core liquid crystal. J. Appl. Phys.**2007**, 101, 034105.

32. Jacobs, D.S. Selected Papers on Liquid Crystals for Optics; SPIE Press: Bellingham, WA, USA, 1992; Volume MS46.

33. Zhuang, Z.; Suh, S.; Patel, J.S. Polarization controller using nematic liquid crystals. Opt. Lett.**1999**, 24, 694–696.

34. Bueno, J.M. Polarimetry using liquid-crystal variable retarders: theory and calibration. J. Opt. A Pure Appl. Opt.**2000**, 2, 216–222.

35. Calvez, L.A.; Montant, S.; Freysz, E.; Ducasse, A.; Zhuang, X.W.; Shen, Y.R. Ultrashort orientational dynamics of liquid crystals in the smectic-A phase. Chem. Phys. Lett.**1996**, 258, 620–625.

36. Meadowlark Optics. Swift Liquid Crystal Variable Retarders; Meadowlark Optics: Frederick, CO, USA, 2008; pp. 48–50. Available online:http://www.lamdapacific.com/Webupdate/upload/200942418 2642734.pdf accessed May 8, 2009.

37. Chen, X.Y.; Li, H.; Qiu, Y.S.; Wang, Y.F.; Ni, B. Tunable Photonic Crystal Mach–Zehnder Interferometer Based on Self-collimation Effect. Chinese Phys. Lett.**2008**, 25, 4307–4310.

38. Brzd, K.A.; Karpierz, M.A.; Fratalocchi, A.; Assanto, G.; Nowinowski-Kruszelnicki, E. Nematic liquid crystal waveguide arrays. Opto-Electron. Rev.**2005**, 13, 107–112.

39. Peccianti, M.; Assanto, G. Signal readressing by steering of spatial solitons in bulk nematic liquid crystals. Opt. Lett.**2001**, 26, 1690–1692.

40. Sirleto, L.; Coppola, G.; Breglio, G. Optical multimode interference router based on a liquid crystal waveguide. J. Opt. A Pure Appl. Opt.**2003**, 5, S298–S304.

41. Beeckman, J.; Neyts, K.; Haelterman, M. Patterned electrode steering of nematicons. J. Opt. A: Pure Appl. Opt.**2006**, 8, 214–220.

42. Fratalocchi, A.; Asquini, R.; Assanto, G. Integrated electro-optic switch in liquid crystals. Opt. Expr.**2005**, 13, 32–37.

43. Fratalocchi, A.; Assanto, G.; Brzdakiewicz, K.A.; Karpierz, M.A. Discrete propagation and spatial solitons in nematic liquid crystals. Opt.Lett.**2004**, 29, 1530–1532.

44. Fratalocchi, A.; Assanto, G.; Brzdakiewicz, K.A.; Karpierz, M.A. Optical multiband vector breathers in tunable waveguide arrays. Opt. Lett.**2005**, 30, 174–176.

45. Fratalocchi, A.; Assanto, G. All-optical switching in a liquid crystalline waveguide. Appl. Phys. Lett.**2005**, 86, 051109.

46. Fratalocchi, A.; Assanto, G.; Brzdkiewicz, K.A.; Karpierz, M.A. All-optical switching and beam steering in tunable waveguide arrays. Appl. Phys. Lett.**2005**, 86, 051112.

47. McPhail, D.; Min, G. Use of polarization sensitivity for three-dimensional optical data storage in polymer dispersed liquid crystals under two-photon illumination. Appl. Phys. Lett.**2002**, 81, 1160–1162.

48. Li, F.; Mukohzaka, N.; Yoshida, N.; Igasaki, Y.; Toyoda, H.; Inoue, T.; Kobayashi, Y.; Hara, T. Phase modulation characteristics analysis of optically-addressed parallel-aligned nematic liquid crystal phase-only spatial light modulator combined with a liquid crystal display. Opt. Rev.**1998**, 5, 174–178.

49. Hayashi, T.; Shibata, T.; Kawashima, T.; Makino, E.; Mineta, T.; Masuzawa, T. Photolithography system with liquid crystal display as active gray-tone mask for 3D structuring of photoresist. Sens. Actuat. A: Phys.**2008**, 144, 381–388.

50. Hands, P.J.W.; Tatarkova, S.A.; Kirby, A.K.; Love, G. Modal liquid crystal devices in optical tweezing: 3D control and oscillating potential wells. Opt. Expr.**2006**, 14, 4525–4537.

51. Żmija, J.; Kłosowicz, S.J.; Kędzierski, J.; Nowinowskt-Kruszelnicki, E.; Zieliński, J.; Raszewski, Z.; Walczak, A.; Parka, J. Application of liquid crystals in optical processing of optical signals. Opto-Electron. Rev.**1997**, 5, 93–106.

52. Ibáñez-López, C.; Muñoz-Escrivá, L.; Saavedra, G.; Martínez-Corral, M. Manufacture of pupil filters for 3D beam shaping. Opt. Commun.**2007**, 272, 197–204.

53. Roggemann, M.C.; Welsh, B.M. Imaging Through Turbulence; CRC: Boca Raton, FL, USA, 1996.

54. Hardy, J.W. Active optics: A new technology for the control of light. Proc. IEEE**1978**, 66, 651–697.

55. Roddier, F. Adaptive Optics in Astronomy; Cambridge University Press: Cambridge, UK, 1999.

56. Dayton, D.C.; Browne, S.L.; Sandven, S.P.; Gonglewski, J.D.; Kudryashov, A.V. Theory and laboratory demonstrations on the use of a nematic liquid-crystal phase modulator for controlled turbulence generation and adaptive optics. Appl. Opt.**1998**, 37, 5579–5589.

57. Bonaccini, D.; Brusa, G.; Esposito, S.; Salinari, P.; Stefanini, P.; Biliotti, V. Adaptive optics wave front corrector using addressable liquid crystal retarders II. Proc. SPIE**1991**, 1543, 133–143.

58. Vasiliev, A.A.; Kompanets, I.N.; Parfenov, A.V. Progress in the development and applications of optically controlled liquid crystal spatial light modulators. Sov. J. Quantum Electron.**1983**, 13, 689–695.

59. Amako, J.; Miura, H.; Sonehara, T. Wave-front control using liquid-crystal devices. Appl. Opt.**1993**, 32, 4323–4329.

60. Bold, G.; Barnes, T.; Gourlay, J.; Sharples, R.; Haskell, T. Practical issues for the use of liquid crystal spatial light modulators in adaptive optics. Opt. Commun.**1998**, 148, 323–330.

61. Chen, J.; Hirayama, T.; Ishizuka, K.; Tonomura, A. Spherical aberration correction using a liquid-crystal spatial-light modulator in off-axis electron holography. App. Opt.**1994**, 33, 6597–6602.

62. Dale, S.; Love, G.; Myers, R.; Naumov, A. Wavefront correction using a self-referencing phase conjugation system based on a Zernike cell. Opt. Commun.**2001**, 191, 31–38.

63. Dou, R.; Giles, M.K. Closed-loop adaptive-optics system with a liquid-crystal television as a phase retarder. Opt. Lett.**1995**, 20, 1583–1585.

64. Love, G.D.; Fender, J.S.; Restaino, S.R. Adaptive wavefront shaping with liquid crystals. Opt. Photon. News**1995**, 6, 16–21.

65. Love, G.D. Wavefront control using a high quality nematic liquid crystal spatial light modulator. In Proceedings of Society of Photo-optical Instrumentation Engineers - Advanced imaging technologies and commercial applications, San Diego, CA, USA, July 10-12, 1995.

66. Mu, Q.; Cao, Z.; Li, D.; Hu, L.; Xuan, L. Open-loop correction of horizontal turbulence: system design and result. Appl. Opt.**2008**, 47, 4297–4301.

67. Dou, R.; Giles, M.K. Closed-loop adaptive-optics system with a liquid-crystal television as a phase retarder. Opt. Lett.**1995**, 20, 1583–1585.

68. Liesener, J.; Reicherter, M.; Tiziani, H. Determination and compensation of aberrations using SLMs. Opt. Commun.**2004**, 233, 161–166.

69. Love, G. Liquid-crystal phase modulator for unpolarized light. Appl. Opt.**1993**, 32, 2222–2223.

70. Kelly, T.; Love, G. White-light performance of a polarization-independent liquid-crystal phase modulator. Appl. Opt.**1999**, 38((10)), 1986–1989.

71. Kirby, A.; Love, G. Fast, large and controllable phase modulation using dual frequency liquid crystals. Opt. Expr.**2004**, 12, 1470–1475.

72. Riza, N.; Jorgesen, D. Minimally invasive optical beam profiler. Opt. Expr.**2004**, 12, 1892–1901.

73. Manzanera, S.; Prieto, P.M.; Ayala, D.B.; Lindacher, J.M.; Artal, P. Liquid crystal adaptive optics visual simulator: Application to testing

and design of ophthalmic optical elements. Opt. Expr.**2007**, 15, 16177–16188.

74. Quan, W.; Wang, Z.Q.; Mu, G.; Ning, L. Correction of the aberrations in the human eyes with SVAG1 thin-film transistor liquid-crystal display. OPTIK**2003**, 114, 467–471.

75. Simonov, A.N.; Vdovin, G.; Loktev, M. Liquid-crystal intraocular adaptive lens with wireless control. Opt. Expr.**2007**, 15, 7468–7478.

76. Awwal, A.; Bauman, B.; Gavel, D.; Olivier, S.; Jones, S.; Silva, D.; Hardy, J.; Barnes, T.; Werner, J.; Tyson, R.; Lloyd-Hart, M. Characterization and operation of a liquid crystal adaptive optics phoropter. Proc. SPIE**2003**, 5169.

77. Iwasaki, T.; Kubota, T.; Tawara, A. The tolerance range of binocular disparity on a 3D display based on the physiological characteristics of ocular accommodation. Displays**2009**, 30, 44–48.

78. Mu, Q.; Cao, Z.; Li, D.; Hu, L.; Xuan, L. Liquid Crystal based adaptive optics system to compensate both low and high order aberrations in a model eye. Opt. Expr.**2007**, 15, 1946–1953.

79. Naumov, A.; Yu, M.; Loktev, I.; Guralnik, R.; Vdovin, G. Liquid-crystal adaptive lenses with modal control. Opt. Lett.**1998**, 23, 992–994.

80. Durán, V.; Climent, V.; Tajahuerce, E.; Jaroszewicz, Z.; Arines, J.; Bará, S. Efficient compensation of Zernike modes and eye aberration patterns using low-cost spatial light modulators. J. Biomed. Opt.**2007**, 12, 014037.

81. Arines, J.; Durán, V.; Jaroszewicz, Z.; Ares, J.; Tajahuerce, E.; Prado, P.; Lancis, J.; Bará, S.; Climent, V. Measurement and compensation of optical aberrations using a single spatial light modulator. Opt. Expr.**2007**, 15, 15287–15292.

82. Shirai, T. Liquid-Crystal Adaptive Optics Based on Feedback Interferometry for High-Resolution Retinal Imaging. Appl. Opt.**2002**, 41, 4013–4023.

83. Kitaguchi, Y.; Bessho, K.; Yamaguchi, T.; Nakazawa, N.; Mihashi, T.; Fujikado, T. In vivo measurements of cone photoreceptor spacing in myopic eyes from images obtained by an adaptive optics fundus camera. Jpn. J. Opht.**2007**, 51, 456–461.

84. Pal, D.; Misra, P.; Misra, A.K.; Manohar, R.; Shukla, J.P. Effect of dichroic dye on dielectric and optical properties of a nematic liquid crystal. Res. J. Phys.**2007**, 1, 10–18.

85.   Teng, W.; Jeng, S.; Kuo, C.; Lin, Y.; Liao, C.; Chin, W. Nanoparticles-doped guest-host liquid crystal displays. Opt. Lett.**2008**, 33, 1663–1665.

Chapter 1

Embedded Adaptive Optics for Ubiquitous Lab-on-a-Chip Readout on Intact Cell Phones

Pakorn Preechaburana [1,2,*], Anke Suska [1] and Daniel Filippini [1]

[1]*Optical Devices Laboratory, Division of Applied Physics, IFM–Linköping University, SE-58183 Linköping, Sweden; E-Mails: anksu@ifm.liu.se (A.S.); danfi@ifm.liu.se (D.F.)*

[2]*Department of Physics, Faculty of Science and Technology, Thammasat University, Pathum Thani 12121, Thailand*

ABSTRACT

The evaluation of disposable lab-on-a-chip (LOC) devices on cell phones is an attractive alternative to migrate the analytical strength of LOC solutions to decentralized sensing applications. Imaging the micrometric detection areas of LOCs in contact with intact phone cameras is central to provide such capability. This work demonstrates a disposable and morphing liquid lens concept that can be integrated in LOC devices and refocuses micrometric features in the range necessary for LOC evaluation using diverse cell phone cameras. During natural evaporation, the lens focus varies adapting to different type of cameras. Standard software in the phone commands a time-lapse acquisition for best focal selection that is sufficient to capture and resolve, under ambient illumination, 50 μm features in regions larger than 500 × 500 μm^2. In this way, the present concept introduces a generic solution compatible with the use of diverse and unmodified cell phone cameras to evaluate disposable LOC devices.

Keywords: adaptive optics; lab-on-a-chip readout; optical chemical sensing; ubiquitous sensing; cell phones

1. INTRODUCTION

Disposable lab-on-a-chip (LOC) devices are an attractive platform for the implementation of compact and robust analytical tests, which minimize sample volumes and simplify the handling of the measurements, both important factors in distributed analyses [1,2]. Among the diverse possibilities existing for LOC readout, optical methods [3] are those relevant to this work.

Disposable LOC devices have been demonstrated for numerous sensing and clinical applications [3], however, their dissemination is restricted by the instrumentation required for readout. LOC solutions for point of care (POC) or other distributed detections [2] are typically associated with dedicated and specific off-chip readers [4,5].

Thus, although LOC devices can be disposable and deployable at a large scale, the availability of readers and their specific characteristics restrict the dissemination of analyses based on this technology. Therefore, if disposable LOC devices could be evaluated using generic and common platforms, such as cell phones, the benefits of this technology could be made ubiquitous.

On the other hand, dedicated instruments for chemical sensing, both compact or sizable, which make use of regular cell phones for imaging [6,7] and communication purposes [8] have been demonstrated in recent years. In some cases the cell phones are embedded within the instrument [8] and permanently modified, whereas in other examples there is only a temporary connection [6] to the instrument and the phone remains usable for its natural purpose. In both cases, the components additional to the phone are not common and restrict the ubiquity of the combined solution.

In contrasts, ubiquitous chemical sensing approaches have been developed during the past ten years to take advantage of mass-produced consumer electronic devices such as flatbed scanners [9], DVD/CD drives [10], computer sets [11], and also cell phones [12]. In these examples, components are sensibly combined to minimize additional interfacing elements that could restrict ubiquity. In this work we follow these principles aiming at solution that can be integrated in disposable LOC devices for ubiquitous sensing.

Here we investigate off-chip readout of disposable LOC devices on cell phones without additional accessories and using adaptive optics integrated in the same disposable LOC that will be evaluated. The device sits on the camera surface, which provides a standard mechanical support for the device, optical coupling and a compact configuration. The device temporarily sticks on the camera during evaluation and is disposed afterwards.

Intact cell phone cameras cannot focus at the short distances required by this concept, and the LOC must incorporate a refocusing element to image its micrometric detection area. Simple fixed lenses can be implemented for a particular camera type [13]; however, different brands and models have slightly different optical designs, and a generic solution to this problem demands to adapt to all of these conditions with a unified concept. Adaptive optics is central for autofocusing and can be implemented in different ways [14–17] as dedicated components, but in this work we seek a solution that can be embedded in disposable LOC devices, such as a sessile drop [18–20] complemented by data analysis.

Here we demonstrate a disposable morphing lens concept that can be integrated in the LOC device, and operates on different phone and computer cameras, rendering these platforms capable to image the micrometric detection regions necessary for the evaluation of LOC devices.

2. EXPERIMENTAL SECTION

2.1. Lens Supporting Device

The lens supporting part of the LOC device was made from Dow Corning Sylgard 184 PDMS (polydimethylsiloxane [21]) with a base/curing agent ratio of 10:1, as negative replicas of a SU-8 (10) template (Microchem Corp., Newton, MA, USA) created with a refined micro projection lithography system (MPLS) described elsewhere [22]. The template created a 30 μm deep circular depression that confines the liquid lens.

To fabricate the PDMS substrate, 10 g of Dow Corning Sylgard 184 base and 1 g of curing agent were mixed and stirred in a cup for 5 min. The mixture was degased in a desiccator connected with a rotary pump for 45 min, and afterwards poured on a SU-8(10) template. The PDMS film was then cured at 65 °C for 2 h in an environmental oven. The result was an adhesive microstructured 150 μm thick film that was cut with a blade and placed on phone cameras, serving as support of the liquid lens, and the PDMS elements that hold the LOC substrate (Figure 1(c)).

2.2. Test Structures

The LOCs devices used for testing are 3D chambers created using a refined MPLS approach [22]. In order to make closed chambers with their own roof, the exposure depth of a SU-8/S1818 mixture is precisely controlled, as described in a previous work [22]. The mixture of SU-8 (50) (Microchem Corp., Newton, MA, USA) with 10% in volume of S1818 G2 (Rohm and Hass) enhances light absorption for the spectral radiance used in this mask less MPLS platform, which utilizes a DMD slides projector (2500 ANSI lumens Optoma EP1690) as controlled light source set on the epi-fluorescence channel of a routine microscope (Zeiss Axiovert 40 CFL). The mixture behaves as a negative photoresist with higher absorption in the blue region of the visible spectrum.

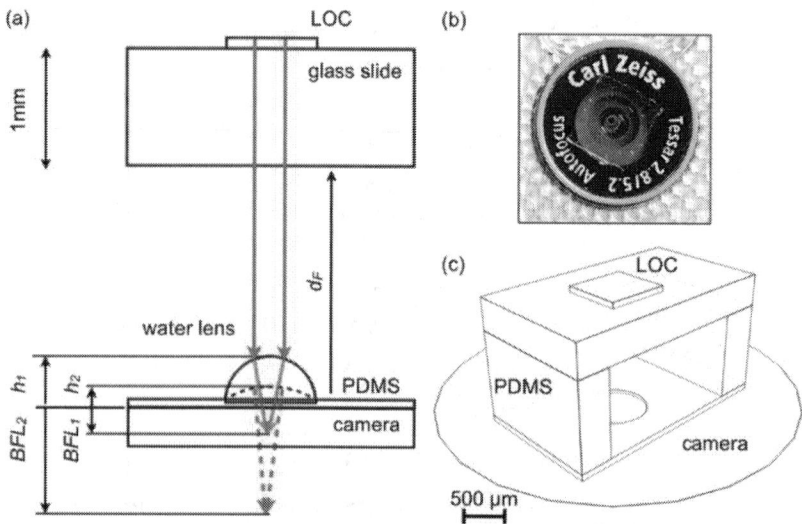

Figure 1. (a) Cross-section of the measuring device. The liquid lens is represented at two different volumes with the associated back focal lengths (BFL). The forward distance (d_F) is 1 and 2 mm in the experiments and it is defined by PDMS structural elements; (b) Nokia 6720 cell phone rear camera with a PDMS lens substrate adhered to its surface; (c) 3D scheme of the device including the integrated adaptive focusing element.

SU-8(50)/S1818 was spin coated on a clean glass slide (Menzel-Glaser, Braunschweig, Germany) producing a processed film thickness of 200 μm. Two steps of soft baking at 65 °C for 5 min and 85 °C for 25 min, on a hotplate, were applied before sealing the walls of the micro chambers and inner structures to the glass substrate by exposing the film with a patterned source at 81 mJ·cm^{-2} through the glass slide. The samples were then flipped over and exposed from the photoresist side at 27 mJ·cm^{-2} to configure the monolithic roofs and ceilings. The samples were post-baked at 65 °C for 3 min and 85 °C for 15 min on a hotplate, developed by immersion in SU-8 developer (mr-dev 600, Microresist technology) for 2 h at room temperature and dried in N_2 afterwards.

For the demonstration of color detection experiments, the chambers were filled with 0.1% solution of resazurine in water (Chroma-Gesellschaft, Münster, Germany), delivered with a 5 mL syringe set with a 0.5 mm diameter needle. While video imaged with the Nokia 6720 front camera, pH3 buffer solution

(CertiPUR, Merck, Darmstadt, Germany) was delivered to the second reservoir. The acquired video stream was visually inspected to select few frames showing the progress of the color change.

2.3. Imaging

Three different cameras were used to test the adaptive lenses. The front and rear cameras of a Nokia 6720 classic cell phone and the frame embedded camera of a MacBook Pro Apple computer.

The front camera in the Nokia cell phone is a QVGA camera (320 × 240 pixels, 8 bit color channels for still images and 176 × 144 pixels in video mode) and it was operated with the software provided by the phone manufacturer, which produces .jpg pictures and .3gp format videos.

Nokia 6720's rear camera is a 5 MP, 2,592 × 1,944 pixels still camera with Carl Zeiss optics and autofocus for minimum focal distances of several centimeters. This camera is also capable of VGA video (640 × 480 pixels, 8 bits color channels) recording at 15 fps. The software provided with the phone has additional modes for this camera, which for still images was set in macro mode, maximum resolution and time-lapse acquisition at 6 fpm.

The MacBook Pro computer camera is a VGA device (640 × 480 pixels, 8 bits color channels) recording H.264 encoded videos in .mov format controlled by QuickTime Player Version 10.0 running on Mac OSX 10.6.7. The device was positioned at the center of the camera lens and the PDMS surface inherently adheres to the camera surface providing sufficient mechanical and optical coupling.

Liquid lenses were made of a drop of distilled water delivered with a 5 mL syringe (set with a 0.5 mm diameter stainless steel needle, Microlance 3 from BD Medial Systems, Drogheda, Ireland) directly on the PDMS reservoir.

All cameras lay horizontally and pointing upwards during the measurements, which proceeded immediately after creating the liquid lens. Measurements were carried out at 20 °C room temperature and normal indoors illumination of about 1,000 Lm/m^2.

LOC samples were imaged in a routine inverted fluorescence microscope, Zeiss Axiovert 40 CFL (Carl Zeiss AG, Germany), with white light illumination and through a 10× (0.25) A-plan objective lens and a 2.5× objective. Images were acquired with a 5 MP, 8 bits color channels, Cannon PowerShot A95, mounted to the camera channel of the microscope.

Sessile drop evaporation was characterized by imaging the process with an Olympus SZ60 stereomicroscope at 6.3× magnification using an Au mirror on a 45° PDMS support. A distilled water drop was delivered on a PDMS surface and imaged at a 15 s interval with a Canon EOS 500D DSRL camera, acquiring 15 MP raw format images.

2.4. Data Processing

Acquired image sequences and video streams were manually inspected to determine best focus. Autofocusing code and plug-ins are available for automatic focus selection [23], however, for the purposes of this study the visual inspection of the captured image sequences was sufficient.

Image processing was carried out in 64-bit Image J 1.44014 and data collected and edited in Keynote 5.05 and Matlab 2008b.

Image composition was made by layer stacking in Photoshop CS4. 26 time-lapse images of the sessile drop evaporation were registered and manually masked with a 60% opacity brush to recover the drop surface reflections. Quantification of the drop chord and height was carried out in Image J for individual time-lapse acquisitions of the drop evaporation.

For the characterization of color reactions the images were collected in a single collage, split in color channels and the blue channel spatially averaged in Image J before profiling.

Modeling was carried out in Raytrace 2.21 (IME Software) for 2D ray tracing conceptual analysis and Atmos-Optical Design and Analysis Software [24] for lens design and characterization.

3. RESULTS AND DISCUSSION

Adaptive optics is central in autofocusing cameras and can be implemented in different ways [14–17], but in all cases they are conceived as permanent components. In contrast, this work aims at a solution coherent with the evaluation of disposable LOC devices on intact cell phones. Accordingly, the adaptive optics concept we study in this work must be compatible with integration within the disposable LOC device.

Autofocus cameras in cell phones cannot focus at the short range required for LOC evaluation and an additional optical element should provide this function; however a permanent focusing element would introduce an extra component, which would restrict the ubiquity of the solution. Contrarily, the morphing lenses used in this work are versatile to adapt to different cell phone cameras, whereas they can be integrated in the LOC tests, and disposed together after use.

Figure 1(a) shows the schematics of the disposable morphing lens design. The LOC device is represented as mounted on a 1 mm thick substrate (e.g., standard glass slides) and separated from the lens-supporting element by a forward distance (d_F) shorter than 2 mm, which defines a compact configuration for the entire arrangement. Ambient light is used for illumination.

The adaptive lens is created on a 150 μm thick PDMS layer that mildly adheres to the camera surface (Figure 1(b)) producing a reliable optical coupling between these elements. After the measurement the device is easily removed and disposed leaving the camera intact. The PDMS element is patterned with a 30 μm deep circular depression (800 μm diameter) that confines the lens to a fixed position (Figure 1(c)), and holds the LOC devices at a fixed distance from the camera.

The lens consists of a drop of distilled water delivered through a 0.5 mm needle. At its maximum volume in the operating regime the thickness of the lens along the optical axis is h_1 and for this condition the curvature is maximum, thus producing the shortest back focal length (BFL_1).

Lens curvature is determined by the water drop volume and the contact angle with the substrate, which can be widely controlled using PDMS [25] as substrate. During natural evaporation the drop loses volume reducing its curvature and consequently increasing the BFL (BFL_2 in Figure 1(a)). During this process the BFL is scanned in a range sufficient to refocus the LOC device on diverse types of cameras.

At room temperature the 0.134 μL lens considered in Figure 1(a) takes several minutes to completely evaporate. This mechanism provides enough time for time-lapse image acquisitions or video recordings that capture different focusing conditions. In a post-processing step the best-focused images for each particular platform can be selected. Exactly the same procedure, and the same configuration, can be used on any cell phone or computer set, since the focal range that can be scanned with this design is large enough to accommodate optical differences across models and brands. In this work we corroborated this assumption on two different cameras of a standard cell phone (Nokia 6720 Classic) and the frame embedded camera of a MacBook Pro Apple computer.

Figure 2 collects the optical simulation of the lens shown in Figure 1(a). Figure 2(a) shows the behavior of the BFL for h values between 50 and 400 μm at a fixed d_F of 2 mm. A small variation of 350 μm in h controls a BFL change of more than 4 mm. In the same way, the aperture of the system decreases with the curvature, implying a greater depth of field at longer BFLs (and smaller h values).

The spot diagram in Figure 2(b) characterizes the 2° distortions of the system for green light (λ = 560 nm), an intermediate range in the visible spectrum where these cameras operate. Not surprisingly, this simple configuration [26] suffers from multiple types of aberrations. The spot diagram in Figure 2(b) indicates spherical and coma aberrations increasing with the curvature (and thickness h), and the performance degrade towards shorter wavelengths.

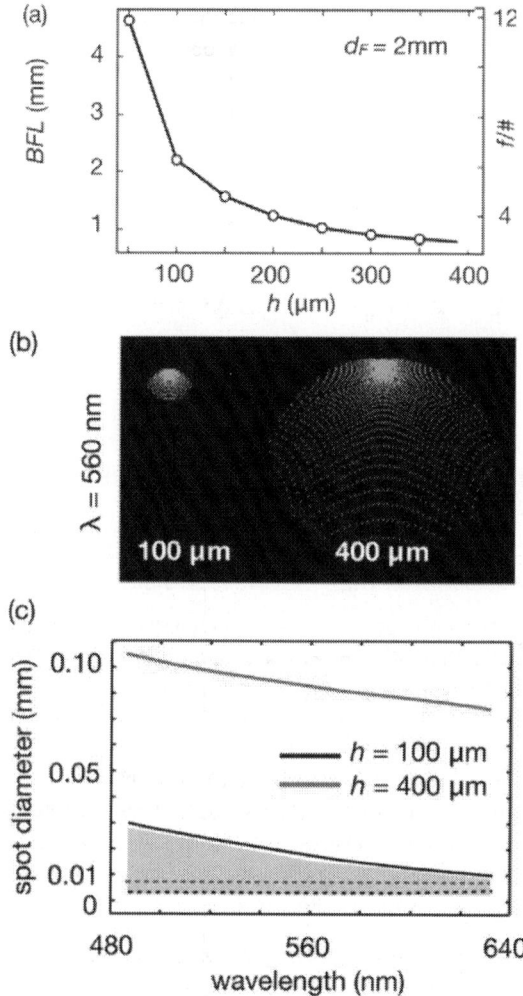

Figure 2. (a) Calculated back focal length (BFL) and aperture (f/#) vs. lens thickness along the optical axis (h); (b) Calculated spot diagram for λ = 560 nm and 2° angle for h = 100 μm and 400 μm; (c) RMS spot diameter in mm vs. wavelength for h = 100 μm in black solid lines and h = 400 μm in red solid lines. The diffraction-limited Airy diameter is indicated in dashed lines for the same two conditions.

The root mean square (RMS) spot diameter [27], the diameter of a circle containing approximately 68% of the focused energy, is shown in Figure 2(c), for the same curvatures as in Figure 2(b) and for different wavelengths. The dashed lines correspond to the diffraction-limited system (Airy diameters [28]), for the two previous curvatures.

As can be seen, the expected performance increases at longer wavelengths almost approaching the diffraction limit for 630 nm and h = 100 μm. Overall, the spot diameter is between 10 and 100 μm depending on the curvature and illuminating conditions, and under 30 μm for systems that operate with BFLs longer than 2 mm.

Airy diameters are between 4 and 8 μm for green light and above typical pixel sizes in standard cell phones, which can be between ~1.23 μm width for 5 MP in 1/4 "format and ~2 μm in 1/2.5".

The adaptive lens was experimentally tested imaging LOC microstructures of known dimensions. Figure 3(a) shows a 25× optical microscopy image of a SU-8/S1818 device fabricated with a refined 3D mask less micro prototyping MPLS method [22].

These structures have two parts: two open lateral flow service channels connected to 1 mm diameter reservoirs made of SU-8 (10) (30 μm thick) and a SU-8/S1818 3D chamber fused to the service channels. The chamber is 200 μm high and the whole system is a monolithic unit sealed to the glass slide substrate. The images were taken from the glass side.

Figure 3(b,c) are 100× magnifications of the chamber and display two different models tested in this work, an empty chamber (Figure 3(b)) and a chamber with a 6 × 6 array of 50 μm diameter pillars connecting the glass surface with the chamber ceiling.

Figure 3(d) shows selected full frames (174 × 144 pixels, 8 bits per channels color images) from a 1 min video stream acquired with the front camera of a standard Nokia 6720 cell phone. This video is acquired 2 min after the liquid lens was created and it was measured with a d_F = 2 mm configuration. The video acquisition was controlled with the software provided with the phone.

The camera was focused in the background of the scene, which shows the ceiling of the room and a fluorescent lamp, which provided the illumination. No other source was employed in the experiments. Simultaneously, the area of the camera lens occupied by the liquid lens refocuses the test microstructure at different BFLs, while changing curvature due to evaporation. As can be seen, the sharpest images are obtained, with this camera, between 48 s and 63 s. It is also possible to observe that at the best focus the system maximized the depth of field, bringing together in focus the more than 100 μm high chamber ceiling and the 30 μm thick service channel, which became visible only after 48 s.

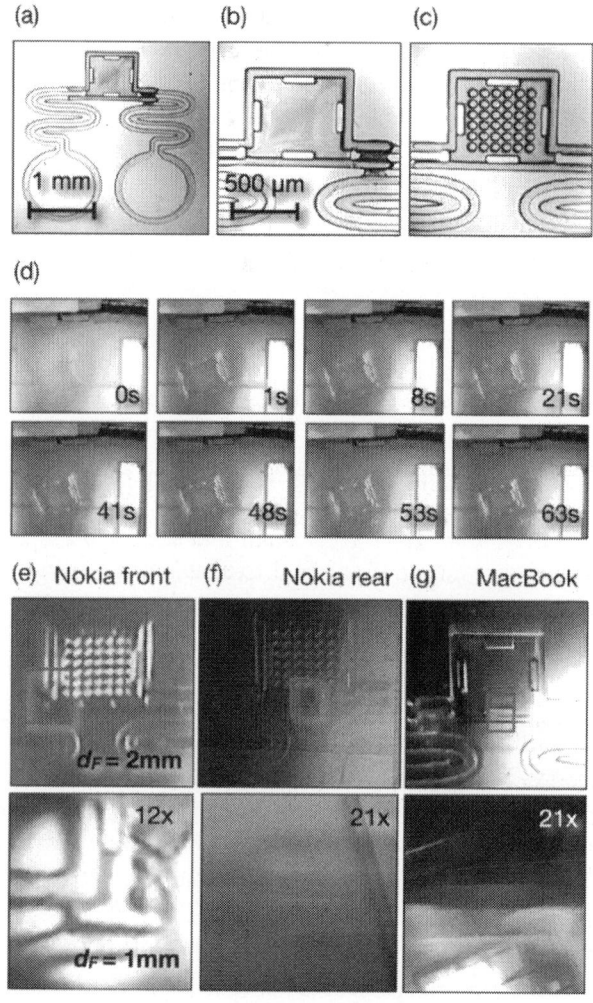

Figure 3. (a) Microscopy image of a 3D test microstructure at 25× magnification; (b) Detail at 100× magnification of one type of micro chamber used in this work; (c) Detail at 100× magnification of the second type of micro chamber used in this work; (d) Selected full frames (176 × 144 pixels, 8 bits per colour channel) from a 1 min video stream recorded with a Nokia 6720 cell phone front camera. The focus changes while the liquid lens evaporates and between 41 s and 53 s the best focus and depth of field is achieved; (e) Selected area of a microstructure imaged with a Nokia 6720 cell phone front camera operating in video mode for d_F = 2 mm and d_F = 1 mm in the lower panel, where an estimated 12× zoom factor is indicated; (f) Selected area of a microstructure imaged with a Nokia 6720 cell phone rear camera operating in 5 MP time-elapsed still image acquisition at 6 fpm for d_F = 2 mm and d_F = 1 mm in the lower panel, where an estimated 21× zoom factor is indicated; (g) Selected area of a microstructure imaged with a MacBook Pro camera operating in VGA (640 × 480 pixels) video mode for d_F = 2 mm and d_F = 1 mm in the lower panel, where an estimated 21× zoom factor is indicated.

Depending on the location of the illumination, high contrast ratios exceeding the camera dynamic range, can occur in the region where the microstructure is imaged. Repositioning the cell phone is one possible supervised solution, whereas implementing high dynamic range (HDR) recording, which has been demonstrated for this same type of cell phone [12], could provide unsupervised imaging in future developments.

Figure 3(e–g) illustrate the performance of the three different cameras at imaging the microstructures using two different optical magnifications defined by the distance d_F = 2 mm in the upper panels and d_F = 1 mm, in the three lower panels. These images are the same selected areas cropped from full frames of different resolutions.

Figure 3(e) is taken from a 176 × 144 pixels frame video acquired with the Nokia 6720 front camera in white ambient light. Even at this low resolution, perhaps the most demanding condition one can propose, it was possible to resolve the 50 μm pillars with an acceptable depth of field. Shifting to d_F = 1 mm provided a more than 12× optical zoom (red square in upper panel taken as reference) with enough depth of field to simultaneously capture the base and ceiling of the chamber. Note that the rectangular openings in the chamber were in the ceiling, more than 100 μm from the substrate surface where walls were imaged, and both could be identified.

Figure 3(f) is a still image taken from the rear 5 MP camera (2,592 × 1,944 pixels) of the Nokia 6720 cell phone. In order to find the right focus a time elapsed acquisition at 6 fpm was run. This acquisition mode is also available with the native software installed in the phone. In the d_F = 2 mm range the edges were sharp, the 50 μm pillars could be better resolved and there was enough depth of field to capture channels and chambers in focus. One advantage of the 5 MP images is that they can tolerate digital zooming and image enhancement.

In the d_F = 1 mm range a zoom factor of more than 21× could be estimated from the figure, although the depth of field was severely reduced in this condition, as illustrated by the focused image only at the edge of one the ceiling's rectangular openings.

Figure 3(g) shows the best-focused frames acquired with the MacBook Pro camera. In this case, a 640 × 480 pixels video was acquired at 15 fps with QuickTime Player, a free application from Apple.

The selected image for d_F = 2 mm range showed an intermediate resolution between the two Nokia cameras, but with good depth of field and clarity. As in the case of the 5 MP Nokia camera the depth of field deteriorated at d_F = 1 mm, and in this case the estimated zoom factor was also about 21×.

Sessile drop evaporation is a complex non-stationary phenomenon displaying more alternatives than the particular regime of operation considered so far.

The drop geometry is defined by its contact angle with the host substrate, in this case a PDMS slab fabricated as detailed in the experimental section. As reported in [25] the contact angle with PDMS can be widely varied after UV irradiation if necessary to serve different design requirements.

During the evaporation of a liquid drop heat and mass transfer processes occur simultaneously, which involve associated heat transfer to the substrate, effect of the surface tension on convection within the drop, effect of contaminants and environmental conditions. Advanced modeling and detailed experiments on sessile drops are reported in the literature [18–20] and are beyond the scope of this work. Here, instead, drop evaporation was characterized for the conditions they have been used for optical detection.

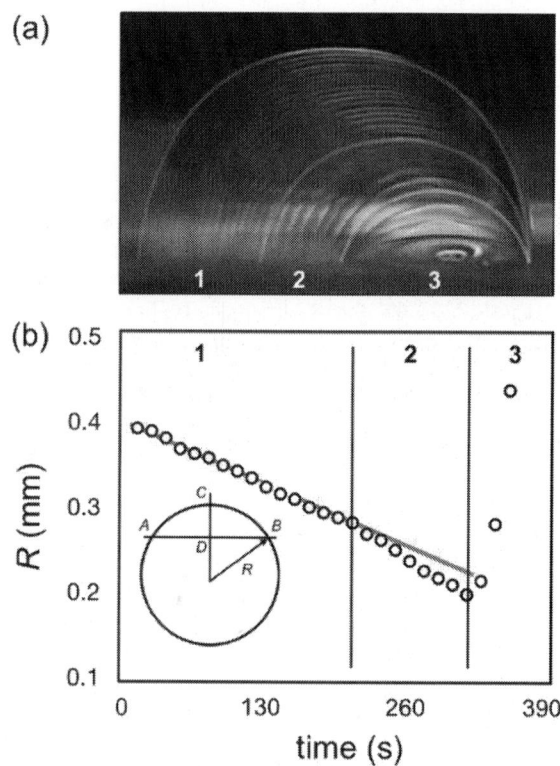

Figure 4. (a) Collection of sessile drop images captured with a stereomicroscope at a 15 s interval. The evaporation regimes (indicated as 1, 2 and 3) are highlighted; **(b)** Characterization of the drop evaporation by the drop radius of curvature, calculated from measured chord and height distances at 15 s intervals.

Figure 4(a) illsustrates the transition of a water drop on a PDMS substrate at room temperature as measured at 15s interval. The figure shows three different regimes in the behavior of the drop. In the first regime (1 in Figure 4(a)), occurring for about half of the recorded time, the contact angle remains practically constant and the evaporation changes the drop curvature by reducing the chord and height. In the second regime (2 in Figure 4(a)) the process accelerates, which can be noticed by the larger spacing between captures, and a noticeable change in contact angle, although the drop chord and height keep changing. Finally in the third regime (3 in Figure 4(a)) the changes accelerate even further, but the drop is pinned [18,19] at a constant chord and only the contact angle and associated height changes.

From measurements of chord and height taken from the individual pictures is possible to calculate the radius of curvature R as:

$$R = \frac{\left(\dfrac{d_{AB}}{2}\right)^2 + \left(d_{CD}\right)^2}{2 \cdot d_{CD}}$$

(1)

where d_{AB} is the chord and d_{CD} is the height h. The evolution of R (Figure 4(b)) shows a linear decrease in the first regime (1 in Figure 4(b)), in what is characterized as an unpinned drop [19], where the contact line recedes, while the contact angle remains constant. In the second regime (2 in Figure 4(b)) the noticeable changes of contact angle in the still unpinned drop are reflected on a faster decrease of R. In the third regime (3 in Figure 4(b)) the drop is pinned and changes curvature at a faster rate producing a large range of increasing Rs (and correlated BFLs), resulting in the conditions we exploit to adapt LOC imaging to diverse phone cameras.

Although the mechanism of de-pinning is beyond the scope of this work, the deviation from equilibrium of the contact angle is a correlated phenomenon [19]. On the other hand, the pinning drops are related to the surface roughness and eventual contamination, thus the pinning effect is expectable in most cases [19] and reliably present for the operating conditions of our devices.

Although our morphing lenses operate in the third time of regime, the results in Figure 4 suggests that other lens designs are possible, which could exploit slower focal changes to evaluate assays that demand longer response times.

Variations in ambient temperature and relative humidity can certainly affect the evaporation rate and regime of the focusing element. In the present conditions the systems operates satisfactory for normal air-conditioned environments, but if required the evaporation rate can be reduced by partially enclosing the lens environment.

In recent years a number of distributed microscopy principles associated with cell phones for imaging and communication have been demonstrated [6–8]. These examples employ dedicated instruments [7], reusable additional devices [6], operate with a single type of camera and require accurate sample positioning [13] or imply permanent modifications to the cell phones [8].

In contrast with these approaches the present principle is a disposable element integrated in the LOC device operating on intact cell phone cameras, and uses available ambient illumination as light source.

It is worth noticing that the performance of the adaptive lens is enough for the purpose it has been conceived: enables to image LOCs in the range of dimensions and resolutions that are required, it makes it adaptively and as part of a disposable device and operates in a compact and fixed configuration. The merit of the present solution is not the ultimate performance in each of these areas but the ability to collect all these aspects in a single disposable element, and establishing the compromises that make the off-chip LOC readout feasible on regular phones.

The capabilities of the conceived device to support sensing uses were tested for the detection of a transient chemical reaction within a LOC device.

To assess the limits of performance, the simpler front camera of the Nokia phone was used in video mode. The micro chamber in Figure 2(a,b) was imaged in this experiment at d_F = 2 mm and 176 × 144 pixels color video resolution at 15 fps.

A 0.1% solution of resazurine in water (pH 7, blue colour) was delivered to one reservoir of the micro chamber, after the adaptive lens was in the focusing region and the video acquisition had started. Once the solution reached the chamber and turned it blue, a pH3 buffer solution drop was delivered to the second reservoir and a colour change front (from blue to orange) crossed the measuring chamber. At 15 fps the reaction front could be captured in multiple frames, four of which are collected in Figure 5. These pictures show a blue chamber with a yellow front advancing from the right of the image, and completely crossing the chamber in about 3 s.

The color change can be more quantitatively rendered by profiling the blue channel of these images along a line (Figure 5, indicated in the t_0 panel). This result confirmed that a simple camera could follow a fast chemical reaction in the confined area (0.6 mm^2) of a regular LOC element.

Summarizing, the results collected in Figure 3 show the ability of the considered concept to image micrometric features from 3D microstructures in the range representative of LOC detection areas, making the technique a feasible alternative for the evaluation of LOC devices on intact cell phones and computer sets without additional instrumentation or accessories, only using native software for acquisition and under available illumination.

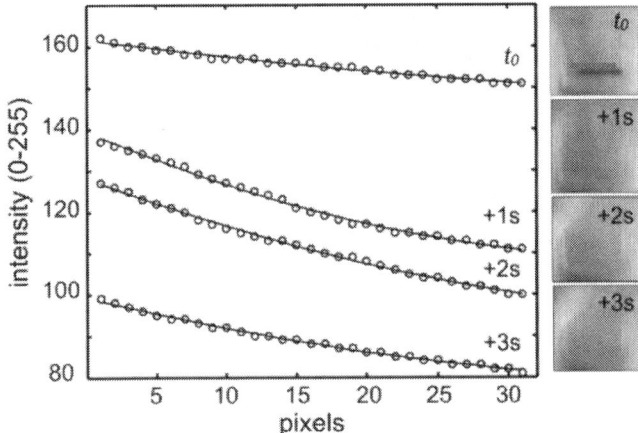

Figure 5. Color changing reaction of resazurine solution from pH7 to pH3 within a microchamber, captured at 15 fps with a Nokia 6720 front camera in video mode and analyzed through the blue camera channel.

These results demonstrate that a simple and generic device, integrating a morphing lens, can operate across diverse brands, models and types of compact cameras delivering the required performance for LOC sensing experiments.

The proposed concept is forgiving to imperfections, and delivers consistent results by complementing the versatile imaging and acquisition capabilities of intact cell phones. Further magnification is possible using thinner LOC substrates and approaching the specimen to the lens, such it has been demonstrated for fixed lenses with micro-positioned samples [13], however, the development in that direction certainly depends on the detection target, for general LOC evaluation the present range is sufficient and enables the use of robust substrates and classical LOC configurations, which are important aspects to enable ubiquitous LOC usage.

In this context, further progress would involve the incorporation of HDR acquisition to secure results in arbitrary illuminating conditions and the refinement of the fluidics to create the focusing element. Nevertheless, already at the present stage the possibility to detect transient chemical reactions within microstructures, using the simplest imaging configuration and the most common hardware and software resources readily available in cell phones, has been demonstrated.

4. CONCLUSIONS

Imaging of LOC micrometric features within regions larger than 500 × 500 μm^2 has been demonstrated using ambient illumination and consumer cameras on intact cell phones and computer sets. The proposed method is generic and was conceived to adapt to diverse models and brands of pervasive consumer imagers using disposable devices that could be deployed in large numbers. The use of a morphing focusing element integrated in disposable LOC devices permitted their evaluation without introducing permanent accessories or specialized sample positioning, which would limit the ubiquity of cell phones as measuring platforms. The present concept highlights the possibility to materialize decentralized sensing relying on classical LOC technologies by temporarily co-opting intact cell phones as universal off-chip readers; In this case, tested with the monitoring of a transient chemical reaction using the simplest camera configuration.

ACKNOWLEDGMENTS

This work has been supported by a grant from the Linköping Centre for Life Science Technologies (LIST), Sweden, and a PhD scholarship from Thammasat University of Thailand for Pakorn Preechaburana.

REFERENCES

1. Yager, P.; Edwards, T.; Fu, E.; Helton, K.; Nelson, K.; Tam, M.R.; Weigl, B.H. Microfluidic diagnostic technologies for global public health. Nature**2006**, 442, 412–418.

2. Martinez, A.W.; Phillips, S.T.; Whitesides, G.M. Three-dimensional microfluidic devices fabricated in layered paper and tape. Proc. Natl. Acad. Sci. USA**2008**, 105, 19606–19611.

3. Kuswandi, B.; Nuriman, B.; Huskens, J.; Verboom, W. Optical sensing systems for microfluidic devices. Anal. Chim. Acta**2007**, 601, 141–155.

4. Whitesides, G.M. The origins and the future of microfluidics. Nature**2006**, 442, 368–373.

5. Chin, C.D.; Linder, V.; Sia, S.K. Lab-on-a-chip devices for global health: Past studies and future opportunities. Lab Chip**2007**, 7, 41–57.

6. Zhu, H.; Yaglidere, O.; Su, T.; Tseng, D.; Ozcan, A. Cost-effective and compact wide-field fluorescent imaging on a cell-phone. Lab Chip**2011**, 11, 315–322.

7. Breslauer, D.; Maamari, R.; Switz, N.; Lam, W.; Fletcher, D. Mobile phone based clinical microscopy for global health applications. PLoS One**2009**, 4, e6320.

8. Aoki, P.M.; Honicky, R.J.; Mainwaring, A.; Myers, C.; Paulos, E. Common Sense: Mobile Environmental Sensing Platforms to Support

Community Action and Citizen Science. Proceedings of Tenth International Conference on Ubiquitous Computing (Ubicomp), Seoul, Korea, 21–24 September 2008; Paper 201. pp. 59–60.

9. Rakow, N.; Suslick, K. A colorimetric sensor array for odour visualization. Nature**2000**, 406, 710–712.

10. Potyrailo, R.A.; Morris, W.G.; Leach, A.M.; Hassib, L.; Krishnan, K.; Surman, C.; Wroczynski, R.; Boyette, S.; Xiao, C.; Shrikhande, P.; et al. Theory and practice of ubiquitous quantitative chemical analysis using conventional computer optical disk drives. Appl. Opt.**2007**, 46, 7007–7017.

11. Filippini, D.; Alimelli, A.; Natale, C.D.; Paolesse, R.; D'Amico, A.; Lundström, I. Chemical sensing with familiar devices. Angew. Chem. Int. Ed.**2006**, 45, 3800–3803.

12. Preechaburana, P.; Macken, S.; Suska, A.; Filippini, D. HDR imaging evaluation of a NT-proBNP test with a mobile phone. Biosens. Bioelectron.**2011**, 26, 2107–2113.

13. Smith, Z.J.; Chu, K.; Espenson, A.R.; Rahimzadeh, M.; Gryshuk, A.; Molinaro, M.; Dwyre, D.M.; Lane, S.; Matthews, D.; Wachsmann-Hogiu, S. Cell-phone-based platform for biomedical device development and Education Applications. PLoS One**2011**, 6, e17150.

14. Yu, H.; Zhou, G.; Chau, F.; Sinh, S. Tunable electromagnetically actuated liquid-filled lens. Sens. Actuat. A**2011**, 167, 602–607.

15. Zhang, W.; Aljasem, K.; Zappe, H.; Seifert, A. Completely integrated, thermo-pneumatically tunable microlens. Opt. Express**2011**, 19, 2347–2362.

16. Song, C.; Nguyen, N.; Yap, Y.; Luong, T.; Asundi, A. Multi-functional, optofluidic, in-plane, bi-concave lens: Tuning light beam from focused to divergent. Microfluid. Nanofluid.**2011**, 10, 671–678.

17. Malouin, B., Jr.; Vogel, M.; Olles, J.; Cheng, L.; Hirs, A. Electromagnetic liquid pistons for capillarity-based pumping. Lab Chip**2011**, 11, 393–397.

18. Erbil, H.Y. Evaporation of pure liquid sessile and spherical suspended drops: A review. Adv. Colloid Interface Sci.**2012**, 170, 67–86.

19. Sefiane, K. The coupling between evaporation and absorbed surfactant accumulation and its effect on the wetting and spreading behaviour of volatile drops on a hot surface. J. Pet. Sci. Eng.**2006**, 51, 238–252.

20. Hu, H.; Larson, R.G. Evaporation of sessile droplet on a substrate. J. Phys. Chem. B**2002**, 106, 1334–1344.

21. Lötters, J.; Olthuis, W.; Veltink, P.; Bergveld, P. The mechanical properties of the rubber elastic polymer polydimethylsiloxane for sensor applications. J. Micromech. Microeng.**1997**, 7, 145–147.

22. Preechaburana, P.; Filippini, D. Fabrication of monolithic 3D microsystems. Lab Chip**2011**, 11, 288–255.

23. Image Processing and Analysis in Java. Plugins. Available online: http://rsbweb.nih.gov/ij/ (accessed on 28 April 2012).

24. ATM Optical Design and Analysis Software. Lens Design and Characterization Available online: http://www.atmos-software.it/Atmos.html (accessed on 28 April 2012).

25. Graubner, V.M.; Jordan, R.; Nuyken, O.; Kötz, R.; Lippert, T.; Schnyder, B.; Wokaun, A. Wettability and surface composition of poly(dimethylsiloxane) irradiated at 172 nm. Polym. Mater.: Sci. Eng.**2003**, 88, 488–489.

26. Escudero-Sanz, I.; Navarro, R. Off-axis aberrations of a wide angle schematic eye model. J. Opt. Soc. Am. A**1999**, 16, 1881–1891.

27. Fischer, R.E.; Tadic-Galeb, B.; Yoder, P.R. Computer Performance Evaluation. In Optical System Design, 2nd ed.; SPIE Press, McGraw-Hill: London, UK, 2008; Chapter 10.

28. Fischer, R.E.; Tadic-Galeb, B.; Yoder, P.R. Diffraction, Aberrations, and Image Quality. In Optical System Design, 2nd ed.; SPIE Press, McGraw-Hill: London, UK, 2008; Chapter 3.

Chapter 8

The Influence of Filter Slit on the Imaging Performance of the Solar Grating Spectrometer Based on Adaptive Optics

Lianhui Zheng[1,2,3,4], Changhui Rao[1,4], Naiting Gu[1,4], Qi Qiu[2]

[1]*Key Laboratory of Adaptive Optics, Chinese Academy of Sciences, Chengdu, China*

[2]*School of Optoelectronic Information, University of Electronic Science and Technology of China, Chengdu, China*

[3]*University of Chinese Academy of Sciences, Beijing, China*

[4]*Institute of Optics and Electronics, Chinese Academy of Sciences, Chengdu, China*

ABSTRACT

The solar grating spectrometer is an essential tool to study the thermodynamics properties of the solar atmosphere with different height distributions, but its imaging performance will be degraded by the wavefront aberration generalized by the atmospheric turbulence. On the other hand, the narrow slit of the grating spectrometer will filter the wavefront aberration to some extent. The influence of the filter slit on the wavefront aberration and the correction requirement of the adaptive optics are analyzed theoretically and experimentally. We demonstrate that the influence of filter slit on the different types and magnitudes of wavefront aberration is different, and the RMS value of the wavefront aberration less than 0.3λ is down to below almost 60% after the filter slit, and it can lower the correction range requirement of the adaptive optics. The numerical simulation and experiment results show that: after the adaptive optics correction, the influence of the wavefront aberration on the spectral resolution is neglected, and the energy utilization is considerably improved; both numerical simulation and experiment results are in good agreement.

Keywords: Solar Atmosphere, Aberration, Adaptive Optics, Prototype, Spectral Resolution

1. INTRODUCTION

It is well known that the evolving magnetic field causes the solar atmosphere eruption, and it produces most of the sometimes spectacular visible phenomena, such as the sunspots, prominences, flares, coronal mass ejections and so on [1] [2] . Real-time observation and forecast of the solar atmosphere activities are very important since the violent activities of the solar atmosphere and period variation will influence the living environment of human being, and the imaging grating spectrometer is an important tool to achieve this goal [3] - [5] . It not only can perform the imaging observation, but also can perform the spectral observation and identify the thermodynamics parameters of the solar atmosphere by the spectral information, such as the magnetic field, pressure and element abundance. As a result, the imaging grating spectrometer is broadly applied in the solar atmosphere observation.

However, the imaging performance of the grating spectrometer installed in the ground-based solar telescope is limited by the wavefront aberration [6] [7] , i.e. the dynamic wavefront aberration generalized by the atmospheric turbulence and the static wavefront aberration of the optical system. The wavefront aberration not only will degrade the spatial resolution, but also will degrade the spectral resolution and energy utilization. Fortunately, after development for several years, Adaptive Optics (AO) technique has made great progress, and it has become an important tool to reduce the influence of the atmospheric aberration and static aberration [8] -[10] . Therefore, the Adaptive Optics system is integrated in the solar telescope to compensate the atmospheric turbulence influence on the imaging performance of the solar telescope.

On the other hand, unlike the general optical system, the used slit of the grating spectrometer is narrow enough, and it will filter the wavefront aberration to some extent. Quantitatively studying the influence of the filter slit on the wavefront aberration with different types and magnitudes becomes a necessity. Since the filter slit will filter the wavefront aberration, the traditional AO system cannot be directly applied to compensate the wavefront aberration. Our goal is to demonstrate the influence of the filter slit on the wavefront aberration and the influence of the Adaptive Optics theoretically and experimentally. To the best of our knowledge, it is an innovative work.

This paper is organized as follows: in Section 2, the influence factors of the imaging grating spectrometer are analyzed; in Section 3, the numerical simulation is performed and is validated experimentally; in Section 4, the conclusions are summarized.

2. THE INFLUENCE FACTORS ON THE IMAGING PERFORMANCE OF THE GRATING SPECTROMETER

The spectral resolution and energy utilization are the key parameters of the imaging grating spectrometer, and they are mainly influenced by the wavefront aberration, slit width and the spectral sampling of the CCD camera. Among them, the influence of the slit width and the spectral sampling of the CCD camera on the spectral resolution cannot be avoided, and can be represented by the net spectral resolution [4] . On the other hand, the quantitative influence of the filter slit on the spectral resolution and energy utilization of the grating spectrometer is still rare understood, and we will discuss it in the following paragraphs.

The optical layout of the grating spectrometer is depicted in the Figure 1. The field of view of the imaging grating spectrometer is limited by the slit since the slit is the field stop. Generally, the slit is rectangle, and the slit length is long enough. Hence, the field of view is limited by the slit width.

The spectral purity [4] is used to represent the influence of the slit width on the spectral resolution, given by

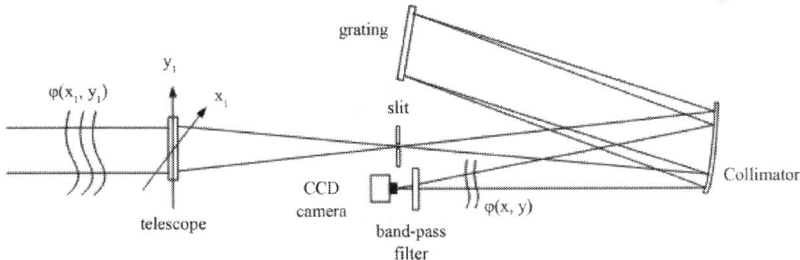

Figure 1. The optical layout of the grating spectrometer.$\phi(x_1, y_1)$ is the wavefront aberration of the solar telescope and the atmospheric turbulence, and $\phi(x, y)$ is the wavefront aberration of the grating spectrometer.

$$\Delta\lambda_{sp} = w_s \frac{d\cos\alpha}{mf}$$

$$(1)$$

where the w_s is the slit width of the grating spectrometer, α is the incidence angle of the grating, m is the diffraction order, f is the focal length of the collimator, d is grating constant.

After considering the influence of the slit width, the spectral resolution [4] is given by

$$\Delta\lambda = \sqrt{\Delta\lambda_{gr}^2 + \Delta\lambda_{sp}^2}$$

(2)

where

$$\Delta\lambda_{gr} = \frac{\lambda}{w_g \sigma m}$$

(3)

where the w_g and the σ is the illuminated width and the groove density of the grating, respectively.

It is apparent that the energy utilization increases with the slit width, but at the cost of the spectral resolution based on Equation (2). When the slit width is equal to the airy disk diameter of the solar telescope $(1.22\lambda f/D)$, the influence of the slit width on the spectral resolution is minimized, and the energy utilization is maximized [7].

Actually, the real spectral resolution $\Delta\lambda_{real}$ often is determined by the Rayleigh Criterion [11] [12]. That is, the spectral line s1 and s2 with equal intensity can be distinguished, when $|\lambda1 - \lambda2|$ is equal to the full width at half maximum (FWHM) of the spectral line (s1 or s2). On the other hand, the spectral resolution will be degraded by the aberration, and the s1 and s2 cannot be identified as distinct spectral line again, as depicted in the Figure 2.

In order to directly measure the relative spectral resolution degradation, the K is given by

$$K = \left| \frac{\Delta\lambda_{real} - \Delta\lambda}{\Delta\lambda_{real}} \right| \times 100\%$$

(4)

where the $\Delta\lambda_{real}$ is the real spectral resolution influenced by the aberration, and $\Delta\lambda$ is the spectral resolution without the influence of the aberration. The K represents the variable of the $\Delta\lambda$ caused by the aberration.

For the point source, the energy utilization η is given by

$$\eta = \iint \frac{PSF(x,y)_{\varphi \neq 0}}{PSF(x,y)_{\varphi = 0}} dxdy \times 100\%$$

(5)

where the ϕ is the wavefront aberration, the $PSF(x,y)_{\varphi \neq 0}$ and $PSF(x,y)_{\varphi = 0}$ are the point spread function

with and without the influence of the aberration, respectively.

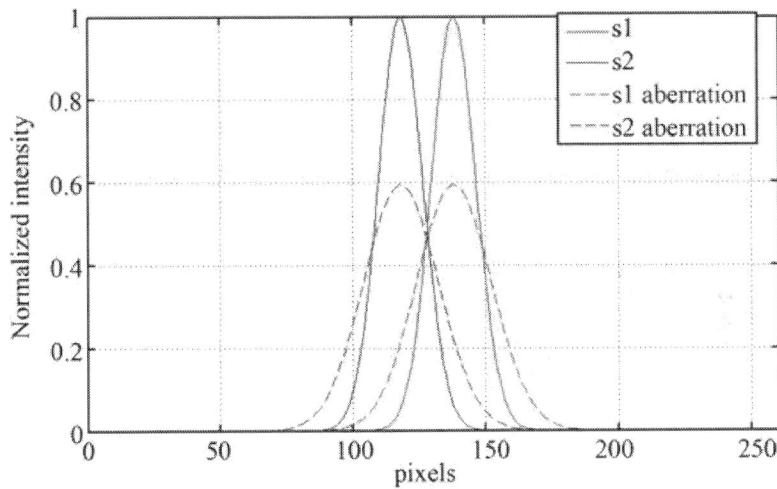

Figure 2. The spectral line influenced by the aberration. The red and the blue solid lines are the spectral line s1 and s2, respectively; the red (s1aberration) and blue (s2 aberration) dot lines are the spectral lines s1 and the s2 influenced by the aberration, respectively.

The observed spectral line [11] is given by

$$O(\lambda) = I(\lambda) \otimes PSF(\lambda)$$

(6)

where the $O(\lambda)$ is the observed spectral line, $I(\lambda)$ is the real spectral line, the $PSF(\lambda)$ denotes the modulation of the system, and the $\otimes$ denotes the convolution operator.

Unlike the general optical system, the slit of the grating spectrometer will filter the wavefront aberration to some extent. According to the optics theory, the point spread function after the filter slit is given by

$$PSF_s(x,y) = PSF_{tel}(x,y) \, sinc^2(x/w_s) \, sinc^2(y/l_s) \tag{7}$$

where the w_s and l_s are the slit width and length, respectively; The $PSF_{tel}(x,y)$ is the modulation of the solar telescope and the atmospheric turbulence, given by

$$PSF_{tel}(x,y) = \left| FFT\left\{ A_0 e^{i\varphi(x_1,y_1)} \right\} \right|^2$$

where the FFT{●} denotes the Fourier transform operator, the $\varphi(x_1,y_1)$ is the wavefront aberration of the solar telescope and the atmospheric turbulence, the $\varphi(x,y)$ is the wavefront aberration of the optical system of the grating spectrometer. The $sinc(x/w_s)$ and $sinc(y/l_s)$ is given by

$$sinc(x/w_s) = \begin{cases} 1 & -0.5w_s \leq \overset{\cdot}{x} \leq 0.5w_s \\ 0 & \text{others} \end{cases}$$

$$sinc(y/l_s) = \begin{cases} 1 & -0.5l_s \leq y \leq 0.5l_s \\ 0 & \text{others} \end{cases}$$

It can be inferred from the Equation (7) that the slit will filter the wavefront aberration due to the diffraction of the slit, and the high frequencies of the wavefront aberration will be blocked by the slit.

The point spread function at the focal plane of the grating spectrometer is given by

$$PSF_{gf}(x,y) = PSF_s(x,y) \otimes PSF_{sp}(x,y) \tag{8}$$

where the $PSF_{sp}(x,y)$ is the modulation of the grating spectrometer, given by

$$PSF_{sp}(x,y) = \left| FFT\left\{ Ae^{i\varphi(x,y)} \right\} \right|$$

Supposing the dispersion direction is along the x-axis, according to the linear dispersion definition [12] , the relation between the spatial coordinate x and the spectral coordinate λ is given by

$$x = \frac{mf}{d \cos \beta} \lambda \tag{9}$$

where the β is the diffraction angle of the grating.

And the $PSF(\lambda)$ is given by

$$PSF(\lambda) = \int PSF_{gf}(x,y)dy \Big|_{x=\frac{mf}{d\cos\beta}\lambda}$$

(10)

Substituting the Equation (10) into the Equation (6), the observed spectral line is given by

$$O(\lambda) = I(\lambda) \otimes \int PSF_{gf}(x,y)dy \Big|_{x=\frac{mf}{d\cos\beta}\lambda}$$

(11)

Supposing the value P is the maximum value of the observed spectral line, and the spectral resolution influenced by the wavefront aberration can be obtained by solving the equation O(λ) = 0.5P. Assuming the λ+ and λ- are the real solutions, and λ+> λ-, hence the real spectral resolution influenced by the wavefront aberration is given by

$$\Delta\lambda_{real} = \lambda_+ - \lambda_-$$

(12)

And the relative spectral resolution K can be expressed as follows:

$$\kappa = \left| \frac{\lambda_+ - \lambda_-}{\Delta\lambda} - 1 \right| \times 100\%$$

(13)

3. NUMERICAL SIMULATION AND EXPERIMENT VALIDATION

3.1. System Description

The AO system is integrated in the solar grating spectrometer, and it consists of a Hartmann Shack Wavefront sensor (HS WFS) and an electrically addressed phase-only liquid-crystal spatial light modulator (LC SLM), as depicted in the Figure 3. The polarized HeNe laser with 632.8 nm wavelength is used as the light source for all the experiments, its FWHM is 1 pm. Our goal is to demonstrate the influence of the wavefront aberration on the spectral resolution, hence the experiment conclusions are also validated for the broad solar spectrum.

The RMS value of the initial wavefront aberration of the optical elements is 0.15λ, detected by the HS WFS with 28 × 28 sub-apertures. After the AO correction, the RMS value of the residual wavefront aberration is roughly 0.025λ, and a nearly diffration-limited focal spot is attained, the result of the experiment is as depicted in the Figure 4.

In the following experiments, an original wavefront aberration will firstly be generated by the LC SLM and will exist throughout the whole correction procedure. The generated aberration will be detected by the HS WFS. Based on the reconstructed wavefront generalized by the HS WFS, the control computer calculates the conjugated wavefront at each position with the response matrix, and it is applied to control the corresponding pixel of the LC SLM and then the aberrated wavefront is corrected. In the experiment, the tiny slit widths we used are 50 μm and 100 μm, respectively, which are roughly equal to one airy disk diameter of the optical system $(1.22\lambda f/D)$ and two airy disk diameter $(2.44\lambda f/D)$, respectively, where f is the focal length of L1, λ is the wavelength, D is the input aperture diameter.

3.2. The Influence of the Filter Slit on the Wavefront Aberration

To demonstrate the influence of the filter slit on the wavefront aberration with different types and magnitudes,

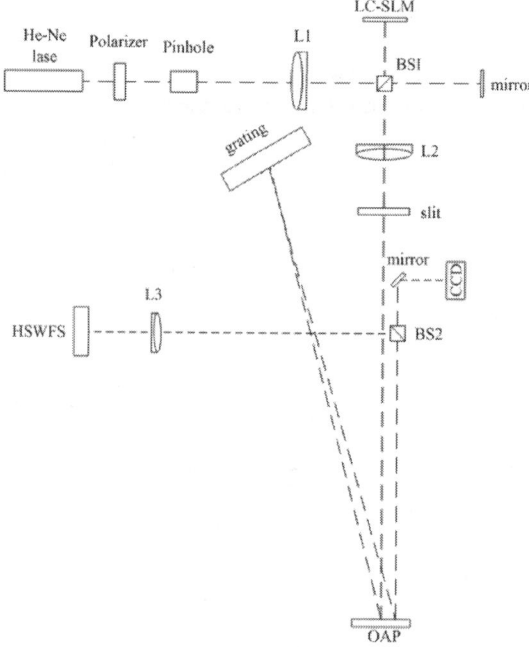

Figure 3. The optical layout of the grating spectrometer. L1 - L3: lenses; BS1 - B2: 50/50 non-polarizing beam splitters; OAP: off axis parabola mirror.

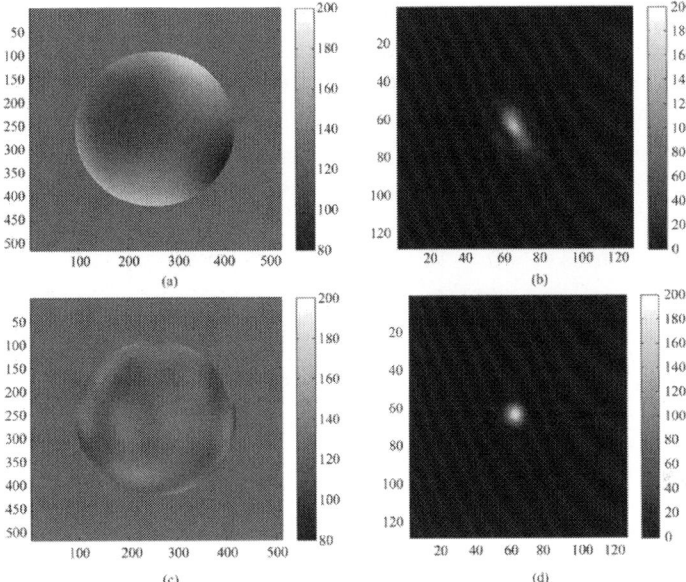

Figure 4. The wavefront aberration of the optical system before and after the AO correction. (a) and (b) represent the wavefront aberration and the corresponding focal spot before the AO correction, respectively; (c) and (d) represent the residual wavefront aberration and the focal spot after the AO correction, respectively.

the experiment validation is performed. The wavefront aberrations described by the Zernike polynomials with order number from 3rd to 15th. After the initial aberration of the optical system is corrected by the AO, and a new wavefront aberration is superimposed in the initial wavefront aberration. The RMS values of the wavefront aberrations are 0.1λ, 0.2λ and 0.3λ, which is generalized by the LC SLM. Zernike tip and tilt terms have been eliminated. To directly demonstrate the influence of the filter slit, ΔRMS is given by

$$\Delta RMS = \frac{RMS - RMS_s}{RMS} \times 100\%$$

(14)

where, the RMS and RMS_s represent the RMS value before and after the filter slit, respectively. The bigger the ΔRMS is, the stronger the influence of filter slit on the wavefront aberration is. The experimental results are illustrated in Figure 5.

We demonstrate that: when the same slit is used, the influence of the filter slit on the different order Zernike aberration is different. The bigger the magnitude of the Zernike aberration is, the stronger the influence of the filter slit on the wavefront aberration is. On the other hand, to the same Zernike aberration, the smaller the slit is, the stronger the influence of the filter slit on the wavefront aberration is. Generally, when the wavefront aberration less than 0.3λ, the $\Delta RMS \leq 60\%$ is attained after the filter slit. It is apparent that the wavefront aberration needed to be corrected becomes smaller after the filter slit, hence the filter slit can lower the requirement of the AO correction.

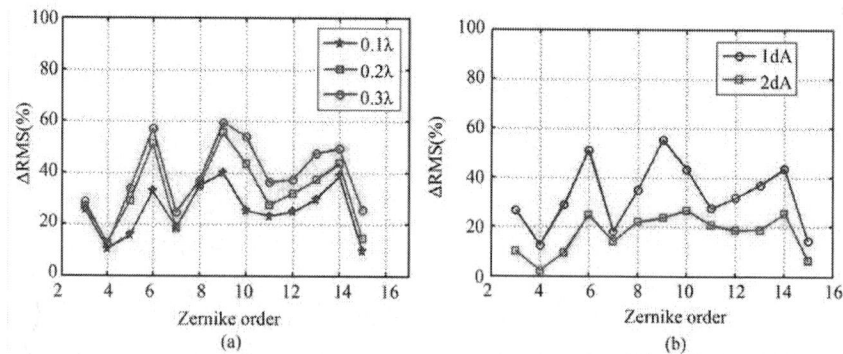

Figure 5. The influence of the filter slit on the wavefront aberration. (a) When the slit width is 1dA, the influence of the filter slit on the different Zernike order aberration, the RMS values of the aberration are 0.1λ, 0.2λ and 0.3λ; (b) When the RMS value of the aberration is 0.2λ, the influence of the different slit widths on the wavefront aberration, slit are 1dA and 2dA, respectively.

3.3. Single Zernike Order Aberration Correction

To investigate the influence of the Adaptive Optics on the κ and η, a series of closed-loop corrections of aberrated wavefronts described by Zernike polynomials with order number from 3rd to 15th are carried out. Since the residual wavefront aberration of the optical system roughly is 0.025λ after the AO correction, then a new wavefront aberration is superimposed into the corrected wavefront aberration. The RMS value of the wavefront aberration is 0.2λ, which is generalized by the LC SLM. Zernike tip and tilt terms are eliminated. Close-loop correction example for the 9th order Zernike aberrations is illustrated in the Figure 6. Apparently, a nearly diffraction focal spot is obtained after the AO correction. The influence of the AO correction on the κ and η is depicted in the Figure 7. The results show an effective wavefront correction, the κ is less than 2%, and η is better than 95%.

3.4. Kolmogorov Phase Screen Correction

The validation experiments of the influence of AO correction on the κ and η are performed. The atmospheric phase screen is consistent with Kolmogorov's theory. The phase screen consists of 3rd to 15th Zernike orders, and the D/ro = 5, D/ro = 7 and D/ro = 10 are used, where the D is the input aperture diameter, and ro is the Fried parameter. Close-loop correction examples for the D/ro = 5 is illustrated in the Figure 8. The influence of the AO correction on the κ and η is illustrated in the Figure 9. The numerical simulation and experiment results show that an effective wavefront correction is obtained. The influence of the wavefront aberration generalized by the atmospheric turbulence on the κ and η can be neglected after the AO correction. The κ is less than 2%, and the energy utilization η is considerably improved, better than 95%.

4. CONCLUSIONS

The solar grating spectrometer with high spectral resolution is an important tool to study the characteristics of the solar atmosphere. Otherwise, the energy utilization is import to the signal-to-noise and temperal resolution. However, the spectral resolution and energy utilization will be influenced by the wavefront aberration. Unlike the general optics system, the slit of the grating spectrometer is narrow enough, and it will filter the wavefornt aberration to a certain extent. Hence, when the AO system is applied to compensate the wavefront aberration, the influence of the filter slit on the wavefront aberration should be considered.

In this paper, the influence of the filter slit on the wavefront aberration is analyzed theoretically and experimentally. The results show that the RMS value of the aberration less than 0.3λ is down to below almost 60% after the filter slit. It is apparent that the wavefront aberration needed to be corrected becomes smaller after the filter slit. Besides, the closed-loop of the AO corrections is conducted. We demonstrate that diffraction-limited focal spot is attained. The influence of the residual wavefront aberration on the spectral resolution is neglected, and the energy utilization is considerably improved.

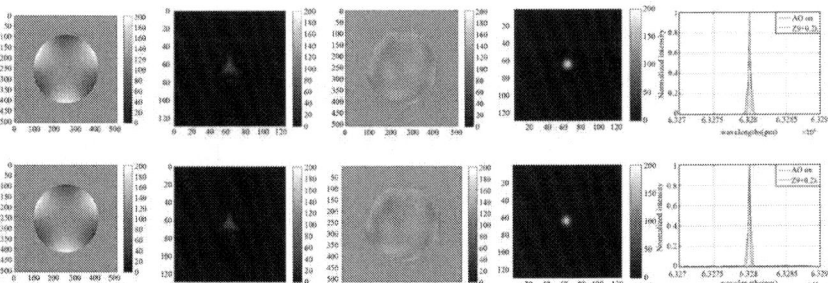

Figure 6. Close-loop correction result of the 9th order aberration. The upper and the lower rows represent the numerical simulation and experiment results respectively, from left to right, including original wavefront, focal spot before AO correction, residual wavefront after AO correction, focal spot after AO correction and the spectral line before and after the AO correction.

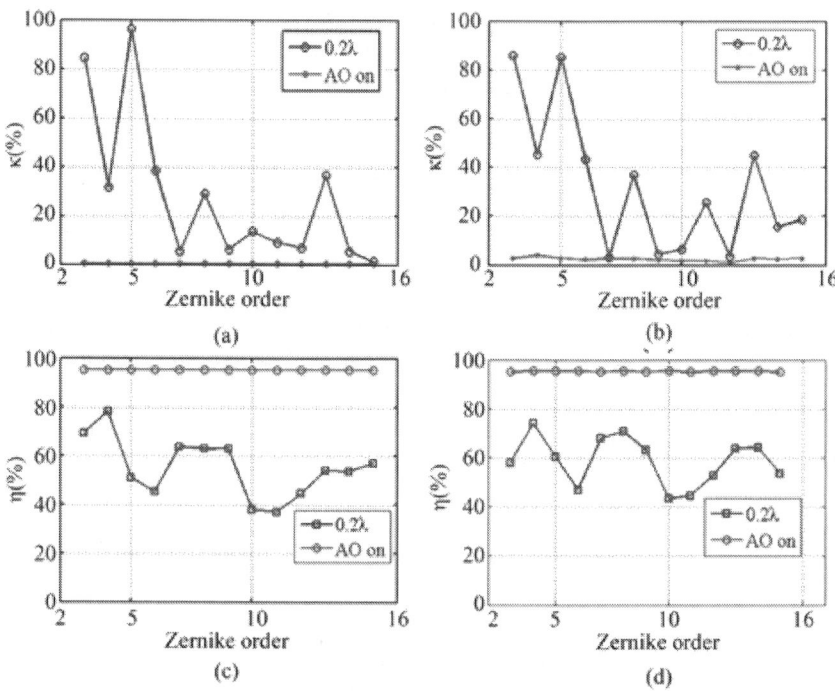

Figure 7. The influence of the AO correction on the κ and η. (a) and (b) are the numerical and the experiment on the κ, respectively. (c) and (d) are the numerical and the experiment results on the η, respectively.

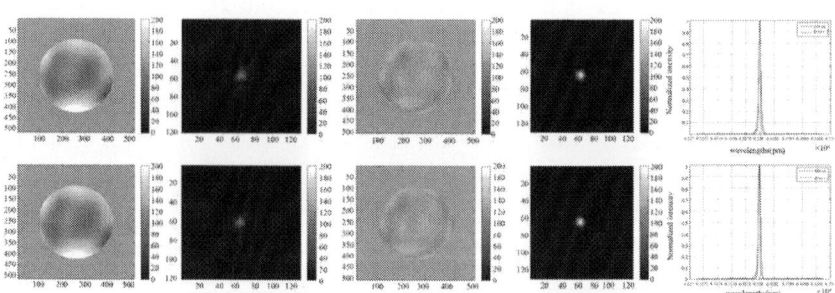

Figure 8. Close-loop correction results of the Kolmogorov phase screen with D/ro = 5. The first row represent the numerical simulation results, including original wavefront, focal spot before AO correction, residual wavefront after AO correction, focal spot after AO correction and the spectral line before and after the AO correction. The second row represents the experiment results corresponding to first row.

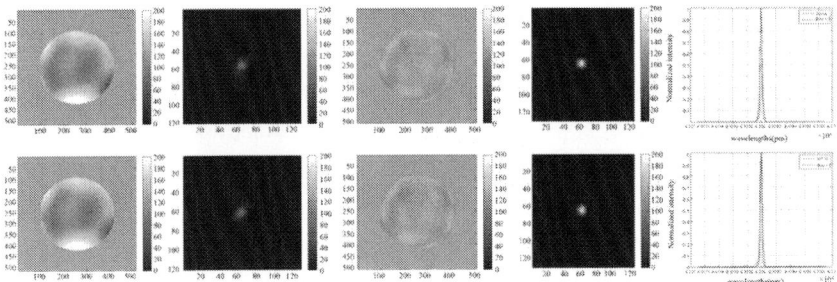

Figure 9. The influence of the AO correction on the κ and η. From left to right represent the influence of the AO correction on the κ and η, respectively.

ACKNOWLEDGEMENTS

This work was funded by the Joint Funds of the National Natural Science Foundation of China under No. 11178004.

REFERENCES

1. Volkmer, R., von der Lühe, O., Denker, C., et al. (2010) GREGOR Solar Telescope: Design and Status. Astronomische Nachrichten, 331, 624-627.

2. van Noort, M.J. and Rouppe van der Voort, L.H.M. (2006) High-Resolution Observations of Fast Events in the Solar Chromosphere. Astrophysical Journal, 648, L67-L70. http://dx.doi.org/10.1086/507704

3. Collados, M., Calcines, A., Díaz, J.J., et al. (2008) A High-Resolution Spectrograph for the Solar Telescope GREGOR. SPIE Proceedings, Ground-Based and Airborne Instrumentation for Astronomy II, Marseille, 10 July 2008, 70145Z.

4. Chae, J., Park, M., Ahn, K., et al. (2012) Fast Imaging Solar Spectrograph of the 1.6 Meter New Solar Telescope at Big Bear Solar Observatory. Solar Physics, 288, 1-22.

5. Mein, P., Mein, N. and Bommier, V. (2009) Fast imaging Spectroscopy with MSDP Spectrometers. Vector Magnetic Maps with THEMIS/MSDP. Astronomy & Astrophysics, 507, 531-539.

6. Roddier, F. (1981) The Effects of Atmospheric Turbulence in Optical Astronomy. Progress in Optics, 19, 281-376.

7. Wu, G. (1978) Design of Optical Spectrometer. Science Press, Beijing.

8.   Rao, C., Zhu, L., Gu, N., et al. (2010) Performance of the 37-Element Solar Adaptive Optics for the 26 cm Solar Fine Structure Telescope at Yunnan Astronomical Observatory. Applied Optics, 49, G129-G135.

9.   Rao, C., Zhu, L., Gu, N., et al. (2013) Solar Adaptive Optics System for 1-M New Vacuum Solar Telescope. Third AO4ELT Conference-Adaptive Optics for Extremely Large Telescopes, Florence, May 2013. http://ao4elt3.sciencesconf.org/13295/document

10.   Rimmele, T.R. (2004) Recent Advances in Solar Adaptive Optics. SPIE Proceedings, 5490, 34-46.

11.   Huang, Y.R., Xu, A.A. and Tang, Y.Y. (1987) Immediate Measurement for Astrophysics. Science Press, Beijing.

12.   Palmer, C.A., Loewen, E.G. and Thermo, R. (2005) Diffraction Grating Handbook. Newport Corporation, Springfield.

Chapter 9

Optical Power Allocation for Adaptive Transmissions in Wavelength-Division Multiplexing Free Space Optical Networks

Hui Zhou[a, b], Shiwen Mao[a,], Prathima Agrawal[a,]

[a] Department of Electrical and Computer Engineering, Auburn University, Auburn, AL 36849-5201, USA

[b] Amazon Web Service, Amazon Inc., 345 Boren Ave N, Seattle, WA 98109, USA

ABSTRACT

Attracting increasing attention in recent years, the Free Space Optics (FSO) technology has been recognized as a cost-effective wireless access technology for multi-Gigabit rate wireless networks. Radio on Free Space Optics (RoFSO) provides a new approach to support various bandwidth-intensive wireless services in an optical wireless link. In an RoFSO system using wavelength-division multiplexing (WDM), it is possible to concurrently transmit multiple data streams consisting of various wireless services at very high rate. In this paper, we investigate the problem of optical power allocation under power budget and eye safety constraints for adaptive WDM transmission in RoFSO networks. We develop power allocation schemes for adaptive WDM transmissions to combat the effect of weather turbulence on RoFSO links. Simulation results show that WDM RoFSO can support high data rates even over long distance or under bad weather conditions with an adequate system design.

Keywords : Free space optics, Radio over free space optics, Wavelength-division multiplexing

1. INTRODUCTION

Recently, demand for multimedia service with high quality of service (QoS) requirements has been drastically increasing. To cater to the increasing demand on capacity, optical fibers have been utilized for years to deliver high volume of data, often between central and remote universal stations. Due to the relatively high cost of deploying optical fibers, Free Space Optics (FSO) has been developed as a cost-effective alternative wireless access technology for multi-Gigabit rate communication networks. FSO provides an excellent alternative to optical fiber systems for last-mile access networks [1], ranging from local area network (LAN)-to-LAN connection for enterprise/campus, high capacity military communications, to disaster recovery and emergency response, among others.

Attracting considerable attention from the research community, the FSO technology has also been developed for carrying various wireless services, as known as Radio on Free Space Optics (RoFSO). FSO systems can operate on wavelengths in the 1520–1600 nm range, which makes the development of wavelength-division multiplexing (WDM) RoFSO systems feasible. Advanced Dense Wavelength Division Multiplexing (DWDM) RoFSO systems have been developed to support the simultaneous transmission of multiple wireless signals [1]. Despite of its great potential of supporting data intensive communications for various applications, a line-of-sight (LOS) path is required in any FSO system. Consequently, FSO is highly susceptible to the atmospheric environment due to the inhomogeneity of air temperature and pressure, or flying objects [2]. To harvest the high potential of RoFSO, fading-mitigation techniques should be employed to mitigate atmospheric turbulence-induced intensity fluctuations.

To this end, topology control [3], [2], [4] and [5], load-balancing [6], and spatial diversity techniques [7] have been studied and proved to be effective in maintaining a good system performance. Adaptive transmissions have been recently introduced into FSO systems and is emerging as a potential solution to mitigate the effect of atmospheric turbulence [8]. In FSO systems, channels are usually slow-fading and FSO transceivers have full-duplex capabilities. With negligible effect on data rates, a small portion of the bandwidth can be used for feedback of channel state information (CSI). In some hybrid RF/FSO systems, the RF channel can be used for CSI feedback [9]. Thus reliable CSI could be available in FSO systems, which will be highly useful for designing adaptive transmission schemes.

In this paper, we propose optical power allocation schemes for an adaptive WDM RoFSO system in which variable wavelengths are adopted to mitigate the effect of weather turbulence [10]. Proposed optical power allocation schemes optimally allocate transmit power to achieve maximum capacity and enhance the performance of the WDM RoFSO system. Nowadays, the bandwith wavelengths between 1520 nm and 1600 nm have already been used in FSO systems. Furthermore, the emerging quantum cascade laser (QCL) technology can offer great flexibility on adjusting an RoFSO transceiver to operate on the optimal transmit wavelengths [11]. Under a total power constraint, different optical powers can be allocated to the chosen wavelengths in a WDM RoFSO system, to achieve further enhanced system performance.

We investigate the problem of optical power allocation under the total power budget and eye safety power constraints for adaptive WDM transmission in RoFSO systems. To achieve capacity gain, we first analyze a conventional FSO system and develop a simple water-filling based algorithm to derive the optimal power allocation for the chosen wavelengths. For WDM RoFSO systems, a near-optimal RoFSO power allocation algorithm is developed based on the reformulation–linearization technique (RLT) [12], which can provide a linear programming (LP) relaxation of the complex problem. A computationally efficient scheme is also developed based on an approximation of the channel model. Finally, we investigate the diversity gain in the WDM RoFSO system. The performance of the proposed schemes is evaluated with simulations, and is demonstrated to be highly effective for achieving high system capacity under various scenarios.

The remainder of this paper is organized as follows. The related work is discussed in Section 2. The system model is presented in Section 3, while the three power allocation schemes are developed in Section 4 to fully utilize DWDM RoFSO systems. Simulation results are analyzed in Section 5. Section 6 concludes this paper.

2. RELATED WORK

Attracting significant interest both in academia and industry, the FSO technology has been recognized as a promising solution for high capacity, long distance communications. The performance of FSO networks is highly depend on the availability and reliability of the LOS path since an FSO transmitter is highly directional. Weather turbulence strongly affects FSO communication links. Weather effects on the connectivity of FSO networks were studied in [13]. The influence of turbulence-accentuated interchannel crosstalk on WDM FSO system performance has been studied in [14].

Many fading-mitigation techniques have recently been employed to maintain a good FSO system performance. In [15], a multipath fading resistant FSO communication system architecture was introduced to combat adverse weather conditions. Spatial diversity techniques, extensively studied in conventional RF communication systems [16], can also be applied in FSO systems to improve system performance. A multiple-input multiple-output (MIMO) FSO system can achieve significant diversity gain in the presence of atmospheric fading by deploying multiple transmit or receiver apertures [7] and [17]. A cooperative diversity technique [18], [19] and [20] is also a cost-effective alternative to maintain system performance. In a recent work [21], a one-relay cooperative diversity scheme was proposed for combating turbulence-induced fading, while cooperative diversity was analyzed for non-coherent FSO communications.

Adaptive transmission technology has been introduced into FSO systems to mitigate weather turbulence. Djordjevic in [22] applied the conventional wireless adaptive modulation and coding method in an FSO system and further studied adaptive low-density-parity-check (LDPC) coded modulation to compensate performance degradation when turbulence is strong. Karimi and Uysal in [8] designed transmission algorithms with consideration of the number of bits carried per chip time (BpC) in an FSO link, in which intensity modulation/direct detection (IM/DD) with M-ary pulse position modulation (M-PPM) was employed. Several other adaptive schemes have also been proposed and studied. In [11], the authors proposed using variable wavelength to combat the effects of atmospheric interference. Varying wavelength becomes feasible as the QCL technology becomes more mature. In [23], the authors proposed an adaptive transmission scheme to satisfy the requirements of various wireless services. A WDM power allocation method considering Optical Modulation Index (OMI) was proposed in their adaptive RoFSO system design. The authors in [24] studied the potential of the MIMO channel for combating link fading. However, the performance of the FSO MIMO system is constrained by the thermal noise limited receivers and thus Avalanche photodiodes (APDs) [25] were studied and commonly used in FSO systems.

In this paper, we propose power allocation schemes for adaptive WDM transmissions to achieve capacity gain or diversity gain. WDM has been employed in FSO transmission systems and has been shown to be capable of supporting very high data rate transmission in [26]. RoFSO technology makes it possible to transmit multiple RF signals using WDM. RoFSO provides a promising alternative to optical fiber systems. In [27], the authors designed and evaluated an RoFSO system as an universal platform for the integration of optical fiber and FSO networks. In [28], optical fading in FSO Channels was statistically analyzed, while [1] provides a comprehensive study of RoFSO and the satisfactory results confirmed that the effect of scintillation on RoFSO performance can be estimated by an analytical model.

In the adaptive WDM RoFSO system we studied, a variable number of wavelengths are adopted to mitigate weather turbulence and optical transmit power is optimally allocated to achieve maximum capacity gain. Although turbulence may not change significantly with wavelength in some weather conditions, considering the huge bandwidth that the DWDM RoFSO system can support, it is still non-trivial to study the problem of utilizing wavelength properly. Moreover, since QCL technology can offer great flexibility on adjusting wavelengths, more wavelengths can be utilized as FSO systems advance. Alternatively, the multiple wavelengths used in the WDM RoFSO system can be utilized to achieve robustness of the system.

3. SYSTEM AND CHANNEL MODEL

In this section, we will introduce the channel model and system models. We summarize the notation used in this paper in Table 1.

Table 1. Table of notation.

Symbol	Definition
h_l	FSO link attenuation
h_f	Atmospheric turbulence
A_{TX}	Aperture area of the transmitter
A_{RX}	Aperture area of the receiver
α	Atmospheric attenuation coefficient
V	Visibility in kilometers
Λ	Wavelength
q	A parameter related to visibility
CNR	Carrier to noise ratio
OMI	Optical modulation index
RIN	Relative intensity noise
K	Boltzman's constant
T	Temperature
m	Photodiode gain
e	Electrical charge
F	Excess noise factor
G_f	Photodiode output conductance
h	Channel state of a FSO link
B	Bandwidth
N	Number of available wavelength bands
M	Number of used wavelength bands
h_i	Channel gain for channel i
P_i	Optical power allocated to channel i
P	Peak power bound for transmitted pulse
P_{max}	Power budget
λ, λ_i	Lagrange multipliers

3.1. Channel model

FSO transceivers are highly directional, but FSO links are prone to degradation due to weather turbulence. We consider both effects of path loss and turbulence-induced fading over FSO links [18]. The optical channel state h is modeled as a product of two factors

$$h = h_l \cdot h_f,$$

(1)

where h_l denotes the attenuation and h_f represents the atmospheric turbulence. Attenuation h_l is a function of optical wavelength Λ and link distance d, as

$$h_l = \frac{A_{TX} A_{RX} \cdot e^{-\alpha d}}{(\Lambda \cdot d)^2},$$

(2)

where A_{TX} and A_{RX} are the aperture areas of transmitter and receiver, respectively. The atmospheric attenuation coefficient α is given by

equation(3)

$$\alpha = \left(\frac{3.91}{V}\right) \cdot \left(\frac{\Lambda}{55}\right)^{-q},$$

(3)

where V is the visibility in kilometers and q is a parameter related to the visibility as [8]

$$q = \begin{cases} 0.585 V^{1/3}, & V \leq 6\,\text{km} \\ 1.3, & 6\,\text{km} \leq V \leq 50\,\text{km}. \end{cases}$$

(4)

For fading h_f, we assume atmospheric turbulence can be modeled as a log-normal distribution [29]. The log-normal model is a widely used fading model, especially under weak-to-moderate turbulence conditions.

The channel model for an FSO link can be written as

$$y = h \cdot P_t \cdot x + n,$$

where x and y are transmitted and received signal respectively; P_t is the power of the transmitted pulse; and n is the additive Gaussian noise.

3.2. System Model

RoFSO is a new technology that provides high data rate and reliable transmission. The RoFSO system using WDM allows simultaneous transmission of multiple data streams consisting of various wireless and wireline services at very high rates. An advanced DWDM RoFSO system is illustrated in Fig. 1[1].

Figure1. Illustration of an advanced DWDM RoFSO system.

We assume a WDM FSO system that is capable of operating in the wavelength band from 1520 nm to 1600 nm. Apart from the data-transmitting antenna, we assume that an atmospheric influence measurement antenna or weather measurement device is equipped at the FSO BS. Thus, we can estimate atmospheric loss for channels using different wavelengths. Alternatively, CSI can be obtained by using a small portion of the bandwidth to provide channel information without tangibly affecting the data rate. In some hybrid RF/FSO systems, the RF channel can be used for CSI feedback. Since atmospheric turbulence is a major degrading factor in FSO systems, we propose power allocation schemes for the adaptive RoFSO system, in which both wavelength and transmit power will be adaptively allocated according to channel conditions.

An important parameter to evaluate the performance of RF-FSO is the carrier to noise ratio (CNR). The CNR of an RoFSO system using an APD photo detector is given as [1]

$$CNR = \frac{0.5(OMI \cdot mP_r)^2}{RIN \cdot P_r^2 + 2em^{(2+F)}P_r + 4KT \cdot G_f},$$ (6)

where OMI is the optical modulation index, m is the photodiode gain, T is the temperature, K is Boltzman's constant, e is the electrical charge and G_f is the photodiode output conductance. In the denominator, $4KTG_f$ is thermal noise; F is the excess noise factor; m is the photodiode gain; e is the electrical charge; $2em^{(2+F)}P_r$ is the optical short noise; and $RINP_r^2$ is the relative intensity noise from Laser diode (LD). The numerator represents the received signal power. $P_r = rP_{pd}$, where r is the photodiode responsivity and P_{pd} is the received power at the detector and is given asthe product of transmit power and channel gain. Without loss of generality, we assume all tones are modulated with the same OMI that will not introduce intermodulation distortion.

4. ADAPTIVE WDM TRANSMISSION

To mitigate the effect of weather turbulence, we adapt wavelength and adjust power allocation to achieve better system performance. First, we consider an RoFSO system using wavelength Λ in the range of [1520,1600] nm. The available wavelengths are divided into N parts, which are non-overlapping with adequate spacing. We assume the FSO channel is slow varying [8]. For each channel with wavelength Λ_i, we can estimate its channel state h_i or obtain the channel state through a feedback channel.

In different weather conditions, it is desirable to choose the wavelengths that have the best channel conditions. We assume that the WDM FSO system will use at most M different wavelengths. Among the N available wavelengths, we will first choose M wavelengths that have the greatest channel gains. The next step is to allocate transmit powers to different wavelengths such that the system capacity is maximized.

4.1. Conventional FSO system

We first consider conventional modeling of wireless channels under white Gaussian noise. For the sake of simplicity, we define channel gain over noise for channel i by abusing of notation as

$$h_i = \frac{|h_i \cdot h_f|^2}{N_0}.$$

(7)

According to the estimated channel gains, we choose M wavelengths that can offer the greatest channel gains from the N available wavelengths. Here M is a constant jointly determined by the channel conditions and the capacity need of the system. It can be set to N, for example, if all the wavelengths are to be used.

Let P_i be the optical power allocated to channel i. Due to fixed power budget P_{max} for each FSO base station, we have the following total power constraint:

$$\sum_{i=1}^{M} P_i \leq P_{max}.$$

(8)

In FSO systems, eye safety should always be taken into consideration in the system design. Thus we have the additional power constraint

$$0 \le P_i \le \overline{P} \quad \text{for all } i. \tag{9}$$

where $\overline{P}$ is the peak power bound for the transmit powers. Thus, we formulate the following capacity maximization problem:

$$\max \quad \sum_{i=1}^{M} \log(1 + P_i \cdot h_i) \tag{10}$$

$$\text{s.t.} \quad \sum_{i=1}^{M} P_i \le P_{max} \tag{11}$$

$$0 \le P_i \le \overline{P}, \quad \text{for all } i. \tag{12}$$

By applying Karush–Kuhn–Tucker (KKT) theorem, we can find that optimal power allocation satisfies

$$\begin{cases} P_i = \min\left\{ \overline{P}, \left[\dfrac{1}{\lambda} - \dfrac{1}{h_i} \right]^+ \right\} \\ \sum_{i=1}^{M} P_i = P_{max}, \end{cases} \tag{13}$$

where λ is the Lagrange multiplier. The inverse of λ is often regarded as the water level.

The algorithm to solve the capacity maximization problem is to first sort the channels according to their channel gains. We then find the number of channels n, which are allocated with a nonzero power, as in Steps 2–3. The water volume H_n that is required to fill n channels can be calculated as

$$\sum_{i=1}^{n} i \cdot \left(\frac{1}{h_{i+1}} - \frac{1}{h_i} \right),$$

and H_n should not be greater than P_{max}. Then we can calculate the water level and allocate power to each selected channel according to (13). This procedure is based on the assumption that a feasible power should also satisfy the eye safety power constraint. If the allocated power $(1/\lambda - 1/h_i)$ is greater than $\bar{\bar{P}}P^-$, we need to adjust the water level accordingly. The detailed water-filling algorithm is presented in Algorithm 1.

Algorithm 1. The water-filling algorithm.

1 Sort channels according to channel gains as:
$h_1 \geq h_2 \geq ... \geq h_M$;
2 Calculate $H_n = \sum_{i=1}^{n} i(1/h_{i+1} - 1/h_i)$;
3 Find n such that $H_{n+1} \geq P_{max} \geq H_n$;
4 Determine Water level:
$(1/\lambda) = (1/n)(P_{max} + \sum_{i=1}^{n}(1/h_i))$;
5 Allocate power to each channel: $P_i = [1/\lambda - 1/h_i]^+$, for all i ;
6 **if** $P_i \leq \bar{P}$, for all i **then**
7 | Terminate with the power allocation ;
8 **end**
9 Set $i = 1$;
10 **while** $P_i > \bar{P}$ **do**
11 | Set $P_i = \bar{P}$ and $P_{max} = P_{max} - \bar{P}$;
12 | Delete channel i in the next power allocation round
13 | $i++$;
14 **end**
15 Go to Step 2 ;

4.2. DWDM RoFSO system

Next, we develop the model for the RoFSO channels as described in Section 3, which is more suitable for RoFSO using APD photo-detectors. CNR defined in Section 3.2 will be an important parameter to evaluate the RoFSO performance.

4.2.1. Reformulation and relaxation based approach

To simplify notation, we denote the constant 0.5(mOMI)^{2}0.5(mOMI)2 by a , the relative intensity noise level RINRIN by b , the optical short noise $2em^{(2+F)}$2em(2+F) by c , and the thermal noise $4KTG_f$4KTGf by d. Thus, our adaptive power allocation problem becomes

$$\max \sum_{i=1}^{M} \log\left(1 + \frac{a_i(P_ih_i)^2}{b_i(P_ih_i)^2 + c_iP_ih_i + d_i}\right)$$

(14)

$$\text{s.t.} \sum_{i=1}^{M} P_i \leq P_{max}$$

(15)

$$0 \leq P_i \leq \bar{P} \quad \text{for all } i. \tag{16}$$

If we denote CNR by

$$\gamma_i = \frac{a_i(P_i h_i)^2}{b_i(P_i h_i)^2 + c_i P_i h_i + d_i}, \tag{17}$$

the optimization problem, termed Problem OPT-RoFSO, can be rewritten as

$$\max \sum_{i=1}^{M} \log(1+\gamma_i) \tag{18}$$

$$\text{s.t} \sum_{i=1}^{M} P_i \leq P_{max} \tag{19}$$

$$0 \leq P_i \leq \bar{P} \quad \text{for all } i \tag{20}$$

$$\gamma_i = \frac{a_i(P_i h_i)^2}{b_i(P_i h_i)^2 + c_i P_i h_i + d_i} \quad \text{for all } i. \tag{21}$$

It is challenging to solve this problem due to its complexity and nonlinear nonconvex properties. In the following, we adopt the RLT technique to obtain an LP relaxation of Problem OPT-RoFSO and derive a feasible near-optimal solution [12].

The RLT relaxation is as follows. Letting

$$c_i = \log(1+\gamma_i),$$

the objective function $\sum_{i=1}^{M} c_i$ will now be linear and new constraints $c_i = \log(1+\gamma_i)$ are introduced. We first linearize the logarithmic terms in the new constraints using the polyhedral outer approximation as follows.

From (21), it can be seen that γ_i is a monotone increasing function of P_i. Letting P_i be 0 and $\bar{P}$, we obtain the lower and upper bounds of γ_i, respectively. We denote the upper bound of γ_i by $\bar{\gamma}_i$, while the lower bound is 0. We use the four-point approximation and obtain the following new linear constraints:

$$\begin{cases} c_i \geq \dfrac{\gamma_i}{\bar{\gamma}_i} \cdot \log(1+\bar{\gamma}_i) \\[2ex] c_i \leq \log(1+\gamma_i^k) + \dfrac{\gamma_i - \gamma_i^k}{1+\gamma_i^k}, \end{cases} \tag{22}$$

where

$$\gamma_i^k = \frac{1}{3} \cdot k\bar{\gamma}_i \quad \text{for } k = 0, 1, 2, 3.$$

The first equation in (22) is for the segment connecting the two end points of the logarithm function and the second equation in (22) is for the tangent lines at the four points on the logarithm function respectively. A four-point approximation for log(1+γ)log(1+γ) is illustrated in Fig. 2. The corresponding convex envelope is formed by four tangent lines and a chord connecting the two end points.

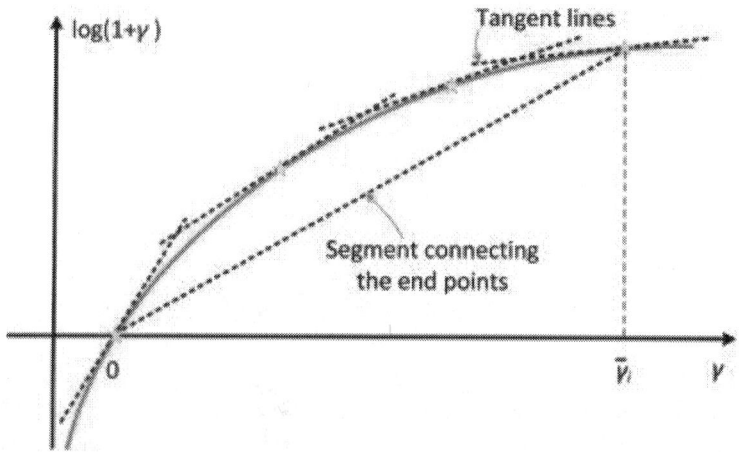

Figure 2. Four-point polyhedral outer approximation for log(1+γ)log(1+γ).

Problem OPT-RoFSO now becomes a polynomial programming problem. We next introduce substitution variables and the corresponding RLT bound-factor product constraints to remove the quadratic terms and to obtain an LP relaxation. Specifically, constraint (21) contains quadratic terms. We can rewrite (21) as

$$b_i(P_i h_i)^2 \gamma_i + c_i h_i P_i \gamma_i + d_i \gamma_i - a_i(P_i h_i)^2 = 0.$$

To remove quadratic terms, we define substitution variables

$$\begin{cases} u_i = P_i^2 & \text{for all } i \\ v_i = \gamma_i P_i & \text{for all } i \\ w_i = \gamma_i u_i & \text{for all } i \\ \beta_i = c_i v_i & \text{for all } i. \end{cases}$$

$$(24)$$

Thus constraint (21) becomes

$$b_i h_i^2 w_i + h_i \beta_i + d_i \gamma_i - a_i h_i^2 u_i = 0. \tag{25}$$

Since P_i is bounded, it can easily show that $u_i \in (0, \overline{P}^2)$ $u_i \in (0, P^-2)$. For variables v_i, with $0 \le \gamma_i \le \overline{\gamma}_i$ $0 \le y_i \le y^-i$ and $0 \le P_i \le \overline{P}$ $0 \le P_i \le P^-$, we can obtain the following RLT bound-factor product constraints:

$$\begin{cases} (\gamma_i - 0)(P_i - 0) \ge 0 \\ (\overline{\gamma}_i - \gamma_i)(P_i - 0) \ge 0 \\ (\gamma_i - 0)(\overline{P} - P_i) \ge 0 \\ (\overline{\gamma}_i - \gamma_i)(\overline{P} - P_i) \ge 0. \end{cases} \tag{26}$$

Substituting $v_i = \gamma_i P_i$ $vi = yiPi$, we obtain the following four linear constraints for variable v_i:

$$\begin{cases} v_i \ge 0 \\ \overline{\gamma}_i P_i - v_i \ge 0 \\ \overline{P} \gamma_i - v_i \ge 0 \\ \overline{P} \overline{\gamma}_i - \overline{\gamma}_i P_i - \overline{P} \gamma_i + v_i \ge 0. \end{cases} \tag{27}$$

For variable w_i, with $0 \le \gamma_i \le \overline{\gamma}_i$ $0 \le y_i \le y^-i$ and $0 \le u_i \le \overline{P}^2$ $0 \le u_i \le P^-2$, we obtain the following four linear constraints in the same manner:

$$\begin{cases} w_i \ge 0 \\ \overline{\gamma}_i u_i - w_i \ge 0 \\ \overline{P}^2 \gamma_i - w_i \ge 0 \\ \overline{P}^2 \overline{\gamma}_i - \overline{\gamma}_i u_i - \overline{P}^2 \gamma_i + w_i \ge 0. \end{cases} \tag{28}$$

We deal with the variables β_i in the same manner, with $0 \le c_i \le \overline{c}_i$ $0 \le c_i \le c^-i$ and $0 \le v_i \le \overline{\gamma}_i \overline{P}$ $0 \le v_i \le y^-i P^-$, and obtain the following four linear constraints for variable β_i.

$$\begin{cases} \beta_i \ge 0 \\ \overline{c}_i v_i - \beta_i \ge 0 \\ \overline{\gamma}_i \overline{P} c_i - \beta_i \ge 0 \\ \overline{c}_i \overline{\gamma}_i \overline{P} - \overline{\gamma}_i \overline{P} c_i - \overline{c}_i v_i + \beta_i \ge 0, \end{cases} \tag{29}$$

where $\overline{c}_i = \log(1 + \overline{\gamma}_i)$ $c^-i = \log(1 + y^-i)$.

Now the original problem is relaxed to an LP problem with the additional constraints and variables as follows:

$$\max \sum_{i=1}^{M} c_i \tag{30}$$

$$\text{s.t. } \sum_{i=1}^{M} P_i \leq P_{max} \tag{31}$$

$$0 \leq P_i \leq \overline{P} \quad \text{for all } i \tag{32}$$

$$0 \leq u_i \leq \overline{P}^2 \quad \text{for all } i \tag{33}$$

$$b_i h_i^2 w_i + h_i \beta_i + d_i \gamma_i - a_i h_i^2 u_i = 0,$$
$$\text{for all } i \tag{34}$$

$$\text{New linear constraints (22) for all } i \tag{35}$$

$$\text{RLT bound} - \text{factor constraints (27),}$$
$$\text{(28) and (29) for all } i. \tag{36}$$

The relaxed problem can be solved in polynomial time with an LP solver. Note that during the procedure of reformulation and linearization, we preserve the original power constraints of Problem OPT-RoFSO, i.e., (19) and (20). Hence the optimal transmit power allocation policy obtained for the LP relaxation is also feasible to the original problem OPT-RoFSO. The feasibility of the LP solution is summarized in the following proposition:

Proposition 1.

The optimal transmit power allocation policy to the LP relaxation of Problem OPT-RoFSO is a feasible solution to the original problem.

4.2.2. A faster near-optimal algorithm

Due to the complexity of the RLT-based method, it may not be suitable when the channels vary quickly. We next develop a more computationally cost-effective scheme in the following. If we ignore the relative intensity noise and optical short noise, CNR can be approximated by $0.5P_r^2 (mOMI)^2 / 4KTG_f$ $0.5Pr2(mOMI)2/4KTGf$. To simplify notation, we denote constant value $0.5(mOMI)^2/4KTG_i 0.5(mOMI)2/4KTGf$ by a. We obtain the following optimization problem:

$$\max \sum_{i=1}^{M} \log\left(1 + a_i (P_i h_i)^2\right) \tag{37}$$

$$\text{s.t. } \sum_{i=1}^{M} P_i \leq P_{max} \tag{38}$$

$$0 \leq P_i \leq \overline{P} \quad \text{for all } i. \tag{39}$$

According to Karush–Kuhn–Tucker (KKT) theorem, if $\boldsymbol{P^*} = [P_1^*, P_2^*, \cdots, P_M^*]$ P*=[P1*,P2*,···,PM*] is a local maximizer for the above optimization problem, there exists $\lambda \in \mathbb{R}$λ∈R and $\lambda_i \in \mathbb{R}$λi∈R, for i∈{1,2,···,M}i∈{1,2,···,M}, such that

$$
\begin{cases}
\dfrac{\partial[\log(1 + a_i(P_i^* h_i)^2)]}{\partial P_i} - \lambda - \lambda_i = 0 & \text{for all } i \\[2mm]
\lambda\left(\displaystyle\sum_{i=1}^{M} P_i^* - P_{max}\right) = 0 \\[2mm]
\lambda_i(P_i^* - \overline{P}) = 0 & \text{for all } i \\[1mm]
\lambda \geq 0, \\[1mm]
\lambda_i \geq 0 & \text{for all } i.
\end{cases}
\tag{40}
$$

According to (40), if $\lambda_i > 0$λi>0, then $P_i^* = \overline{P}$ Pi*=P⁻; and if $\lambda_i = 0$λi=0, we can solve (40) to have

$$P_i^* = \frac{1}{\lambda} + \sqrt{\frac{1}{\lambda^2} - \frac{1}{a_i h_i^2}}. \tag{41}$$

Thus we find that the optimal power allocation satisfies

$$
\begin{cases}
P_i = \min\left\{\overline{P}, \dfrac{1}{\lambda} + \sqrt{\dfrac{1}{\lambda^2} - \dfrac{1}{a_i h_i^2}}\right\}, & \text{if } \lambda^2 \leq a_i h_i^2 \\[3mm]
P_i = 0, \text{otherwise} \\[1mm]
\displaystyle\sum_{i=1}^{M} P_i = P_{max}.
\end{cases}
\tag{42}
$$

Regarding the inverse of λ as some kind of "water level," we find that the greater the channel gain, the larger the deviation of the power allocated to this channel from the water level. Usually we cannot directly solve from (42) for the optimal power allocation. An iterative algorithm is needed to obtain an appropriate λ and solve this optimization problem. The detailed algorithm is presented in Algorithm 2.

Algorithm 2.

RoFSO power allocation algorithm.

1 Sort channels according to channel gains as:
$h_1 \geq h_2 \geq ... \geq h_M$;
2 Solve $\sum_{i=1}^{M} P_i = P_{max}$ numerically according to (42) ;
3 Obtain λ and calculate P_i for $i \in [1, M]$ according to (41) ;
4 **if** $P_i \leq \bar{P}$, *for all i* **then**
5   Terminate with the power allocation ;
6 **end**
7 Set $i = 1$;
8 **while** $P_i > \bar{P}$ **do**
9   Set $P_i = \bar{P}$;
10     $i{+}{+}$;
11 **end**
12 Go to Step 2 ;

5. PERFORMANCE EVALUATION

In this section, we evaluate the performance of the proposed algorithms with MATLAB simulations. We calculate channel gains as shown in Section 3.1 and CNR are calculated according to (6) for the evaluated schemes. The simulation parameters are the same from prior work and are listed in Table 2[1] and [23].

Table 2. Simulation parameters.

Symbol	Value	Definition
D_t	15 mm	Tx. aperture diameter
D_r	0.1 m	Rx. aperture diameter
B	1 GHz	Bandwidth
d	1 km	Distance
P_{max}	0.5 W	Power budget
$\bar{P}$	0.1 W	Peak power constraint
OMI	17.5%	Optical modulation index
m	5	Photodiode gain
RIN	-150 dB/Hz	Relatively intensity noise

We investigate the three algorithms introduced in the previous section for power allocation in the WDM RoFSO system. Wavelengths in the band 1520–1580 nm are assumed in our WDM RoFSO simulations. We adopt a 5 nm spacing between adjacent wavelengths used in the simulations.

In Fig. 3, we compare the three algorithms introduced in Section 4 and examine the impact of the power budget on the total system capacity. We increase P_{max} from 0.5 to 1 with step-size 0.1 and plot the total capacity. As can be seen in Fig. 3, the RoFSO power allocation algorithm outperforms the other two algorithms with considerable gains. Since the relative intensity noise and the optical short noise are very small in our simulations, the RoFSO power allocation algorithm will achieve the best near-optimal solution. The RLT algorithm, although consumes much running time, can only produce an optimal power allocation for the relaxed LP problem. The power allocation solution obtained from RLT is feasible but achieves the worst performance in terms of capacity gain due to relaxation. We also find that the total capacity increases along with the power budget for both the water-filling algorithm and RoFSO power allocation algorithm, albeit not obviously for the latter. After the power budget becomes even larger, there is much less space for capacity increment due to the eye safety power constraint.

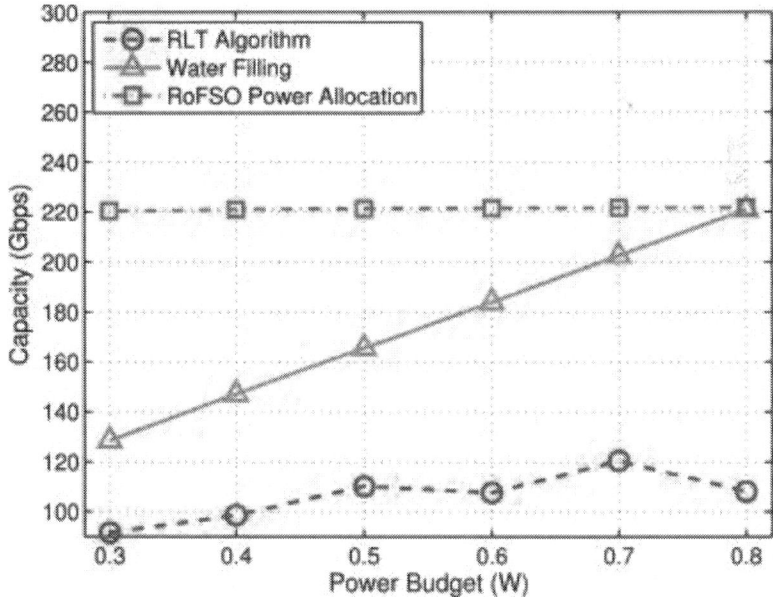

Figure 3. System capacity versus the power budget P_{max}.

Next, we examine the effect of weather conditions on the system capacity in Fig. 4. The distance between FSO BS's is set to 500 m. We change the weather condition from clear to foggy by changing the atmospheric attention coefficient. The coefficient is 0.48 db/km for clear weather, 2.8 db/km for hazy weather, and 15 db/km for foggy weather [30]. As can be seen from Fig. 4, the system capacity decreases as weather gets worse. We also find that the simple water-filling algorithm which is developed for conventional RF systems is not suitable for the RoFSO system when weather condition is severe. The capacity achieved by the simple water-filling algorithm decrease the most as the weather becomes worse. When the weather is foggy, the capacity achieved by the water-filling algorithm is about 10 Gbps.

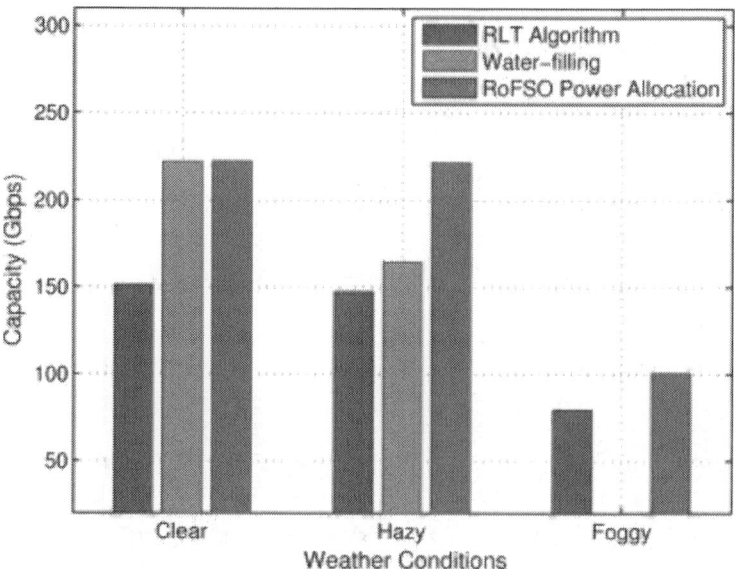

Figure 4. System capacity versus weather condition.

We also plot the total capacity vs. the distance between the FSO BS's in Fig. 5. As expected, the total capacity decreases as the distance is increased. When the distance between the FSO transceivers is relatively small, the difference between the capacities achieved by three algorithms is also small. But as we increase the distance from 500 m to 1 km or even greater, the capacities achieved by simple water filling algorithm and RLT algorithm drop dramatically. The capacity obtained by using the RoFSO power allocation algorithm decreases the least and is always greater than capacities produced by the other two schemes.

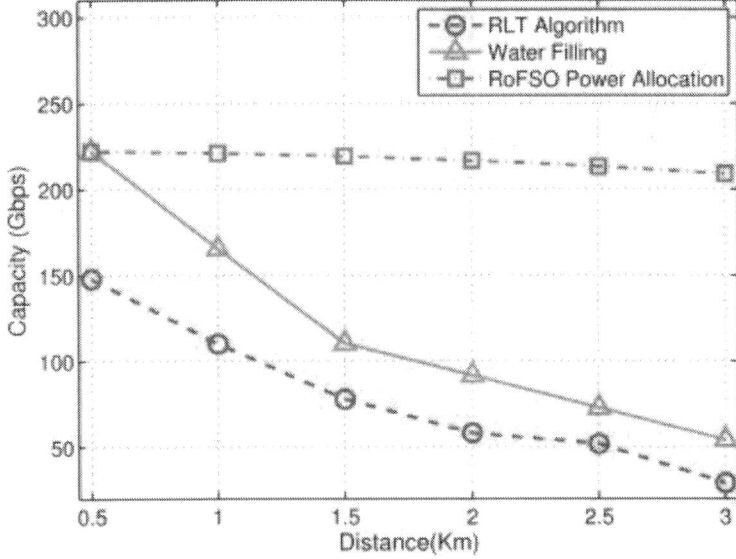

Figure 5. System capacity versus distance.

Finally, we examine the impact of the number of subcarriers used in an adaptive WDM RoFSO system on the system capacity. In the wavelength bands starting from 1520 nm, we adopt a 1 nm spacing between adjacent wavelengths. The simulation results in clear weather are presented in Fig. 6. As we can see in Fig. 6, a system capacity increase if we adopt more subcarriers in the adaptive WDM RoFSO system. When there are 15 channels in the system, the difference between the achieved capacity of the three schemes is not much great. However, as the number of subcarriers increases, the advantage of the RoFSO power allocation scheme becomes greater.

We also run our simulations in hazy and foggy weather and presented in Fig. 7 and Fig. 8, respectively. In Fig. 7, the RoFSO power allocation algorithm achieves greater capacity with more subcarriers. System capacities increased with the number of subcarriers for all three algorithms. However, due to the simplified objective function used in the development of the water-filling algorithm, the water-filling algorithm does not achieve as much capacity as the RLT algorithm when the number of sub-carriers is 35. The RoFSO power allocation algorithm makes better use of available subcarriers to enhance system capacity and thus provide the greatest capacity. When weather conditions become worse, the system capacity become much small and the results are shown in Fig. 8. When the weather is foggy, the capacity achieved by the water-filling algorithm is very small. But the RoFSO power allocation algorithm and the RLT algorithm can maintain system performance in foggy weather. Also when more subcarriers are utilized in the system, more system capacity can be achieved by using these two algorithms.

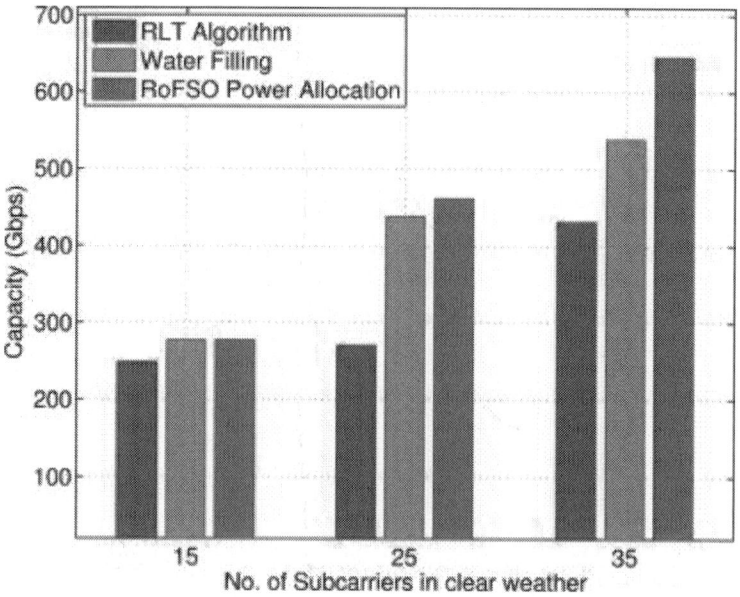

Figure6. System capacity versus the number of subcarriers in clear weather.

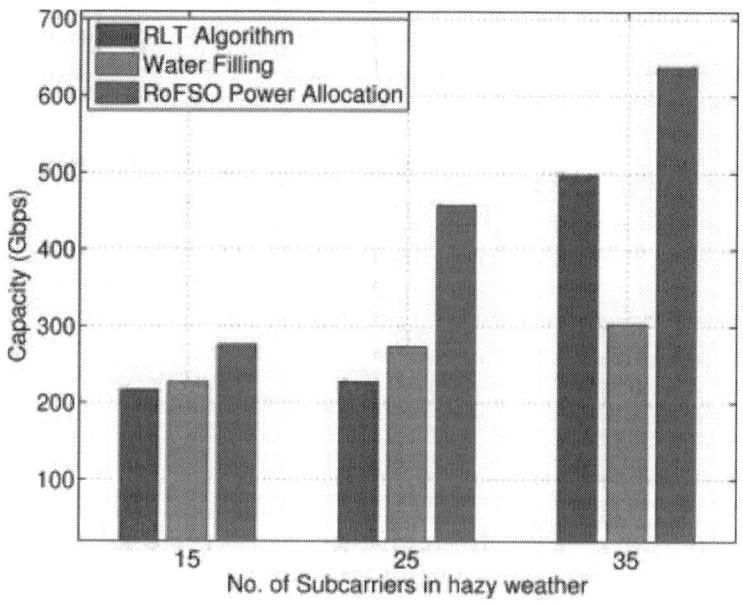

Figure7. System capacity versus the number of subcarriers in hazy weather.

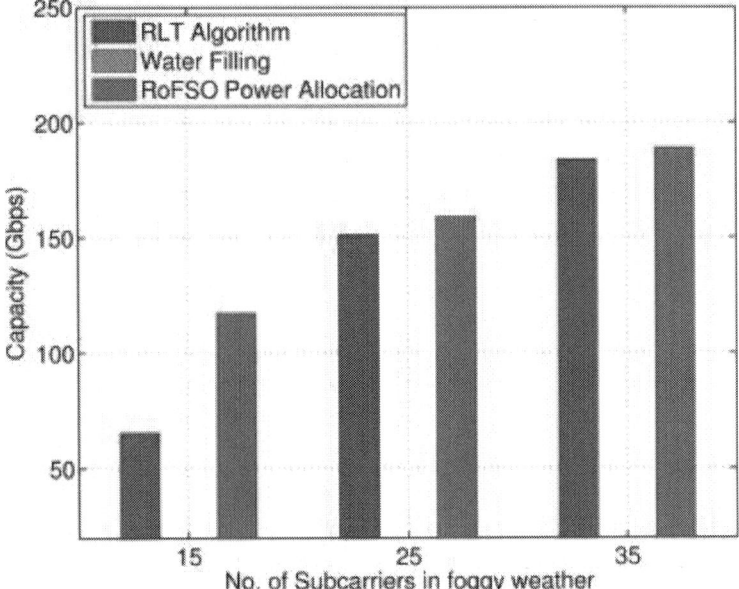

Figure8. System capacity versus the number of subcarriers in foggy weather.

To conclude, the system performance in terms of capacity by using the RoFSO power allocation algorithm is the best in all situations studied here. These simulation results indicate that with proper system design, the RoFSO system can support high data rates even over long distance and under bad weather conditions.

6. CONCLUSIONS

In this paper, we investigated the problem of optical power allocation under a power budget constraint and eye safety power constraint for adaptive WDM transmission to mitigate the effect of weather turbulence, and solution algorithms are developed. For convectional transmission systems, a simple water-filling algorithm can be adopted to allocate power to different wavelengths; for WDM RoFSO systems, an RoFSO power allocation algorithm was demonstrated to achieve the greatest system capacity. It is capable of supporting high data rates even over long distance or under bad weather conditions. Utilizing multiple wavelengths in the WDM FSO system for diversity gain was also investigated in the paper.

ACKNOWLEDGEMENTS

This work was supported in part by the U.S. National Science Foundation (NSF) under Grant CNS-1320664, and through the Wireless Engineering Research and Education Center (WEREC) at Auburn University. Any opinions, findings, and conclusions or recommendations expressed in this material are those of the author(s) and do not necessarily reflect the views of the NSF.

REFERENCES

1. P. Dat, A. Bekkali, K. Kazaura, K. Wakamori, *et al.*Studies on characterizing the transmission of RF signals over a turbulent FSO linkOpt. Express, 17 (2009), pp. 7731–7743

2. H. Zhou, A. Babaei, S. Mao, P. Agrawal, Algebraic connectivity of degree constrained spanning trees for FSO networks, in: Proceedings of IEEE ICC13′, Budapest, Hungary, 2013, pp. 1–6.

3. I.K. Son, S. Kim, S. Mao, Building robust spanning trees in free space optical networks, in: IEEE MILCOM10′, San Jose, CA, 2010, pp. 1857–1862.

4. I.K. Son, S. Mao, Design and optimization of a tiered wireless access network, in: Proceedings of IEEE INFOCOM10′, San Diego, CA, 2010, pp. 1–9.

5. I.-K. Son, S. Mao, S.K. DasOn the design and optimization of a free space optical access networkOpt. Switch.Netw., 11 (Part. A) (2014), pp. 29–43 |

6. I.-K. Son, S. Mao, S.K. DasOn joint topology design and load balancing in FSO networksOpt. Switch.Netw., 11 (Part A) (2014), pp. 92–104 |

7. Farid, S. HranilovicDiversity gain and outage probability for MIMO free-space optical links with misalignmentIEEE Trans. Commun., 60 (2) (2012), pp. 479–487

8. M. Karimi, M. UysalNovel adaptive transmission algorithms for free-space optical linksIEEE Trans. Commun., 60 (12) (2012), pp. 3808–3815

9. H. Moradi, M. Falahpour, H. Refai, P. LoPresti, M. Atiquzzaman, On the capacity of hybrid FSO/RF links, in: Proceedings of IEEE GLOBECOM10′, 2010, pp. 1–5.

10. H. Zhou, S. Mao, P. Agrawal, Optical power allocation for adaptive WDM transmission in free space optical networks, in: Proceedings of IEEE WCNC 2014, Istanbul, Turkey, 2014, pp. 2677–2682.

11. X. LiuFree-space optics optimization models for building sway and atmospheric interference using variable wavelengthIEEE Trans. Commun., 57 (2) (2009), pp. 492–498

12. Y. Huang, S. Mao, Downlink power control for variable bit rate videos over multicell wireless networks, in: Proceedings of IEEE INFOCOM ,11' 2011, pp. 2561–2569.

13. Vavoulas, H. Sandalidis, D. VaroutasWeather effects on FSO network connectivityIEEE/OSA J. Optical Commun. Netw., 4 (10) (2012), pp. 734–740

14. Aladeloba, M. Woolfson, A. PhillipsWDM FSO network with turbulence-accentuated interchannel crosstalkIEEE/OSA J. Opt. Commun. Netw., 5 (6) (2013), pp. 641–651

15. V. Sharma, G. Kaur, Modelling of OFDM-ODSB-FSO transmission system under different weather conditions, in: 2013 Third International Conference on Advanced Computing and Communication Technologies (ACCT), 2013, pp. 154–157.

16. V.V. Sivakumar, D. Hu, P. Agrawal, Relay positioning for energy saving in cooperative networks, in: IEEE 45th Southeastern Symposium on System Theory, 2013, pp. 1–5.

17. Johnsi, V. Saminadan, Performance of diversity combining techniques for FSO-MIMO system, in: 2013 International Conference on Communications and Signal Processing (ICCSP), 2013, pp. 479–483.

18. H. Zhou, D. Hu, S. Mao, P. Agrawal, Joint relay selection and power allocation in cooperative FSO networks, in: Proceedings of IEEE GLOBECOM13', Atlanta, GA, 2013, pp. 1–6.

19. H. Zhou, S. Mao, P. AgrawalOn relay selection and power allocation in cooperative free space optical networksPhoton. Netw.Commun. J. (PNET), 29 (1) (2015), pp. 1–11

20. M. Kashani, M. Safari, M. UysalOptimal relay placement and diversity analysis of relay-assisted free-space optical communication systemsIEEE/OSA J. Opt. Commun. Netw., 5 (1) (2013), pp. 37–47

21. Abou-Rjeily, A. SlimCooperative diversity for free-space optical communications: transceiver design and performance analysisIEEE Trans. Commun., 59 (3) (2011), pp. 658–663

22. DjordjevicAdaptive modulation and coding for free-space optical channelsIEEE/OSA J. Opt. Commun.Netw., 2 (5) (2010), pp. 221–229

23. K.-H. Kim, T. Higashino, K. Tsukamoto, S. Komaki, WDM optical power allocation method for adaptive radio on free space optics system design, in: 2011 International Topical Meeting on Microwave

Photonics & 2011 Asia-Pacific Microwave Photonics Conference (MWP/APMP), Singapore, 2011, pp. 361–364.

24. S.G. Wilson, M. Brandt-Pearce, Q. Cao, J.H. Leveque, Free-space optical MIMO transmission with Q-ary PPM, IEEE Trans. Commun. 53 (8) (2005) 1402–1412.

25. N. Cvijetic, S. Wilson, M. Brandt-PearcePerformance bounds for free-space optical MIMO systems with APD receivers in atmospheric turbulenceIEEE J. Sel. Areas Commun., 26 (3) (2008), pp. 3–12

26. E. Ciaramella, Y. Arimoto, G. Contestabile, M. Presi, A. D'Errico, V. Guarino, M. Matsumoto1.28 Terabit/s (32x40 Gbit/s) WDM transmission system for free space optical communicationsIEEE J. Sel. Areas Commun., 27 (9) (2009), pp. 1639–1645

27. K. Kazaura, K. Wakamori, M. Matsumoto, T. Higashino, K. Tsukamoto, S. Komaki, RoFSO: a universal platform for convergence of fiber and free-space optical communication networks, in: 2009 Innovations for Digital Inclusions, ITU-T Kaleidoscope, 2009, pp. 1–8.

28. K.-H. Kim, T. Higashino, K. Tsukamoto, et al., Statistical analysis on the optical fading in free space optical channel for RoFSO link design, in: Proceedings of SPIE 7620, Broadband Access Communication Technologies IV, 76200G, 2010, pp. 1–10.

29. M. Safari, M. Rad, M. UysalMulti-hop relaying over the atmospheric poisson channel: outage analysis and optimizationIEEE Trans. Commun., 60 (3) (2012), pp. 817–829

30. V. Rajakumar, M. Smadi, S. Ghosh, T. Todd, S. HranilovicInterference management in WLAN mesh networks using free-space optical linksJ. Lightw. Technol., 26 (13) (2008), pp. 1735–1743

Chapter 10

The Human Eye and Adaptive Optics

Fuensanta A. Vera-Díaz[1] and Nathan Doble

[1] *The New England College of Optometry, Boston MA, USA*

1. INTRODUCTION

Scientists have rapidly taken advantage of adaptive optics (AO) technology for the study of the human visual system. Vision, the primary human sense, begins with light entering the eye and the formation of an image on the retina (Fig 1), where light is transformed into electro-chemical impulses that travel towards the brain. The eye provides the only direct view of the central nervous system and is, therefore, the subject of intense interest as a means for the early detectionof a host of retinal and possibly systemic diseases. However, ocular aberrations limit the optical quality of the human eye, thus reducing image contrast and resolution. With the use of AO it is now routinely possible to compensate for these ocular aberrations and image cellular level structures such as retinal cone androd photoreceptors (Liang et al, 1997; Doble et al, 2011), the smaller foveal cones (Putnam et al, 2010), retinal pigment epithelium (RPE) cells (Roorda et al, 2007), leukocyte blood cells (Martin & Roorda, 2005) and the smallest retinal blood vessels (Tam et al, 2010; Wang et al, 2011),*in vivo* and without the aid of contrast enhancing agents.

The chapter begins with a review of the structure of the human eye before describing the challenges and approaches in using AO to study the visual system.

1.1. The Human Eye and Visual System

The human eye behaves as a complex optical structure sensitive to wavelengths between 380 and 760nm. Light entering the eye is refracted as it passes from air through the tear film-cornea interface. It then travels throughthe aqueous humor and the pupil (a diaphragm controlled by the iris) and is further refracted by the crystalline lens before passing through the vitreoushumor and impinging on the retina (Fig 1). The tear film-cornea interface and the crystalline lens are the major refractive components in the eye and act together as a compound lens to project an inverted image onto the light sensitive retina. From the retina, the electrical signals are transmitted to the visual cortex via the optic nerve (Fig 1). A summary of this path is presented in this section, for detailed information on the anatomy and physiology of the eye the reader is directed to the references (Snell & Lemp, 1998; Kaufman & Alm, 2002; Netter, 2006).

1.1.1. Tear Film-Cornea Interface

The tear film–cornea interface (Fig 1) is the most anterior refractive surface of the eye as well as the most powerful due to the difference between its refractive index and that ofair. The anterior radius of the tear film-cornea interface is approximately 7.80mm and the refractive index of the tear film is 1.336, which give a dioptric power of approximately 43.00 diopters. Therefore, small variations in its curvature can cause significant changes in the power of the eye.

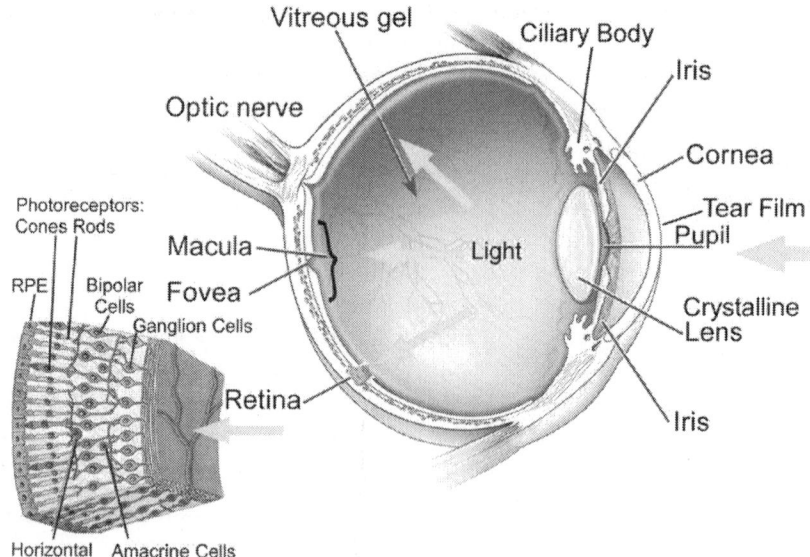

Figure 1. Gross anatomy of the human eye and detail of the retina. The major refractive elements and hence primary sources of aberration are the tear film–cornea interface and the crystalline lens. The incident light on the retina is absorbed by the cone and rod photoreceptors after traversing several retinal layers. Imagemodified from the National Eye Institute, National Institutes of Health.

The cornea is a transparent tissue, achieved by its regular composition of collagen fibers, avascularity and an effective endothelial pump. The cornea in the adult typically measures 10.5 mm vertically and 11.5 mm horizontally and its thickness increases from the center (about 530 µm) to the periphery (about 650 µm). The cornea is more curved than the eyeball and hence protrudes anteriorly. Behind the cornea, the aqueous humor has the same refractive index as the vitreous humor (1.336), whereas the refractive index of the cornea is 1.376.Because the change in refractive index between cornea and aqueous humor is relatively small compared to the change at the air–cornea interface, it has a negligible refractive effect.

1.1.2. Crystalline lens. Accommodation

The crystalline lens is held behind the iris by thin yet strong ligaments, zonules of Zinn, attached to the ciliary processes in the ciliary body (Fig 1). The crystalline lens is flexible and may change its shape using the mechanism of accommodation, by adjusting the ciliary muscle so that the images may be more accurately focused on the retina. It has an ellipsoidal, biconvex shape with the posterior surface being more curved than the anterior. The crystalline lens is typically 10 mm in diameter and has a thickness of approximately 4mm, although its size and shape changes during accommodation, and it continues to grow throughout a person's lifetime. The crystalline lens achieves transparency due to its composition, as 90% of it is formed by tightly packed proteins and there is an absence of organelles such as a nucleus, endoplasmic reticulum and mitochondria within the mature lens fibers.

The intensity of the light reaching the retina is regulated by the diaphragm formed by the iris: the pupil. The pupil is therefore important in regulating the aberrations of the eye, the magnitude of the aberrations increase with larger pupil diameters – section 1.2.1.

1.1.3. Retina

Upon reaching the retinal surface, the light traverses its many layers (Fig 1) before reaching the photoreceptor cells, where the photons are absorbed and transformed into electro-chemical impulses. The gross anatomy of the retina is composed of a macula or central region, with the fovea as the very center. At the fovea the cone photoreceptors have the smallest diameter (1.9-3.4 µm), the highest average density (199,000 cones per mm^2)(Curcio & Allen, 1990) and the eye has the highest resolution (visual acuity, VA). The signals from these photoreceptors are then processed by the many intervening cell types in the retina before exiting towards the brain via the ganglion cells and the optic nerve.

i.Physiology of the Photoreceptors: Rods and Cones

The photoreceptors are photosensitive cells located in the outermost layer of the retina that are responsible for the phototransduction, i.e. they convert photons into electro-chemical signals that can stimulate biological processes.

The proteins (opsins) in the outer segments of these photoreceptors absorb photons and trigger a cascade of changes in the membrane potential; this mechanism is called the signal transduction pathway. In brief, the photoreceptors signal their absorption of photons via a decrease in the release of the neurotransmitter glutamate to the bipolar cells. The photoreceptors are depolarized in the dark, when a high amount of glutamate is being released, and after absorption of a photon they hyperpolarize so less glutamate is released to the presynaptic terminal of the bipolar cells.

The effect of glutamate in the bipolar cells varies depending on the type of receptor imbedded in the bipolar cell's membrane; it may depolarize or hyperpolarize the bipolar cell. This allows one population of bipolar cells to get excited by light whereas another population is inhibited by it, even though all photoreceptors show the same response to light. This complexity is necessary for various visual functions such as detection of colour, contrast or edges. The complexity increases as there are interconnections among bipolar cells, horizontal cells and amacrine cells in the retina. The final result of this complex net is several populations of different classes of ganglion cells that have specific functions in the retina and exit the eye through the optic nerve.

The photoreceptor cells are the rods and cones (Fig 1, 2), named as consequence of their anatomy. Rods are narrower and distributed mostly in the peripheral retina. A third class of lightcells are the photosensitive ganglion cells, discovered in the 1990s (for a review see (Do & Yau, 2010)), which use the photopigment melanopsin and are believed to support circadian rhythm but do not contribute significantly to vision. The human retina contains approximately 120 million rods and 5 million cones, although this amount varies with age and certain retinal diseases. There are also major functional differences between the rods and cones. Rods are extremely sensitive, have more pigment and can be triggered by a very small number of photons. Therefore, at very low light levels (scotopic vision), the visual signal is coming solely fromrods. Rods are almost absent in the fovea, and only a small amount are present in the macular area. Cones, on the other hand, are only sensitive to direct and large amounts of photons; and are used for photopic vision. In humans there are three different types of cone cells that respond approximately to short (S), medium (M) and long (L) wavelengths.

There is a dependence on photoreceptor arrangement with retinal eccentricity, decreasing in regularity and density from the fovea toward the periphery, although the smallest cones are not always located in the center of the fovea (Chui et al, 2008a). At a given retinal location, there is considerable individual variation in cone photoreceptor packing density, although more than 20% of the variance could be accounted for by differences in axial length (Chui et al, 2008b).

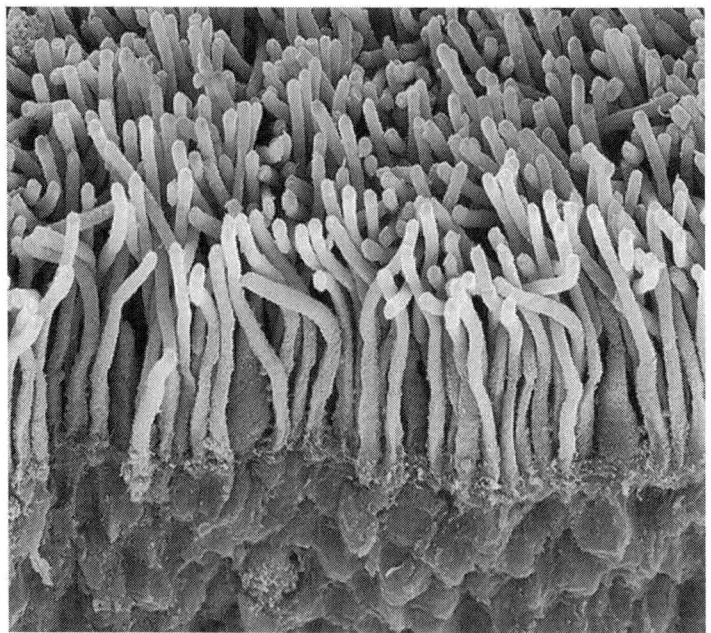

Figure 2. Colored scanning electron micrograph (SEM) of rods (blue) and cones (purple) in the retina of the eye. The outer nuclear layer is brown. Magnification x1800 when printed at 10 centimetres wide.By Steve Gschmeissner.Reproduced with permission from Science Photo Library.

ii Waveguide Properties of the Photoreceptors: The Stiles-Crawford Effect

As mentioned above, cones are sensitive to large amounts of light and only if it is directly incident on them. There is, therefore, a reduction in light sensitivity when its entry point is shifted from the center to the edge of the pupil. This phenomenon, called the Stiles-Crawford Effect (SCE) (Stiles & Crawford, 1933; Westheimer, 2008), plays an important role in vision because unwanted scattered light is rejected. Individual cones have specific waveguiding properties (Enoch, 1963), cone disarray is very small in healthy eyes and ensembles of cones have essentially the same directionality properties as a single cone (Roorda & Williams, 2002). This property of the photoreceptors shows small variations across the retinal field (Westheimer, 1967; Burns et al, 1997). It has been suggested (Vohnsen, 2007) that the photoreceptors may be at least partially adapted to match the average ocular aberrations in order to maximize their light-capturing capabilities.

The study of SCE may provide useful information about subtle structural changes in retinal disease, changes that may not be detected with conventional clinical tests. It has been shown that this property of the photoreceptors is altered in central serous chorioretinopathy (Kanis & van Norren, 2008). Delayed recovery of photoreceptor directionality was found when measuring SCE at a stage of the disease when no abnormalities were found using other common diagnostic techniques such as VA and optical coherence tomography (OCT). Transient changes of the SCE have also been found in the near periphery of myopic eyes with elongated axial lengths (Choi et al, 2004) and in eyes with permanent visual field loss and damage of the inner retinal layers secondary to optic neuropathies (Choi et al, 2008).

iii. Temporal Properties of the Photoreceptors

The photoreceptors outer segments contain discs studded with opsins that capture photons to initiate the phototransduction process. Throughout the day, new discs are added, dozens of discs are shed and phagocytosis occurs at the RPE. *In vivo* detection of disc renewal has only been possible recently with the use of AO. Using an AO flood-illuminated camera, Pallikaris et al (2003) observed changes in cone reflectance over a 24hour period using non-coherent illumination. These changes wereincoherent, not sinusoidal, with both rapid, over minutes, and slow, over hours, changes. They also found these changes to be independent from cone to cone. Hence, they concluded that the changes are not caused by spatiotemporal variation in the optical axes of the cones but were likely caused by changes in the composition of the outer segment-RPE interface due to the migration of melanosomes during disc shedding, or a change in refractive index in the outer segment interface during shedding. If the reflectance changes are related to the renewal process of the receptors, it will be possible to study disruptions in the disc shedding process that occur in diseases such as retinitis pigmentosa.

Other authors have shown faster cone changes. Jonnal et al (2007) showed rapid changes in reflectance in response to visible stimulation of individual photoreceptors. These changes are initiated 5 to 10msec after the onset of the stimulus flash and last 300 to 400msec and are believed to be linked to the process of cone phototransduction. Possible mechanisms for this phenomenon are processes taking place within the cone immediately following stimulation, such as changes in the concentration of G-proteins, hyperpolarization or other changes in the properties of the outer segment membrane, or changes in the physical size of the outer segment secondary to swelling.

Jonnal et al (2010) reported the period for cone reflectance oscillation when using long coherent illumination to range between 2.5 and 3hours, with sinusoidal oscillations occurring during a 24hour period. The power spectra of most cones peaked at a frequency between 0.3 and 0.4 cycles/hour, although this peak varied within a 24hour period. They hypothesized that these oscillations are due to elongation of the cones outer segments(OS) (Jonnal et al, 2010). These rates agree with post-mortem studies in mammals on OS renewal rates on rods (~2µm /day) and cones (~ 1-3µm/day).

1.1.4. The visual system

In brief, the electrical signals at the retina exit each eye via ganglion cells axons through the optic nerve, following a path that crosses at the optic chiasma to later reach the lateral geniculate nucleus (LGN) and from there continue to the primary visual cortex (V1, or striate cortex) first, and to further cortical areas later. The optic chiasm is the point for crossover of information of right and left eyes. The LGN, located at the thalamus, appears to be the first location of feedforward input from higher levels in the brain to the visual input from the eye before most of the visual input travelsto the visual cortex. Note that there is a lateral pathway, that of the superior colliculi, important for eye movement control.

At the visual cortex the signals are processed in V1 and communicated via multiple pathways to numerous visually responsive cortical areas. The visual system comprises a complex network where a cascade of action potentials stream from neuron to neuron forwards, laterally and backwards again. These signals are responsible for our visual perception of the external world, but we are far from understanding how perception of the real world's complex patterns occurs. Visual scientists typically consider that an image can be broken into its components, such as edges, textures, colors, shares, motion, etc. and specialized neurons detect a subset of these components. For a review on receptive field properties of these neurons, retinotopic maps in LGN and V1, orientation and direction selectivity, binocularity and binocular disparity, response timing and other properties of the visual system see online text books (Neuroscience Online, 1997; Webvision, 2011).

1.2. Optical Aberrations of the Eye

In addition to being the main refractive components, the cornea and the crystalline lens are the main sources of aberrations in the human eye. The relative contribution of each of these components can be deduced from total ocular and corneal aberrometry data. The magnitude of the aberration is strongly dependent on individual factors such as age, the state of accommodation or the particular direction through the ocular media. The human eye has monochromatic, longitudinal (up to 2 diopters across the visible spectrum) and transverse chromatic aberrations, the formerbeing significant when using wide bandwidth imaging light sources.

1.2.1. Describing human ocular aberrations

The standard representation of ocular aberrations is in terms of Zernike polynomials (American National Standards Institute (ANSI) – 2010). Zernike polynomials are a mathematical series expansion that are orthogonal over a unit circle. Any wavefront profile can be decomposed into a weighted sum of these polynomials. The low order terms can be translated into the common sphere and cylinder notations used in optometric fields (Porter et al, 2006) and are easily corrected using, for example, spectacles or contact lenses. The higher order Zernike polynomials are traditionally not correctable by such methods, although recently attempts are being made, and require advanced technologies such as AO. Figure3 shows the first 15 Zernike terms and their corresponding far field point spread functions (PSF).

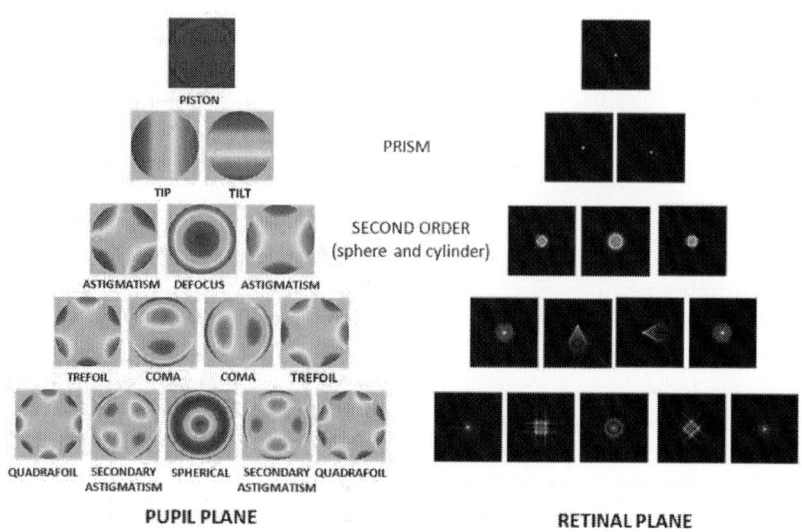

A   B

Figure 3.A) The ocular aberrations can be represented as a weighted sum of Zernike polynomials, each representing a specific aberration. (B) By Fourier transforming and multiplying by the complex conjugate the PSF for each mode can be calculated. Defocus and astigmatism are termed low order modes and are corrected by conventional refractive methods.The higher order modes generally have lower amplitudes but require more elaborate correction technologies.

Porter et al (2001) and Thibos et al (2002) independently measured the wavefront aberration in large human population samples using Shack Hartmann aberrometry. Figure4 shows measured aberrationscoefficients from Porter et al (2001); they measured 109 individuals through a 5.7mm pupil. The majority of the power lies within the low order modes, i.e. defocus (Z_2^0) and astigmatism $(Z_2^{-2})$and(Z_2^2), with these modes accounting for over 92% of the total wavefront aberration variance. Note thatfor this particular study the average defocus coefficient was higher than the general population as they were subjects recruited from a clinic at Bausch & Lomb who were mostly myopic. That said, for high resolution imaging applications where even larger pupilsizes are used, any residual power in the higher order modes can become particularly detrimental.

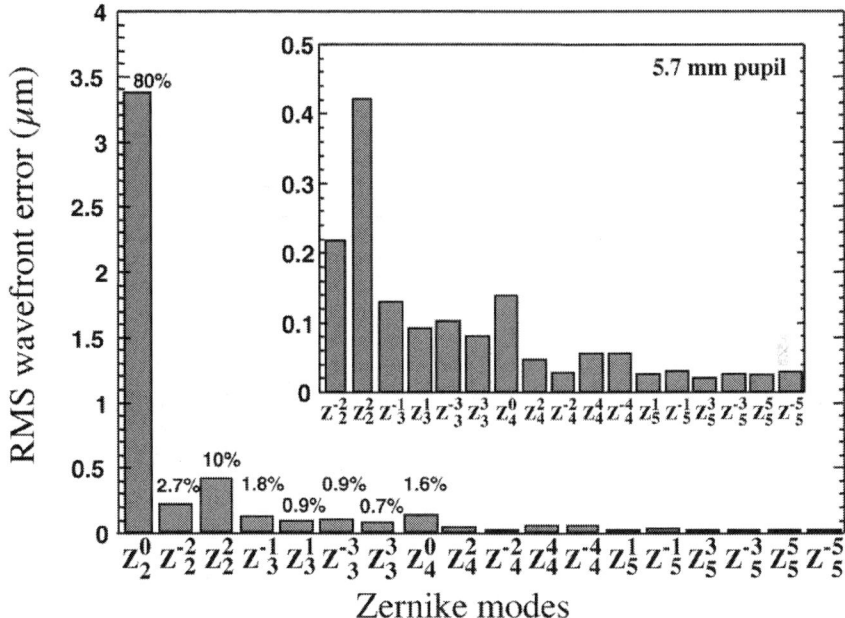

Figure 4. The wavefront aberration decomposed into Zernike polynomials for a large human population (Porter et al, 2001) over a 5.7mm pupil. The majority of the aberration power is found in the low order modes, i.e. defocus (Z_2^0) and astigmatism $(Z_2^{-2}$ and $Z_2^2)$.

The percentages above the first eight modes indicate the percentage of the total wavefront variance. Note: the Zernike order follows that of Noll (1976). Reproduced with permission from the Optical Society of America (OSA).

Doble et al (2007)showed the peak to valley (P-V) wavefront error dependence on pupil size (Fig5) usingaberration data from two human population studies; one comprising of 70 healthy eyes based at the University of Rochester/Bausch & Lomb, and the other consisting of 100 healthy eyes measured at the University of Indiana. Figure 5 shows the wavefront values for each of these populations using different corrective states. Data show that to correct 95% of the normal human population over a 7.5mm pupil upwards of 20μm wavefront correction is required even with the benefit of a second order correction (Fig 5B).

Ocular aberrations also vary with time, mostly due to changes in accommodation (He et al, 2000), although there are other significant contributors such as eye movements. Even when paralyzing accommodation with anticholonergic drugs, the microfluctuations of accommodation can cause significant refractive power changes, up to 0.25 diopters. Hofer et al (2001a; 2001b) and Diaz-Santana et al (2003) have performed detailed measurements on wavefront dynamics and their effect on AO system performance. With a static correction of the higher order aberrations, these dynamic changes can reduce the retinal image contrast by 33% and the Strehl ratio (SR) by a factor of 3 highlighting the need for real time aberration correction (Hofer et al, 2001b). The SR is defined as the ratio of the peak intensity in the aberrated PSF to that of the unaberrated case; an SR greater than 0.8is considered to be diffraction limited.

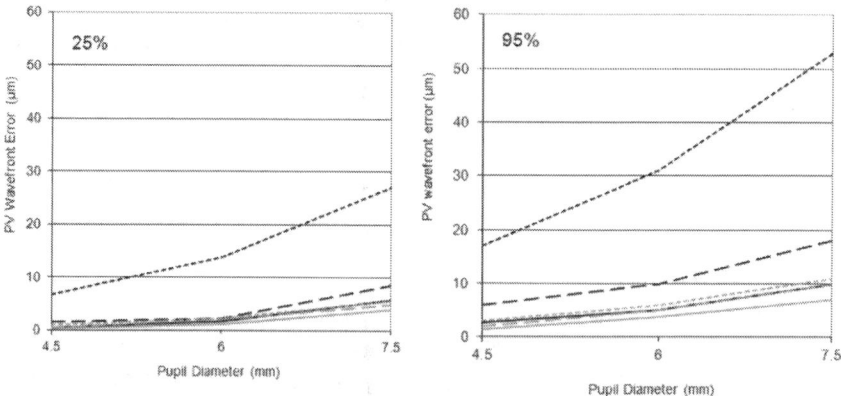

Figure 5. Peak to valley wavefront error that encompasses 25 (left) and 95% (right) of the population in the Rochester (black lines) and Indiana (gray lines) populations. For the Rochester data, three correction states are given: (i) all aberrations present (short dashed lines), (ii) all aberrations present with zeroed Zernike defocus (long dashed lines) and (iii) all aberrations present with zeroed defocus and astigmatism (solid lines). For the Indiana data, the three cases are: (i) residual aberrations after a conventional refraction using trial lenses (short dashed lines), (ii) all aberrations present with zeroed Zernike defocus (long dashed lines) and (iii) all aberrations present with zeroed defocus and astigmatism (solid lines)(Doble et al, 2007). Reproduced with permission from the Optical Society of America (OSA).

1.2.2. The resolution of the human eye
The lateral (transverse) resolution of the eye is given by Eq. 1:

$$r = 1.22\, f\, \lambda\, /\, n\, D$$

(1)

where$r$ is the distance from the center of the Airy diskto the first minima, f is the focal length (of the reduced eye), λ is the wavelength, n is the refractive index and D is the pupil diameter. The maximum resolution would be achieved using the shortest wavelength, λ and the largest possible aperture, D (f being fixed). As an example, for a human eye, with D = 8mm, f = 22.2mm, $n \sim$1.33 and imaging at λ = 550nm, the lateral resolution r is 1.4 μm. In practice, however, ocular aberrations limit this resolution to about 10 μm.

In theory, a lateral resolution of 1.4 μm is sufficient to see the smallest retinal cells.For example, foveal cones have a center to center spacing of 1.9-3.4 μm, and for the rods, the range is 2.2-3.0 μm (Curcio et al, 1990; Jonas et al, 1992). To obtain retinal images with the highest resolution and contrast it is therefore necessary to correct both the low and high order aberrations over a large pupil and moreover track and correct for any associated temporal changes, i.e. we need to employ AO. The ability of AO to dynamically correct (or even induce) higher order spatial modes is becoming increasingly important in the study of the human visual system. The next sections describe how AO is applied to various retinal imaging modalities.

2. THE APPLICATION OF ADAPTIVE OPTICS TO THE HUMAN EYE

The concept of AO was first proposed by the astronomer Horace Babcock in 1953 (Babcock, 1953). However, it was not until the late 1960s/early 70s that the first system was implemented, first by the military followed subsequently by the astronomy community. The first step towards the application of AO to the human eye was the work of Dreher et al (1989) who employed a deformable mirror (DM) to give a static correction of astigmatism in a scanning laser ophthalmoscope (SLO). Later work by Liang et al (1994) saw the first use of a Hartmann-Shack wavefront sensor (HS-WFS) for measurement of the human wavefront aberration who then used a HS-WFS in conjunction with a DM (Liang et al, 1997) to produce some of the first *in vivo* images of the cone photoreceptors. Today, AO has been successfully applied to several retinal imaging modalities employing a variety of DM and WFS technologies.

A detailed discussion of AO is beyond the scope of this chapter and the interested reader is referred to the available reference texts (Hardy, 1998; Porter, 2006; Tyson, 2010).

2.1. Key AO Components

Similar to the AO systems used for other applications such as astronomy and communications, a vision science AO system comprises three main parts:

i.   The Wavefront Sensor (WFS): Most vision science AO systems employ a Hartmann-Shack WFS (Shack & Platt, 1971), althoughcurvature

(Roddier, 1988) and pyramid sensing (Ragazzoni, 1996) have also been employed successfully to the eye (pyramid sensing: Iglesias et al, 2002; curvature sensing: Gruppett et al, 2005). Typically, the ocular wavefront is sampled at 10-20 Hz with closed loop bandwidths of 1-3Hz which is sufficient to correct most of the ocular dynamics (Hofer et al, 2001a). The basic operating principle and design considerations of a WFS are the focus of other chapters in this book and will not be discussed here.

ii. The Wavefront Corrector: These are typically DMs although liquid crystal spatial light modulators (LC-SLMs) have been used in several systems (Thibos & Bradley, 1997; Vargas-Martin et al, 1998; Prieto et al, 2004). Early vision AO systems used large, expensive DMs that were originally designed for military, astronomy or laser applications. These DMs had apertures that were several centimeters in diameter requiring long optical paths to magnify the 6-7mm pupil diameter of the human eye. Today, many systems employ microelectromechanical systems (MEMS) (Fernandez et al, 2001; Bartsch et al, 2002; Doble et al, 2002), electromagnetic (Fernandez et al, 2006) or bimorph type mirrors (Glanc et al, 2004), all of which have much smaller active apertures and a lower cost.

iii. The Control Computer: These take the output of the WFS and converts it to voltage commands that are sent to the wavefront corrector.

There are three main ophthalmic imaging modalities that have successfully employed AO (i) flood illuminated fundus cameras that take a short exposure image of the retina, (ii) confocal laser scanning ophthalmoscopes (cSLOs) that acquire the image by rapidly scanning a point source across the retinal surface and (iii) optical coherence tomography (OCT) which again scans a point source but uses low coherence interferometry to form the image. Each of these modalities are discussed in subsequent sections; however, as the flood illuminated technique is conceptually the simplest it is used here to introduce the application of AO to the human eye.

Figure 6 shows the optical layout of the New England College of Optometry (NECO) AO flood illuminated (flash) fundus camera (Headington et al, 2011). The WFS beacon is used to measure the ocular aberration, a small incident beam (1mm diameter at the cornea) from a superluminescent diode (SLD) at 820nm is focused to a ~10µm diameter spot on the retina. The scattered light exits through the dilated pupil (6mm in diameter) and is redirected by the DM, through the dichroic beamsplitter into the WFS. The aberrations are sampled at 20Hz and the required correction profile is sent to the DM. The system is fast enough to track and correct dynamic ocular aberration changes at a frequency of ~1Hz. As with all AO systems used in vision, the DM and the WFS are placed in optical planes approximately conjugate to the pupil of the eye. Once the aberrations have been corrected, typically below 0.1µm rms over the 6mm diameter pupil, the retinal image is acquired. The imaging source is usually an arclamp and delivers a 4-6msec retinal exposure. The particular imaging wavelength (between 500-800nm) is chosen to highlight a particular retinal feature. The imaging light follows the corrected path through the AO system and is redirected to the science camera via a dichroic beamsplitter. Typical retinal image sizes are 1-3º (0.3-0.9mm diameter).

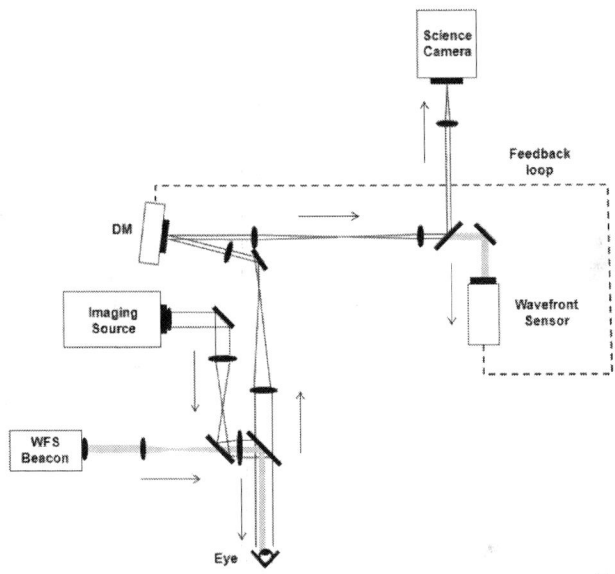

Figure 6. Schematic of the flood illuminated (flash) AO fundus camera (Headington et al, 2011)in use at the New England College of Optometry.

The system described in Figure 6 can be easily modified for functional vision testing. The science camera can be replaced by a visual test pattern, such as a visual acuity or a contrast sensitivity chart. The projected test image is then pre-distorted by the conjugate ocular aberration before being incident on the retina – see later section titled AO for Vision Testing.

2.2. Light Level Considerations

It is essential in the operation of any ophthalmic device that the light levels used are safe. The wide range of imaging modalities, frame rates, wavelengths, field sizes and exposure durations mean that the maximum permissible exposures (MPE) must be calculated on a case by case basis. Several reference standards are used in such calculations (ANSI, 2007; Delori et al, 2007).

2.3. Wavefront Corrector Requirements

Independent of the particular imaging modality, the benefit of AO in ophthalmic systems fundamentally relies on its ability to measure, track and correct the ocular aberrations. It is therefore imperative that the WFS, and in particular the wavefront corrector, have optimal operating characteristics. As DMs are the most commonly used form of wavefront corrector their performance is described in detail here. LC-SLMs (Li et al, 1998) can be modeled as piston-only DMs.

DMs can be divided into two broad categories, continuous surface and segmented (Fig 7). In both cases, there is a set of actuators that physically deform the mirrored surface. Examples of actuation mechanisms can be electrostatic, piezoelectric, magnetic, thermal or voicecoil. Refer to Tyson (2010) for more details on the various types of DMs and their actuation mechanisms. Figure 7A shows a cross section through a continuous surface DM.A two dimensional array of actuators deforms the surface.The greater the number of actuators the higher the spatial frequency correction capability.Light would be incident from the top of the figure. Figure 7B shows a segmented piston/tip/tilt (PTT) DM. In this case, each segment has three degrees of freedom.A common variant is a piston only DM in which an individual segment can only move in the vertical direction. For a continuous surface DM adjusting one actuator causes a deformation of the top mirrored surface and the degree of localization is termed the influence function. Certain DM types, such as membrane and bimorph, have very broad influence functions meaning that activation of one actuator causes a deformation over a large area of the DM. Segmented DMs however have much narrower influence functions; moving a piston only segment only changes that segment's mirror position and not that of its neighbours.The shape of this influence function, along with the number of actuators and the dynamic range, define the corrective ability of a DM.

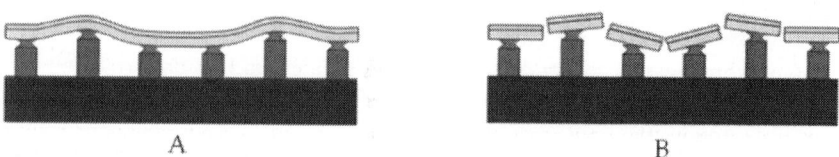

A   B

Figure 7. Deformable mirror (DM) types. (A) Continuous surface mirror, deformed by an underlying array of actuators. (B) Segmented surface - the mirror is composed of a discrete array of segments each of which has three degrees of freedom: piston, tip and tilt (PTT). Piston only segmented DMs are also possible.

Figure 8 shows the DM correction performance as a function of the number of actuators or segments across a 7.5mm pupil for a 0.6μm wavelength. The Rochester population dataset described earlier (Fig 5), was analyzed after zeroing the defocus coefficient. For continuous surface DMs approximately 15 actuators are required to give diffraction limited performance (SR >0.8),with 12 segments giving the same performance for PTT devices(with the caveat that three times as many control voltages are required to move one segment as compared to a single actuator). Piston only DMs require many more segments to achieve good correction with over 100 being necessary; however, these numbers are easily achievable with newer LC-SLMs.

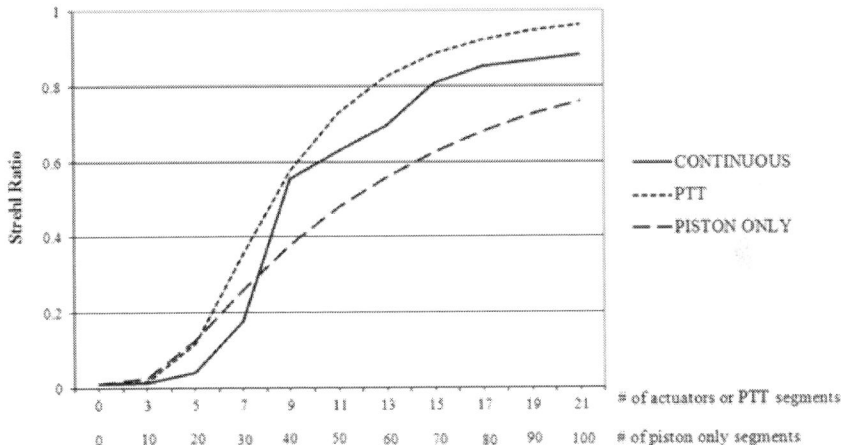

Figure 8.DM correction performance as a function of the number of actuators or segments after zeroing the defocus coefficient. Continuous (solid line) or PTT segmented (short dashed) DMs have comparable performance with 12-15 actuators or segments being required to achieve a SR of 0.8.

3. AO MODALITIES AND APPLICATIONS IN OPHTHALMIC IMAGING

AO improves the capabilities of any ophthalmic instrument where the optics of the eye are involved, from fundus cameras to phoropters. With the high lateral resolution achievable through the use of AO, as described above, it is now possible to detect the earliest changes caused by retinal pathologies. Small structures, such as the smallest microaneurism (early signs of diabetic retinopathy), blood cells, photoreceptor cells, ganglion cells, RPE cells, the smallest capillaries and cells' organelles can now be observed with the high resolution achievable by AO imaging (e.g. Roorda & Williams, 1999; Roorda et al, 2007; Chen et al, 2011; Wang et al, 2011; Zhong et al, 2011). A number of laboratories and clinical centers have begun to evaluate eye diseases using AO imaging. Choi et al (2006) and Wolfing et al (2006) first reportedin vivo images of photoreceptors in a patient with rod-cone dystrophy, which revealed a reduction in cone density. It has also been shown that retinitis pigmentosa and rod-cone dystrophy show a different pattern of cone degeneration (Duncan et al, 2007). Congenital color deficiencies have been studied using AO imaging (e.g. Carroll et al, 2004; Rha et al, 2010)). Choi and colleaguesfound that AO imaging is a reliable technique for assessing and quantifying the changes in photoreceptors in a number of optic neuropathies (Choi et al, 2008) and glaucoma (Choi et al, 2011).AO imaging has also proven to be useful in patients with inherited Stargardt's disease (Chen et al, 2011). Marmor et al (2008) used AO as well as conventional OCT to evaluate the visual significance of the foveal pit and found that it is not required for the specialization of foveal cones. More recently, McAllister et al (2010) have found variation in the degree of foveal hypoplasia and the corresponding variation in foveal cone photoreceptor specialization.

3.1. Flood Illuminated (Flash) AO Fundus Cameras

The first AO retinal imaging systems were flood illuminated designs as depicted in Figure 6.They are extremely versatile and may be configured for a variety of imaging and vision testing experiments. Their disadvantage for imaging is that they are susceptible to ocular scatter as all of the reflected light is imaged onto the science camera, thusreducing imaging contrast, and they have essentially zero axial resolution. In addition, they tend to be slow with sub-Hertz image acquisition rates, although video rate systems have also been built (Rha et al, 2006).

In 1996, Miller et al (1996) obtained the first *in vivo* images of the cone receptors using a high resolution flood illuminated fundus camera (coupled with a precisesecond-order refraction). The introduction ofa full AO system by Liang and Williams (1997) further enhanced the contrast and quality of the cone images. Since then many other flood AO systems have been built (e.g. Hofer, 2001b; Larichev et al, 2002; Glanc et al, 2004; Choi et al, 2006; Rha et al, 2006; Headington et al, 2011). They have utilized improved AO components and imaged a variety of retinal structure and function in both normal (e.g. Roorda & Williams, 1999, 2002; Pallikaris et al, 2003; Putnam et al, 2005; Jonnal et al, 2007; Doble et al, 2011) and diseased eyes (e.g. Carroll et al, 2004; Choi et al, 2006, 2008, 2011; Wolfing et al, 2006; Carroll, 2008).

3.2. Adaptive Optics Confocal Scanning Laser Ophthalmoscopes (AO-cSLO)

In a cSLO, a point of light is scanned rapidly across the retina in a two-dimensional transverse pattern. The reflected light passes through a pinhole that is confocal to a particular retinal layer. This light is then incident on a point detector such as a photomultiplier tube (PMT) or avalanche photodiode (APD). The two dimensional image can then be reconstructed from the detector output. This approach has two major advantages: (i) reduced scattering as only light from a particular point and retinal layer passes through the confocal pinhole (all other light is blocked) and all other light is blocked, and (ii) it allows for video rate imaging of retinal structure and processes.

The two dimensional scan is achieved in modern systems through the use of a fast-mirrored resonant scanner and a slower galvanometric frame scanner; typical frame rates are 20-30Hz. The field of view is similar to flood-based AO systems (1-3º). Standard cSLOs have transverse and axial resolutions of approximately 5 µm and 200 µm respectively, but through the use of AO the resolution is improved to 2.5 µm transversely and <80 µm axially; with these numbers being dependent on the pupil size and imaging wavelength used.

In 1980, Webb and his colleagues(1980) demonstrated the first SLO, which was followed by the work of Dreher et al (1989) who used a DM in conjunction with an SLO. Wade and Fitzke (1998) used an SLO and post-processing to visualize the cone photoreceptors. The first closed-loop AO-cSLO was developed by Roorda et al (2002). Newer systems have pushed the performance even further allowing for the use of dual DMs (Chen et al, 2007a), increased fields of view (Ferguson et al, 2010) and the ability to visualize the rod photoreceptors (Dubra et al, 2011; Merino et al, 2011). For further details the reader is directed to the chapter on AO-cSLOs.

3.3. Adaptive Optics Optical Coherence Tomography (AO-OCT)

Optical coherence tomography (OCT) (Huang et al, 1991; Fercher et al, 1993; Swanson et al, 1993) is a non-invasive imaging modality that exploits the coherence properties of light to form an image. A light source is split into a reference channel and a sample channel via a modified Michelson interferometer. The retinal surface located in the sample channel is rapidly scanned and the reflected light interferes with that of the reference arm. Axial or 'A-scans' are generally acquired first allowing for the subsequent construction of the commonly displayed 'B-scans'. Three-dimensionalvolume scans can then be created. A major advantage of OCT is the axial and lateral resolutions are decoupled. The axial resolution depends on the bandwidth of the light source, the broader the bandwidth the higher the resolution, although compensation of the chromatic aberration of the eye is then required (Fernandez et al, 2005; Zawadzki et al, 2008). Axial resolutions of a few micrometers are possible. The lateral resolution is a function of the pupil size which in turn influences the level of aberration, hence the need for AO.

The first systems combined AO with OCT (Miller et al, 2003; Hermann et al, 2004; Zhang et al, 2005), more recently we are seeing the emergence ofsystems with multimodal imaging capabilities, AO-OCT-SLO systems (Iftimia et al, 2006; Merino et al, 2006; Miller et al, 2011; Zawadzki et al, 2011).

4. AO FOR VISION TESTING

Human visual performance is limited by neural (retinal or at higher visual level) and optical factors. The evaluation of the contribution of each of these factors has proven difficult in the past, since the study of one was confounded by the other and viceversa. Prior attempts to separate the effect of these limiting factors included interferometry and the detection of contrast in images embedded in noise, however both methods have significant procedural limitations.

The use of AO for correcting aberrations in the human eye opened the possibility of evaluating the effect of neural factors sincethe optical effects are compensated for. Soon after the first applications of AO to obtain high-resolution images of the retina (Liang et al, 1997), a number of laboratories began exploring the use of AO to produce aberration-free retinal images to improve visionand evaluate visual function. It was shown that the correction of aberrations improved the contrast sensitivity function (CSF), showing sensitivity at frequencies up to 55 cycles per degree, not possible without AO correction (Liang et al, 1997), and improved the visualsystem's resolution or VA (Yoon & Williams, 2002). AO has also been used to modify the aberrations of the eye to study visual performance(Artal et al, 2004; Chen et al, 2007b). The limiting factor once aberrations are corrected with AO is the sampling of the photoreceptors. Positive effects of the correction of aberrations have been shown in other functions such as face recognition (Sawides et al, 2010) and even some improvement in the periphery of the visual field, where optics do not play such an important role(Roorda, 2011). These results seemed to imply that correcting aberrations, e.g. with refractive surgery, would significantly improve visual performance. However, since the eye is a living organ and the visual system ever changing, there arepotentially significant limitations to the benefit of high order aberration correction to visual function as we describe below.

iAccommodation

Accommodation is the process of changing the power of the eye by modifying the shape and position of the crystalline lens. The amount of accommodation required to form a clear retinal image is controlled by a number of cues, most of which are related to the retinal stimulus quality, such asblur and chromatic aberration, althoughsome, e.g. retinal disparity and proximity, are not.

AO has allowed further evaluation of the stimuli that drive accommodation and disaccommodation, which seem to be non-parallel processes. Among the possible cues that may indicate the sign of defocus to drive accommodation are higher order monochromatic and chromatic aberrations. A number of authors have studied accommodation and disaccommodation with the manipulation of high order aberrations (Hampson et al, 2006; Chin et al, 2009a; Hampson et al,

2010) and suggested that aberrations play a role in the accommodation control of dynamic (stepwise and sinusoidal) stimuli. However, there is controversy as to the role of aberrations in accommodation since other authors have found improvement rather than a reduction of accommodation whencorrecting aberrations (Gambra et al, 2009). Further work is needed to determine the role of higher order aberrations in accommodation. Aberrations may play a role in the time response of accommodation rather than its accuracy (Fernandez & Artal, 2005).

Presbyopia, the decreased ability of the eye to accommodate as it ages, may also benefit from AO aberration correction(see section below on the correction of refractive error using AO).

iiRefraction and Refractive Technologies Using Wavefront Sensing and AO Correction

Autorefractors are computer-controlled instruments that provide objective measurement of the eye's refractive error by measuring the vergence of the light reflected from the retina. They are often used by eye care professionals as a starting point for a subjective refraction. The use of clinically available aberrometers (e.g. Ophthonix Z-View aberrometer; Huvitz HRK-7000AW autorefractor/aberrometer) that measure higher order as well as lower order aberrations instead of the traditional autorefractor are becoming more popular as they have shown greater accuracy (Cooper et al, 2011). A spectacle correction based solely on the measures obtained from these instruments is not appropriate in most patients as they tend to overcorrect astigmatism – and some myopia - and give high errors when determining the axis for low astigmatic magnitudes. A phoropter is an instrument that contains lenses typically used by eye care professionals for subjective refraction of the eye typically used during an eye examination, i.e. correction of the lower aberrations – defocus and astigmatism. Phoroptersincorporating AO would correct for higher order aberrations in addition to defocus and astigmatism using the wavefront pattern obtained with the aberrometer as a base and a subjective refraction as an endpoint.

AO vision simulators may also be useful tools to help finding the best refractive prescription for patients.Rocha et al (2010) used a crx1 AO Visual Simulator (Imagine Eyes SA) to correct and modify the wavefront aberrations in keratoconic eyes and symptomatic postoperative refractive surgery (LASIK) eyes. The AO visual simulator correction improved visual acuity by an average of two lines compared to their best spherocylinder correction. The AO technology may be of clinical benefit when counseling patients with highly aberrated eyes regarding their maximum subjective potential for vision correction.

Lower order aberrations, i.e. defocus and astigmatism, have been measured for hundreds of years and are typically corrected with spectacles, contact lenses, intraocular lenses or refractive surgery. The emerging AO technologies and the improvement in visual performance found with the correction of higher order aberrations has brought excitement to the field of refractive error correction. During the last decade there has been considerable debate concerning the visual impact of correcting the higher order aberrations of the eye.

First attempts to correct higher order aberrations used spectacle lenses and contact lenses. Both of these designs do not benefit significantly from correction of high order aberrations as aberrations are not constant but change with off-axis viewing, movement of the device, pupil size changes and accommodation, among other factors (Lopez-Gil et al, 2007). The use of scleral contact lenses helps with the stability problem of conventional contact lenses. A practical use of these lenses (Sabesan et al, 2007; Katsoulos et al, 2009; Sabesan & Yoon, 2010) for correction of the particularly elevated aberrations found in keratoconus seems plausible. Most of the difficulties found with spectacle lenses and contact lenses may be compensated if other methods of aberration correction, such as refractive surgery or intraocular lenses, are used.

AO has been applied in multifocal intraocular lenses for the correction of presbyopia as itextends the depth of focus by varying the amount of spherical aberration for axial (small pupil) rays for near vision and peripheral rays (larger pupils) for distance. Spherical aberration can be significantly reduced with aspheric intraocular lenses; however, there is a limited reduction in the total high order aberrations, even in perfectly positioned custom aspheric intraocular lenses, which may be influencing the unclear results in the studies assessing the potential benefits on visual performance of these lenses (Einighammer et al, 2009). AO has also been used to show that accommodative intraocular lenses for the correction of presbyopia actually work via pseudoaccommodative rather than accommodative mechanisms (Klaproth et al, 2011).

Refractive surgery has traditionally corrected lower order aberrations.Conventional refractive surgery may disrupt the compensation mechanism of corneal and internal ocular aberrations, creating a larger total amount of high order aberrations (Benito et al, 2009). With the advances in wavefront sensing technology, customized refractive surgery is common nowadays. Compared with conventional treatments, wavefront-guided ablations can achieve a reduction in preexisting higher-order aberrations and less induction of new higher-order aberrations, resulting in improved outcomes for contrast sensitivity and visual symptoms under mesopic and scotopic conditions (Kim & Chuck, 2008). However, concerns regarding the clinical applicability of customized wavefront correction have emerged, and the possibility of achieving supernormal vision in all patients has been challenged (e.g. Yeh & Azar, 2004).

The results of the reviewed studies suggest that many, but not all, observers with normal vision would perceive improvements in their spatial vision with customized (AO) vision correction, at least over a range of viewing distances, particularly when their pupils are larger than 3mm. Keratoconic patients and patients suffering from high spherical aberration, e.g. as a result of conventional refractive surgery, would particularly benefit (Rocha et al, 2010). A recently developed technique combining customized (topography-guided) refractive surgery with riboflavin/UVA cross-linking seems a promising development for the treatment of progressive keratoconus (Krueger & Kanellopoulos, 2010).

iii. Using an AO-cSLO as a High-Frequency Eye Tracker

While the first attempts were made to correct the distortions found in the SLO frames due to eye movements, it was realized that these data are a record of the movements that had occurred during acquisitions. Therefore, the eyes can be tracked with an accuracy and frequency that would not be achieved with the best eye trackers available (Roorda, 2011).

iv. Simultaneous Stimuli Presentation and Image Delivery

With a cSLO the stimulus can be directly encoded in the rastered image, giving a real-time exact position of the stimulus in relation to the surrounding cones. Furthermore, with theincorporation of AO, the stimuli may be delivered to precise regions of the retina, to the level of individual photoreceptors (Sincich et al, 2009). Using a different channel for imaging and stimulus delivery, an infrared light may be used to image the retina while a visible light is used to present the stimuli and be used to record processes occurring at the retinal level as explained in the visual function section below. This technology can also be applied to fluorescence imaging to allow evaluation of sensitivity of non-absorbing structures such as axonal and dendritic structures of primate ganglion cells *in vivo* (Gray et al, 2008). The axial and lateral resolution achieved with AO was high enough to visualize individual dendrites and axons and was able to distinguish between ganglion cell types and function.

Still images and animations can also be delivered into a specific locus at the retina. Surprisingly, moving and stationary targets seem to generate different fixation loci and neither is correlated with the point of maximum cone density (Putnam et al, 2005; Stevenson et al, 2007).

AO-cSLO technology has also the potential for presenting stimuli at the level of a single photoreceptor and performing microperimetry – i.e. measure sensitivity - at an unprecedented degree of retinotopic precision (Makous et al, 2006; Tuten et al, 2011). Such technology would be very useful in studies determining the preferred locus of fixation (Putnam et al, 2005; 2011) and its relation with photoreceptor density. The preferred locus of fixation is an important parameter to obtain in patients with eccentric fixation and in patients with low vision who have central vision loss, e.g. caused by macular degeneration.

v. Visual Acuity and Contrast Sensitivity

By modifying an existing AO imaging device (fundus camera, SLO), stimuli may be imaged onto the retina without aberrations or with a controlled amount/type of aberrations. In addition to work by Liang et al (1997), Yoonand Williams (2002) also used AO to measure VA and contrast sensitivity (CS) through an aberration-corrected eye, with the limiting factor being the sampling of the photoreceptors, and found improved VA and CS at spatial frequencies of 16 and 24 cycles/deg. Most but not all observers showed improvement in VA and CS (e.g. Elliott & Chapman, 2009). Rossi et al (2007) found that myopes do not perform as well after AO correction as emmetropes; it was suggested that this difference is not due to larger cone spacing, as axial myopes (longer eyes) do not show smaller sampling than emmetropes (Li et al, 2010).

viAO Imaging Correlations with Visual Function

A number of reports have shown correlation between AO retinal imaging and visual function tests. Choi et al (2006) first reported that disruption of the cone photoreceptors mosaic in patients with various forms of retinal dystrophies is correlated to functional vision losses (visual fields, contrast sensitivity and and multifocal electroretinography – mfERG) (Fig 9). In toxic maculopathy caused by hydroxychloroquine (antimalarial drug used extensively in the treatment of autoimmune diseases), AO ophthalmoscopy also shows disruption of the cone photoreceptor mosaic in areas corresponding to visual field defects, and shows additional areas of irregularities in cone photoreceptor density in areas with otherwise normal visual field findings, suggesting that AO imaging is detecting changes earlier than visual field tests (Stepien et al, 2009). AO-cSLO imaging of the cone mosaic explained visual performance, a unilateral ring-likeparacentral distortion that could otherwise not be explained using common clinical imaging instruments (Joeres et al, 2008).

More recently, Talcott et al (2011) has found that the use of AO imaging was more useful than the standard of care tests for evaluation of the effect of treatment in three patients with retinal degeneration. Furthermore, AO-cSLO has been used to evaluate cone spacing in familial mitochondrial DNA mutation. Visual function was affected with various levels of severity depending on the cone spacing pattern; it was improved in patients with a contiguous and regular cone mosaic. Patients expressing high levels of the mitochondrial DNA mutation T8993C showed abnormal cone structure, suggesting normal mitochondrial DNA is necessary for normal waveguiding by cones (Yoon et al, 2009). High resolution AO imaging and AO for visual function evaluation of photoreceptors have also been proven useful techniques in selection of patients for therapeutic trials of congenital achromatopsia and for monitoring the therapeutic response in these trials (Genead et al, 2011). AO fundus imaging has also been used to investigate photoreceptor structural changes in eyes with occult macular dystrophy (Kitaguchi et al, 2011).

Furthermore, highresolution imaging with AO-cSLO has contributed significantly to our understanding of Stargardt's disease, a disease that severely affects central vision of young, otherwise healthy, individuals. AO-cSLO imaging showed abnormal cone spacing in regions of abnormal fundus autofluorescence and reduced visual function, although the earliest cone spacing abnormalities were observed in regions of homogeneous autofluorescence and normal visual function (Chen et al, 2011). In addition, visual resolution decreases rapidly outside the foveal center towards the peripheral retina, which seems to be more related to sampling of midget retinal ganglion cells than photoreceptor sampling (Rossi & Roorda, 2010b).

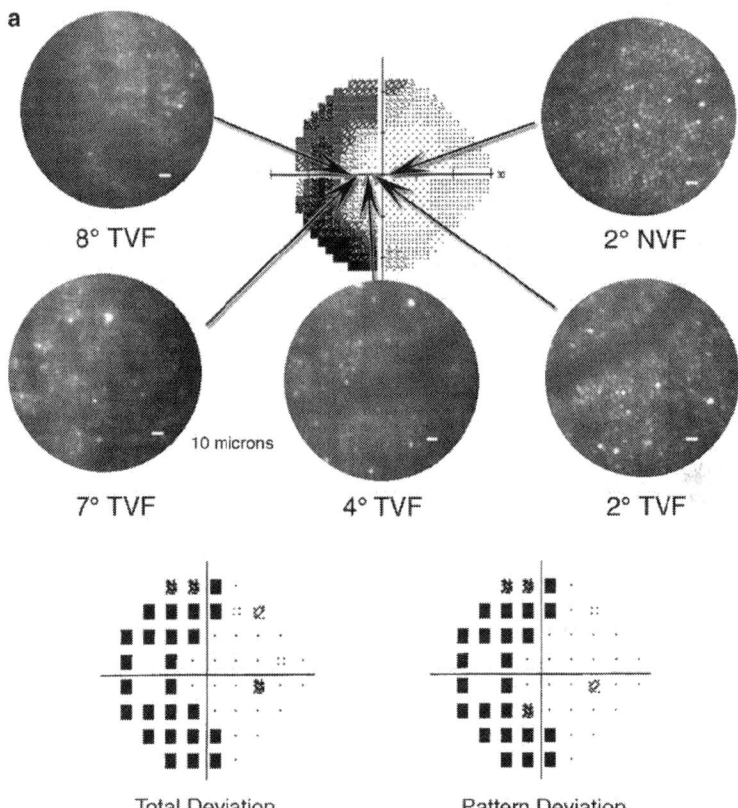

Figure 9. AO-corrected retinal images for a subject with a retinal dystrophy at different locations of the retina shown with the corresponding visual field maps. Reprinted from Choi et al (2006), with permission from the journal Investigative Ophthalmology and Visual Science.

These studies indicate that AO fundus imaging is a reliable technique for assessing and quantifying the changes in the photoreceptor layer as the disease progresses. Furthermore, AO imaging correlates with visual function tests and may be useful in cases where visual function tests provide borderline or ambiguous results, as it allows visualization of individual photoreceptors. Some caution is warranted as there may be greater higher-order wavefront aberrations in eyes with macular disease than in control eyes without disease (Bessho et al, 2009). It has been suggested that portion of the aberration measurementsmay result from irregular or multiple reflecting retinal surfaces.

vii. Visual Perception at the Photoreceptor Level

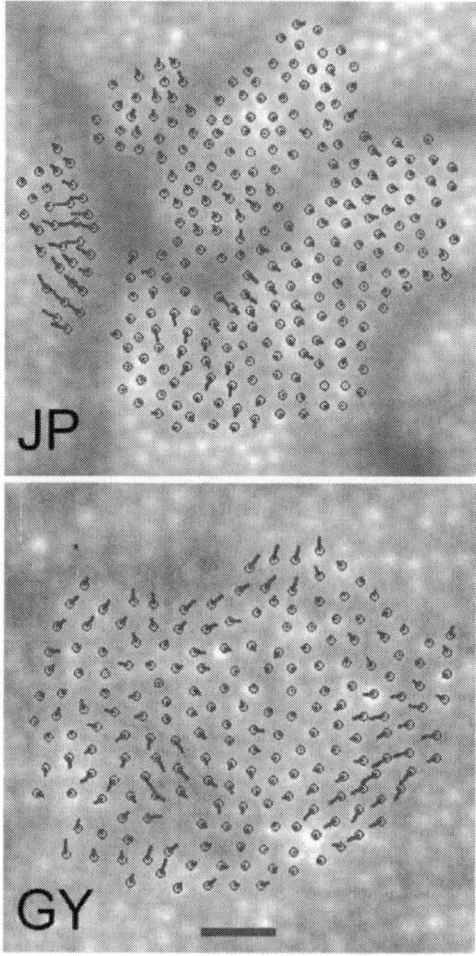

Figure 10. Cone directionality plots for two subjects, from Roorda and Williams (2002). The circles represent cone locations and the lines the direction and magnitude of the departure of each cone's pointing direction in the pupil plane from the average of the ensemble.Reprinted with permission from the Journal of Vision.

AO systems are unique in that they providestructural and functional information of the visual system and therefore allow a new class of experiments with great scientific potential.For example, testing of image perception at the level of an individual photoreceptor is now possible with the use of AO. Roorda and Williams (1999) were the first to show *in vivo* images of the arrangement of the human trichromatic arrangement. By bleaching the retina with three different wavelengths they were able to classify the individual cones according to their photopigment type. The relative numbers of L and M cones varied significantlyamong subjects, even though all had the same color perception.

Using AO-cSLO,Sincich et al (2009) were able to deliver micron-scale spots of light to the centers of the receptive fields of neurons in the macaque LGN and resolve the contribution of single cone photoreceptors to the response of central visual neurons. They imaged and directly stimulated individual cones in the macaque in vivo, while neuron receptive fields were recorded in the LGN. It is therefore possible to now study the properties of different photoreceptors and their influence in visual perception.

As mentioned in the first section of this chapter, Roorda and Williams (2002) imaged the angular tuning properties (SCE) of individual photoreceptors in living human eyes. They found that the disarray in macular cones is very small, implying that the optical waveguideproperties of ensembles of cones are similar to those of the single cones that compose them (Fig 10).

5. ONGOING CHALLENGES IN APPLYING AO FOR VISION

Controlling and correcting aberrations allows the possibility of exploring their effects on human vision and their correction may improve visual performance. However, questions regarding the real benefits of totally eliminating aberrations remain (e.g. Chen et al, 2007b).

The subjective image quality depends on intrinsic optical (e.g. aberrations) and neural factors (Campbell & Green, 1965), as well as the prior experience of the observer. One of the neural limits is adaptation to blur (e.g. Georgeson & Sullivan, 1975; Webster et al, 2002; Vera-Diaz et al, 2010). Adaptation to blur is well known by eye care professionals as their myopic patients often report clearer vision after a period of not wearing their spectacles. Adaptation seems to occur not only for defocus and astigmatism (de Gracia et al, 2011), and other types of induced blur, but the visual system seems to also be adapt to the eye's own high order aberrations profile (Artal et al, 2004; Sawides et al, 2011), somehow removing the effects of blur induced by the optics of the eye. Best subjective image quality is obtained when some amount (~12%) of high order aberrations are left uncorrected (Chen et al, 2007b).Further, observers seem to prefer a small amount of positive spherical defocus for best visual performance

(Werner et al, 2009). Adaptation may have important implications in the correction of aberration with customized refractive surgery or contact lenses, as the benefits of that correction may be overcome with neural adaptation. However, the effect may disappear after adaptation to the new level of aberrations. Rossi and Roorda (2010a) have suggested that adaptation to aberrations does not hinder the correction with AO from givingimmediate visual benefit since AO provideda significant improvement in visual resolution, and visual resolution is a low level task that is not expected to improve with training. It is yet to be determined whether the improvement found in static laboratory measures of VA and CS corresponds to an improvement in real life visual tasks. Furthermore, the improvement in these measures seems to be only found with larger pupil sizes (Elliott & Chapman, 2009) which do not occur in natural viewing. It also seems that the presence of aberrations is utilized as a cue by the visual system for accommodation (Kruger et al, 1993; Chen et al, 2006; Chin et al, 2009b) and perhaps for control of other oculomotor functions of the eye. Therefore, correcting these aberrations may negatively interfere with these visual functions.

In addition to neural limitations, the image quality presented to the retina has other optical dependencies besides aberrations; for example, ocular media density and light scatter, as well as diffraction (the eye is a diffraction-limited optical system at small pupil diameters). These are neither measured nor corrected by AO. Visual function, for example VA and CS at mid- and high-spatial frequencies, is compromised with normal aging. Elliott and Chapman (2009) found that increased high order aberrations that occur with aging (due to coma induced by asymmetric corneas, spherical aberration caused by changes in the lens, etc.) cannot completely account for the decline in spatial vision with aging. A larger role of the aforementioned optical factors and, perhaps more importantly, neural factors, exists in changes in spatial vision found with aging. Elliott et al (2007) and Vera-Diaz et al (2010) found no difference in the strength of adaptation to transient changes in image blur for younger and older observers, suggesting that cortical mechanisms of adaptation remain largely intact with age and could provide a mechanism for long-term adaptation to the increasing degree of high order aberrations with age.

Furthermore, ocular aberrations are dynamic, as explained above, and vary with eye movements, accommodation, etc., which bring subsequent limitations to the benefit of aberration correction.

Other potential limitations to the use of AO for vision science and clinical care are that these devices are not yet user-friendly and are cumbersome, although compact devices are being developed (Mujat et al, 2010). It has also been recently suggested that there are potential differences between AO fundus camera images and AO-cSLO images (Carroll et al, 2010), which requires further investigation. In addition, there are a number of technical limitations of AO systems that require further work. For example, during the time it takes to capture an image with an AO-cSLO device (about 30msecs), the eye moves significantly and in unpredictable ways causing distortions in each frame. These distortions can be corrected for, but this is a work in progress.

6. THE FUTURE OF AO IN VISION SCIENCE

In spite of the limitations described in the previous section, ophthalmic systems that employ AO have the potential to play a crucial role in the clinic. Although the use of AO in clinical settings is still to come, primarily because of the cost and the time-consuming nature of the testing and data processing, the recent substantial and rapid advances of these technologies suggest that their commercialization for ophthalmic clinical use is imminent. A number of prototypes are currently being tested. The AO fundus camera developed by Imagine Eyes © (rtx1™ Adaptive Optics Retinal Camera) is currently commercially available, for research use only, not for sale as a diagnostic device, and clinical trials began in France in 2009 and it may be available soon for clinical use. Physical Sciences Inc.has developed AO imaging devices and an AO-cSLO system, again for research only.

Noteworthy applications of AO imaging are the evaluation of the longitudinal progression of a disease and the evaluation of treatment efficacy. Further insight on the progression of retinal diseases is fundamental for understanding the molecular basics of these diseases. Likewise, the evaluation of the disease response to novel treatments with accurate objective methods such as AO imaging is far more informative than current subjective evaluation methods (e.g. VA and CS), and is critical for determining treatment efficacy. Retinal diseases are in general of slow progression, with some taking several years before changes in functional vision may be appreciated;however, disease progression has been shown with AO imaging techniques. With the use of AO it is now possible to image the smallest structures of the eye *in vivo*; not only retinal cells, but organelles and other microscopic structures are being imaged. As a result of the high transverse resolution of OCT and lateral resolution of SLO with AO, these evaluation measures are quickly being adopted as clinical trials' endpoints.

Talcott et al (2011) has recently completed the first longitudinal study of cone photoreceptors during retinal degeneration (in patients with inherited retinal degeneration, such as retinitis pigmentosa and Usher syndrome) and evaluated the response to a treatment with a ciliary neurotrophic factor. Changes in functional tests such as VA, CS and mfERGwere monitored over 2 years. AO-cSLO images showedreduced cone loss in patients treated compared to the contralateral sham eyes (Sincich et al, 2009).Longitudinal studies on healthy eyes are necessary to create a database of normative data against which to compare disease data.

AO vision simulators may become useful tools to help choose the best refraction for patients.Clinicians will be able to show their patients what their vision would be like if they undergo a particular method of refractive surgery compared to another or compare between various kinds of intraocular lenses. Patient education could benefit from these simulators, particularly with the increased interest in presbyopic surgery, or the use of multifocal contact lenses, as patients may better understand and experiment with the visual benefits of various treatment methods.

7. CONCLUSION

AO is an extremely valuable tool in the study of the human visual system. Through bypassing the limitations of the optics of the eye, AO has enabled scientists to visualize single retinal cells *in vivo* and to probe the limits of human visual performance. Today, the field has moved on from technology development (although challenges still remain) to that of answering fundamental questions onretinal disease development and progression, human visual perception and aiding in the development of a range of new refractive technologies.

8. ACKNOWLEDGEMENTS

This work was supported in part by grantEY020901 from the National Institute of Health, Bethesda, MD, USA.

REFERENCES

1. American National Standard Institute for the Safe Use of Lasers (ANSI). In Standard, A. N. (Ed.), (Vol. ANSI 1361 2007): Laser Institute of America.
2. American National Standard Institute for the Safe Use of Lasers (ANSI). Ophthalmics Methods For Reporting Optical Aberrations Of Eyes. In (ANSI), A. N. S. I. (Ed.), (Z80 28-2010).
3. P. Artal, L. Chen, E. J. Fernandez, B. Singer, S. Manzanera, D. R. Williams, 2004 Neural compensation for the eye's optical aberrations. J Vis, 4 281287 .
4. H. W. Babcock, 1953 The possibility of compensating astronomical
5. D. U. Bartsch, L. Zhu, P. C. Sun, S. Fainman, W. R. Freeman, 2002 Retinal imaging with a low-cost micromachined membrane deformable mirror. J Biomed Opt, 7 451456 .

6. A. Benito, M. Redondo, P. Artal, 2009 Laser in situ keratomileusis disrupts the aberration compensation mechanism of the human eye. Am J Ophthalmol, 147 424431 .

7. K. Bessho, D. U. Bartsch, L. Gomez, L. Cheng, H. J. Koh, W. R. Freeman, 2009 Ocular wavefront aberrations in patients with macular diseases. Retina

8. S. A. Burns, S. Wu, J. C. He, A. E. Elsner, 1997 Variations in photoreceptor directionally across the central retina. J Opt Soc Am A Opt Image Sci Vis, 14 20332040 .

9. F. W. Campbell, D. G. Green, 1965 Optical and retinal factors affecting visual resolution. J Physiol, 181 576593 .

10. J. Carroll, 2008 Adaptive optics retinal imaging: applications for studying retinal degeneration. Arch Ophthalmol, 126 857858 .

11. J. Carroll, M. Neitz, H. Hofer, J. Neitz, D. R. Williams, 2004 Functional photoreceptor loss revealed with adaptive optics

12. J. Carroll, E. A. Rossi, J. Porter, J. Neitz, A. Roorda, D. Williams, J. Neitz, 2010 Deletion of the X-linked opsin gene array locus control region (LCR) results in disruption of the cone mosaic. Vision Res, 50 19891899 .

13. D. C. Chen, S. M. Jones, D. A. Silva, S. S. Olivier, 2007a High-resolution adaptive optics

14. L. Chen, P. Artal, D. Gutierrez, D. R. Williams, 2007b Neural compensation for the best aberration correction. J Vis, 7 91 .

15. L. Chen, P. B. Kruger, H. Hofer, B. Singer, D. R. Williams, 2006 Accommodation

16. Y. Chen, K. Ratnam, S. M. Sundquist, B. Lujan, R. Ayyagari, V. H. Gudiseva, A. Roorda, J. L. Duncan, 2011 Cone photoreceptor abnormalities correlate with vision loss in patients with Stargardt disease. Invest Ophthalmol Vis Sci, 52 32813292 .

17. S. S. Chin, K. M. Hampson, E. A. Mallen, 2009a Effect of correction of ocular aberration dynamics on the accommodation response to a sinusoidally moving stimulus. Opt Lett, 34 32743276 .

18. S. S. Chin, K. M. Hampson, E. A. Mallen, 2009b Role of ocular aberrations in dynamic accommodation control. Clin Exp Optom, 92 227237 .

19. S. S. Choi, N. Doble, J. L. Hardy, S. M. Jones, J. L. Keltner, S. S. Olivier, J. S. Werner, 2006 In vivo imaging of the photoreceptor mosaic in retinal dystrophies and correlations with visual function. Invest Ophthalmol Vis Sci, 47 20802092 .

20. S. S. Choi, J. M. Enoch, M. Kono, 2004 Evidence for transient forces/strains at the optic nerve head in myopia: repeated measurements of the Stiles-Crawford effect of the first kind (SCE-I) over time. Ophthalmic Physiol Opt, 24 194206 .

21. S. S. Choi, R. J. Zawadzki, J. L. Keltner, J. S. Werner, 2008 Changes in cellular structures revealed by ultra-high resolution retinal imaging in optic neuropathies. Invest Ophthalmol Vis Sci, 49 21032119 .

22. S. S. Choi, R. J. Zawadzki, M. C. Lim, J. D. Brandt, J. L. Keltner, N. Doble, J. S. Werner, 2011 Evidence of outer retinal changes in glaucoma patients as revealed by ultrahigh-resolution in vivo retinal imaging. Br J Ophthalmol, 95 131141 .

23. T. Y. Chui, H. Song, S. A. Burns, 2008a Adaptive-optics imaging of human cone photoreceptor distribution. J Opt Soc Am A Opt Image Sci Vis, 25 30213029 .

24. T. Y. Chui, H. Song, S. A. Burns, 2008b Individual variations in human cone photoreceptor packing density: variations with refractive error. Invest Ophthalmol Vis Sci, 49 46794687 .

25. J. Cooper, K. Citek, J. M. Feldman, 2011 Comparison of refractive error measurements in adults with Z-View aberrometer, Humphrey autorefractor, and subjective refraction. Optometry, 82 231240 .

26. C. A. Curcio, K. A. Allen, 1990 Topography of ganglion cells in human retina. J Comp Neurol, 300 525 .

27. C. A. Curcio, K. R. Sloan, R. E. Kalina, A. E. Hendrickson, 1990 Human photoreceptor topography. J Comp Neurol, 292 497523 .

28. P. de Gracia, C. Dorronsoro, G. Marin, M. Hernandez, S. Marcos, 2011 Visual acuity under combined astigmatism and coma: optical and neural adaptation effects. J Vis, 11.

29. F. C. Delori, R. H. Webb, D. H. Sliney, 2007 Maximum permissible exposures for ocular safety (ANSI 2000), with emphasis on ophthalmic devices. J Opt Soc Am A Opt Image Sci Vis, 24 12501265 .

30. L. Diaz-Santana, C. Torti, I. Munro, P. Gasson, C. Dainty, 2003 Benefit of higher closed-loop bandwidths in ocular adaptive optics

31. M. T. , K. W. Yau, 2010 Intrinsically photosensitive retinal ganglion cells. Physiol Rev, 90 15471581 .

32. N. Doble, S. S. Choi, J. L. Codona, J. Christou, J. M. Enoch, D. R. Williams, 2011 In vivo imaging of the human rod photoreceptor mosaic. Opt Lett, 36 3133 .

33. N. Doble, D. T. Miller, G. Yoon, D. R. Williams, 2007 Requirements for discrete actuator and segmented wavefront correctors for aberration

compensation in two large populations of human eyes. Appl Opt, 46 45014514 .

34. N. Doble, G. Yoon, L. Chen, P. Bierden, B. Singer, S. Olivier, D. R. Williams, 2002 Use of a microelectromechanical mirror for adaptive optics

35. A. W. Dreher, J. F. Bille, R. N. Weinreb, 1989 Active optical depth resolution improvement of the laser tomographic scanner. Appl Opt, 28 804808 .

36. A. Dubra, Y. Sulai, J. L. Norris, R. F. Cooper, A. M. Dubis, D. R. Williams, J. Carroll, 2011 Noninvasive imaging of the human rod photoreceptor mosaic using a confocal adaptive optics

37. J. L. Duncan, Y. Zhang, J. Gandhi, C. Nakanishi, M. Othman, K. E. Branham, A. Swaroop, A. Roorda, 2007 High-resolution imaging with adaptive optics

38. J. Einighammer, T. Oltrup, E. Feudner, T. Bende, B. Jean, 2009 Customized aspheric intraocular lenses calculated with real ray tracing. J Cataract Refract Surg, 35 19841994 .

39. D. B. Elliott, G. J. Chapman, 2009 Adaptive gait changes due to spectacle magnification and dioptric blur in older people. Invest Ophthalmol Vis Sci, 51 718722 .

40. S. L. Elliott, J. L. Hardy, M. A. Webster, J. S. Werner, 2007 Aging and blur adaptation. J Vis, 7, 8 EOF .

41. J. M. Enoch, 1963 Optical properties of the retinal receptors. Journal of the Optical Society of America A, 53 7185 .

42. A. F. Fercher, C. K. Hitzenberger, W. Drexler, G. Kamp, H. Sattmann, 1993 In vivo optical coherence tomography

43. R. D. Ferguson, Z. Zhong, D. X. Hammer, M. Mujat, A. H. Patel, C. Deng, W. Zou, S. A. Burns, 2010 Adaptive optics scanning laser ophthalmoscope with integrated wide-field retinal imaging and tracking. J Opt Soc Am A Opt Image Sci Vis, 27, A265277 .

44. E. J. Fernandez, P. Artal, 2005 Study on the effects of monochromatic aberrations in the accommodation response by using adaptive optics

45. E. J. Fernandez, I. Iglesias, P. Artal, 2001 Closed-loop adaptive optics

46. E. J. Fernandez, L. Vabre, B. Hermann, A. Unterhuber, B. Povazay, W. Drexler, 2006 Adaptive optics with a magnetic deformable mirror: applications in the human eye. Opt Express, 14 89008917 .

47. E. . Fernández, W. Drexler, 2005 Influence of ocular chromatic aberration and pupil size on transverse resolution in ophthalmic adaptive optics

48. E. Gambra, L. Sawides, C. Dorronsoro, S. Marcos, 2009 Accommodative lag and fluctuations when optical aberrations are manipulated. J Vis, 9 41 .

49. M. A. Genead, G. A. Fishman, J. Rha, A. M. Dubis, D. M. Bonci, A. Dubra, E. M. Stone, M. Neitz, J. Carroll, 2011 Photoreceptor Structure and Function in Patients with Congenital Achromatopsia. Invest Ophthalmol Vis Sci.

50. M. A. Georgeson, G. D. Sullivan, 1975 Contrast constancy: deblurring in human vision by spatial frequency channels. J Physiol, 252 627656 .

51. M. Glanc, E. Gendron, F. Lacombe, D. Lafaille, J. F. Le Gargasson, P. Léna, 2004 Towards wide-field retinal imaging with adaptive optics

52. D. C. Gray, R. Wolfe, B. P. Gee, D. Scoles, Y. Geng, B. D. Masella, A. Dubra, S. Luque, D. R. Williams, W. H. Merigan, 2008 In vivo imaging of the fine structure of rhodamine-labeled macaque retinal ganglion cells. Invest Ophthalmol Vis Sci, 49 467473 .

53. S. Gruppetta, L. Koechlin, F. Lacombe, P. Puget, 2005 Curvature sensor for the measurement of the static corneal topography and the dynamic tear film topography in the human eye. Opt Lett, 30 27572759 .

54. K. M. Hampson, S. S. Chin, E. A. Mallen, 2010 Effect of temporal location of correction of monochromatic aberrations on the dynamic accommodation response. Biomed Opt Express, 1 879894 .

55. K. M. Hampson, C. Paterson, C. Dainty, E. A. Mallen, 2006 Adaptive optics system for investigation of the effect of the aberration dynamics of the human eye on steady-state accommodation control. J Opt Soc Am A Opt Image Sci Vis, 23 10821088 .

56. J. W. Hardy, 1998 Adaptive Optics

57. J. C. He, S. A. Burns, S. Marcos, 2000 Monochromatic Aberrations in the Accommodated Human Eye. Vision Res, 40 4148 .

58. K. Headington, S. S. Choi, D. Nickla, N. Doble, 2011 Single Cell, In vivo Imaging of the Chick Retina

59. B. Hermann, E. J. Fernandez, A. Unterhuber, H. Sattmann, A. F. Fercher, W. Drexler, P. M. Prieto, P. Artal, 2004 Adaptive-optics ultrahigh-resolution optical coherence tomography

60. H. Hofer, P. Artal, B. Singer, J. L. Aragon, D. R. Williams, 2001a Dynamics of the eye's wave aberration. J Opt Soc Am A Opt Image Sci Vis, 18 497506 .

61. H. Hofer, L. Chen, G. Y. Yoon, B. Singer, Y. Yamauchi, D. R. Williams, 2001b Improvement in retinal image quality with dynamic correction of the eye's aberrations. Opt Express, 8 631643 .

62. D. Huang, E. A. Swanson, C. P. Lin, J. S. Schuman, W. G. Stinson, W. Chang, M. R. Hee, T. Flotte, K. Gregory, C. A. Puliafito, et al. 1991 Optical coherence tomography

63. N. V. Iftimia, D. X. Hammer, C. E. Bigelow, T. Ustun, J. F. de Boer, R. D. Ferguson, 2006 Hybrid retinal imager using line-scanning laser ophthalmoscopy and spectral domain optical coherence tomography

64. I. Iglesias, R. Ragazzoni, Y. Julien, P. Artal, 2002 Extended source pyramid wave-front sensor for the human eye. Opt Express, 10 419428 .

65. S. Joeres, S. M. Jones, D. C. Chen, D. Silva, S. Olivier, A. Fawzi, A. Castellarin, S. R. Sadda, 2008 Retinal imaging with adaptive optics

66. J. B. Jonas, U. Schneider, G. O. H. Naumann, 1992 Count and density of human retinal photoreceptors. Graefe's Archive of Clinical and Experimental Ophthalmology, 230, 505 EOF510 EOF .

67. R. S. Jonnal, J. R. Besecker, J. C. Derby, O. P. Kocaoglu, B. Cense, W. Gao, Q. Wang, D. T. Miller, 2010 Imaging outer segment renewal in living human cone photoreceptors. Opt Express, 18 52575270 .

68. R. S. Jonnal, J. Rha, Y. Zhang, B. Cense, W. Gao, D. T. Miller, 2007 In vivo functional imaging of human cone photoreceptors. Opt Express, 15 1614116160 .

69. M. J. Kanis, D. van Norren, 2008 Delayed recovery of the optical Stiles-Crawford effect in a case of central serous chorioretinopathy. British Journal of Ophthalmology, 92 292292 .

70. C. Katsoulos, L. Karageorgiadis, N. Vasileiou, T. Mousafeiropoulos, G. Asimellis, 2009 Customized hydrogel contact lenses for keratoconus incorporating correction for vertical coma aberration. Ophthalmic Physiol Opt, 29 321329 .

71. P. L. Kaufman, A. Alm, 2002 Adler's Physiology of the Eye (10 ed.): Elsevier Health Sciences.

72. A. Kim, R. S. Chuck, 2008 Wavefront-guided customized corneal ablation. Curr Opin Ophthalmol, 19 314320 .

73. Y. Kitaguchi, S. Kusaka, T. Yamaguchi, T. Mihashi, T. Fujikado, 2011 Detection of photoreceptor disruption by adaptive optics

74. O. K. Klaproth, C. Titke, M. Baumeister, T. Kohnen, 2011 [Accommodative intraocular lenses- principles of clinical evaluation and current results]. Klin Monbl Augenheilkd, 228 666675 .

75. R. R. Krueger, A. J. Kanellopoulos, 2010 Stability of simultaneous topography-guided photorefractive keratectomy and riboflavin/UVA cross-linking for progressive keratoconus: case reports. J Refract Surg, 26, S827832 .

76. P. B. Kruger, S. Mathews, K. R. Aggarwala, N. Sanchez, 1993 Chromatic aberration and ocular focus: Fincham revisited. Vision Res, 33 13971411 .

77. A. V. Larichev, P. V. Ivanov, N. G. Iroshnikov, V. I. Shmalhauzen, L. J. Otten, 2002 Adaptive system for eye-fundus imaging. Quantum Electronics, 32 902908 .

78. F. H. Li, N. Mukohzaba, N. Yoshida, Y. Igasaki, H. Toyoda, T. Inoue, Y. Kobayashi, T. Hara, 1998 Phase modulation characteristics analysis of optically-addressed parallel-aligned nematic liquid crystal phase-only spatial light modulator combined with a liquid crystal display. Opt Review, 5 174178 .

79. K. Y. Li, P. Tiruveedhula, A. Roorda, 2010 Intersubject variability of foveal cone photoreceptor density in relation to eye length. Invest Ophthalmol Vis Sci, 51 68586867 .

80. C. Liang, D. R. Williams, 1997 Aberrations and retinal image quality of the normal human eye. J Opt Soc Am A Opt Image Sci Vis, 14 28732883 .

81. J. Liang, B. Grimm, S. Goelz, J. F. Bille, 1994 Objective measurement of wave aberrations of the human eye with the use of a Hartmann-Shack wave-front sensor. J Opt Soc Am A Opt Image Sci Vis, 11 19491957 .

82. J. Liang, D. R. Williams, D. T. Miller, 1997 Supernormal vision and high-resolution retinal imaging through adaptive optics

83. N. Lopez-Gil, F. J. Rucker, L. R. Stark, M. Badar, T. Borgovan, S. Burke, P. B. Kruger, 2007 Effect of third-order aberrations on dynamic accommodation. Vision Res, 47 755765 .

84. W. Makous, J. Carroll, J. I. Wolfing, J. Lin, N. Christie, D. R. Williams, 2006 Retinal microscotomas revealed with adaptive-optics microflashes. Invest Ophthalmol Vis Sci, 47 41604167 .

85. M. F. Marmor, S. S. Choi, R. J. Zawadzki, J. S. Werner, 2008 Visual insignificance of the foveal pit: reassessment of foveal hypoplasia as fovea plana. Arch Ophthalmol, 126 907913 .

86. J. A. Martin, A. Roorda, 2005 Direct and noninvasive assessment of parafoveal capillary leukocyte velocity. Ophthalmology, 112 22192224 .

87. J. T. Mc Allister, A. M. Dubis, D. M. Tait, S. Ostler, J. Rha, K. E. Stepien, C. G. Summers, J. Carroll, 2010 Arrested development: high-resolution imaging of foveal morphology in albinism. Vision Res, 50 810817 .

88. D. Merino, C. Dainty, A. Bradu, A. G. Podoleanu, 2006 Adaptive optics enhanced simultaneous en-face optical coherence tomography

89. D. Merino, J. L. Duncan, P. Tiruveedhula, A. Roorda, 2011 Observation of cone and rod photoreceptors in normal subjects and patients using a new generation adaptive optics

90. D. T. Miller, O. P. Kocaoglu, Q. Wang, S. Lee, 2011 Adaptive optics and the eye (super resolution OCT). MillerD. T.KocaogluO. P.WangQ.LeeS. (2011). Adaptive optics and the eye (super resolution OCT). Eye (Lond), . (Lond), 25 321330 .

91. D. T. Miller, J. Qu, R. S. Jonnal, K. Thorn, 2003 Coherence Gating and Adaptive Optics

92. D. T. Miller, D. R. Williams, G. M. Morris, 1996 Images of cone photoreceptors in the living human eye. Vision Res, 36 10671079 .

93. M. Mujat, R. D. Ferguson, A. H. Patel, N. Iftimia, N. Lue, D. X. Hammer, 2010 High resolution multimodal clinical ophthalmic imaging system. Opt Express, 18 1160711621 .

94. F. H. Netter, 2006 Netter's Atlas of Human Anatomy. (4 ed.): Saunders-Elsevier.

95. Neuroscience Online. In Byrne, J. H. (Ed.): The University of Texas Health Science Center at Houston (UTHealth). 1997.

96. R. J. Noll, 1976 Zernike polynomials and atmospheric turbulence

97. A. Pallikaris, D. R. Williams, H. Hofer, 2003 The reflectance of single cones in the living human eye. Invest Ophthalmol Vis Sci, 44 45804592 .

98. J. Porter, A. Guirao, I. G. Cox, D. R. Williams, 2001 Monochromatic aberrations of the human eye in a large population. J Opt Soc Am A Opt Image Sci Vis, 18 17931803 .

99. J. Porter, H. Queener, J. Lin, K. E. Thorn, A. Awwal, 2006 Adaptive Optics

100. P. Prieto, E. Fernandez, S. Manzanera, P. Artal, 2004 Adaptive optics with a programmable phase modulator: applications in the human eye. Opt Express, 12 40594071 .

101. N. M. Putnam, D. X. Hammer, Y. Zhang, D. Merino, A. Roorda, 2010 Modeling the foveal cone mosaic imaged with adaptive optics

102. N. M. Putnam, H. J. Hofer, N. Doble, L. Chen, J. Carroll, D. R. Williams, 2005 The locus of fixation and the foveal cone mosaic. J Vis, 5 632639 .

103. N. M. Putnam, P. Tiruveedhula, A. Roorda, 2011 Characterization Of The Preferred Retinal Locus Of Fixation And The Locus Of Perceived Fixation In Relation To The Photoreceptor Mosaic Association for Research and Vision in Ophthalmology. Fort Lauderdale, Florida.

104. R. Ragazzoni, 1996 Pupil plane wavefront sensing with an oscillating prism. J of Mod. Opt, 43 289293 .

105. J. Rha, A. M. Dubis, M. Wagner-Schuman, D. M. Tait, P. Godara, B. Schroeder, K. Stepien, J. Carroll, 2010 Spectral domain optical coherence tomography

106. J. Rha, R. S. Jonnal, K. E. Thorn, J. Qu, Y. Zhang, D. T. Miller, 2006 Adaptive optics flood-illumination camera for high speed retinal imaging. Opt Express, 14 45524569 .

107. K. M. Rocha, L. Vabre, N. Chateau, R. R. Krueger, 2010 Enhanced visual acuity and image perception following correction of highly aberrated eyes using an adaptive optics

108. F. Roddier, 1988 Curvature sensing and compensation: a new concept in adaptive optics

109. A. Roorda, 2011 Adaptive optics for studying visual function: a comprehensive review. J Vis, 11.

110. A. Roorda, F. Romero-Borja, W. Donnelly Iii, H. Queener, T. Hebert, M. Campbell, 2002 Adaptive optics scanning laser ophthalmoscopy. Opt Express, 10 405412 .

111. A. Roorda, D. R. Williams, 1999 The arrangement of the three cone classes in the living human eye. Nature, 397 520522 .

112. A. Roorda, D. R. Williams, 2002 Optical fiber properties of individual human cones. J Vis, 2 404412 .

113. A. Roorda, Y. Zhang, J. L. Duncan, 2007 High-resolution in vivo imaging of the RPE mosaic in eyes with retinal disease. Invest Ophthalmol Vis Sci, 48 22972303 .

114. E. A. Rossi, A. Roorda, 2010a Is visual resolution after adaptive optics

115. E. A. Rossi, A. Roorda, 2010b The relationship between visual resolution and cone spacing in the human fovea. Nat Neurosci, 13 156157 .

116. E. A. Rossi, P. Weiser, J. Tarrant, Roorda, 2007, Weiser, P., Tarrant, J. & Roorda, A. (2007). Visual performance in emmetropia and low myopia after correction of high-order aberrations. J Vis, 7, 14.

117. R. Sabesan, K. Ahmad, G. Yoon, 2007 Correcting highly aberrated eyes using large-stroke adaptive optics

118.   R. Sabesan, G. Yoon, 2010 Neural compensation for long-term asymmetric optical blur to improve visual performance in keratoconic eyes. Invest Ophthalmol Vis Sci, 51 7 38353839 .

119.   L. Sawides, P. de Gracia, C. Dorronsoro, M. Webster, S. Marcos, 2011 Adapting to blur produced by ocular high-order aberrations. J Vis, 11.

120.   L. Sawides, E. Gambra, D. Pascual, C. Dorronsoro, S. Marcos, 2010 Visual performance with real-life tasks under adaptive-optics ocular aberration correction. J Vis, 10, 19.

121.   R. B. Shack, B. C. Platt, 1971 Production and use of a lenticular Hartmann screen. J Opt Soc Am A Opt Image Sci Vis, 61, 656.

122.   L. C. Sincich, Y. Zhang, P. Tiruveedhula, J. C. Horton, A. Roorda, 2009 Resolving single cone inputs to visual receptive fields. Nat Neurosci, 12 967969 .

123.   R. S. Snell, M. A. Lemp, 1998 Clinical Anatomy of the Eye (2 ed.). Oxford: Backwell Science.

124.   K. E. Stepien, D. P. Han, J. Schell, P. Godara, J. Carroll, 2009 Spectral-domain optical coherence tomography

125.   S. Stevenson, G. Kumar, A. Roorda, 2007 Psychophysical and oculomotor reference points for visual direction measured with the adaptive optics

126.   W. S. Stiles, B. H. Crawford, 1933 The luminous efficiency of rays entering the eye pupil at different points. Proceedings of the Royal Society of London. Series B: Biological Sciences, 112 428450 .

127.   E. A. Swanson, J. A. Izatt, M. R. Hee, D. Huang, C. P. Lin, J. S. Schuman, C. A. Puliafito, J. G. Fujimoto, 1993 In vivo retinal imaging by optical coherence tomography

128.   K. E. Talcott, K. Ratnam, S. M. Sundquist, A. S. Lucero, B. J. Lujan, W. Tao, T. C. Porco, A. Roorda, J. L. Duncan, 2011 Longitudinal study of cone photoreceptors during retinal degeneration and in response to ciliary neurotrophic factor treatment. Invest Ophthalmol Vis Sci, 52 22192226 .

129.   J. Tam, J. A. Martin, A. Roorda, 2010 Noninvasive visualization and analysis of parafoveal capillaries in humans. Invest Ophthalmol Vis Sci, 51 16911698 .

130.   L. N. Thibos, A. Bradley, 1997 Use of liquid-crystal adaptive-optics to alter the refractive state of the eye. Optom Vis Sci, 74 581587 .

131.   L. N. Thibos, X. Hong, A. Bradley, X. Cheng, 2002 Statistical variation of aberration structure and image quality in a normal population of healthy eyes. J Opt Soc Am A Opt Image Sci Vis, 19 23292348 .

132. W. S. Tuten, P. Tiruveedhula, A. Roorda, 2011 Adaptive Optics

133. R. K. Tyson, 2010 Principles of Adaptive Optics

134. F. Vargas-Martin, P. M. Prieto, P. Artal, 1998 Correction of the aberrations in the human eye with a liquid-crystal spatial light modulator: limits to performance. J Opt Soc Am A Opt Image Sci Vis, 15 25522562 .

135. F. A. Vera-Diaz, R. L. Woods, E. Peli, 2010 Shape and individual variability of the blur adaptation curve. Vision Res, 50 14521461 .

136. B. Vohnsen, 2007 Photoreceptor waveguides and effective retinal image quality. J Opt Soc Am A Opt Image Sci Vis, 24 597607 .

137. A. Wade, F. Fitzke, 1998 A fast, robust pattern recognition asystem for low light level image registration and its application to retinal imaging. Opt Express, 3 190197 .

138. Q. Wang, O. P. Kocaoglu, B. Cense, J. Bruestle, R. S. Jonnal, W. Gao, D. T. Miller, 2011 Imaging retinal capillaries using ultrahigh-resolution optical coherence tomography

139. R. H. Webb, G. W. Hughes, O. Pomerantzeff, 1980 Flying spot TV ophthalmoscope. Appl Opt, 19 29912997 .

140. M. A. Webster, M. A. Georgeson, S. M. Webster, 2002 Neural adjustments to image blur. Nat Neurosci, 5 839840 .

141. Webvision: The Organization of the Retina

142. J. S. Werner, S. L. Elliott, S. S. Choi, N. Doble, 2009 Spherical aberration yielding optimum visual performance: evaluation of intraocular lenses using adaptive optics

143. G. Westheimer, 1967 Dependence of the magnitude of the Stiles-Crawford effect on retinal location. J Physiol, 192 309315 .

144. G. Westheimer, 2008 Directional sensitivity of the retina: 75 years of Stiles-Crawford effect. Proc Biol Sci, 275 27772786 .

145. J. I. Wolfing, M. Chung, J. Carroll, A. Roorda, D. R. Williams, 2006 High-resolution retinal imaging of cone-rod dystrophy. Ophthalmology, 113, 1019 e1011.

146. S. I. Yeh, D. T. Azar, 2004 The future of wavefront sensing and customization. Ophthalmol Clin North Am, 17 247260 .

147. G. Y. Yoon, D. R. Williams, 2002 Visual performance after correcting the monochromatic and chromatic aberrations of the eye. J Opt Soc Am A Opt Image Sci Vis, 19 266275 .

148. M. K. Yoon, A. Roorda, Y. Zhang, C. Nakanishi, L. J. Wong, Q. Zhang, L. Gillum, A. Green, J. L. Duncan, 2009 Adaptive optics scanning laser

ophthalmoscopy images in a family with the mitochondrial DNA T8993C mutation. Invest Ophthalmol Vis Sci, 50 18381847 .

149. R. J. Zawadzki, B. Cense, Y. Zhang, S. S. Choi, D. T. Miller, J. S. Werner, 2008 Ultrahigh-resolution optical coherence tomography

150. R. J. Zawadzki, S. M. Jones, S. Pilli, S. Balderas-Mata, D. Y. Kim, S. S. Olivier, J. S. Werner, 2011 Integrated adaptive optics

151. Y. Zhang, J. Rha, R. Jonnal, D. Miller, 2005 Adaptive optics parallel spectral domain optical coherence tomography

152. Z. Zhong, H. Song, T. Y. Chui, B. L. Petrig, S. A. Burns, 2011 Noninvasive measurements and analysis of blood velocity profiles in human retinal vessels. Invest Ophthalmol Vis Sci, 52 41514157 .

Chapter 11

Radiation Hydrodynamics Using Characteristics on Adaptive Decomposed Domains for Massively Parallel Star Formation Simulations

Lars Buntemeyer[, a, *], Robi Banerjee[a], Thomas Peters[b, c], Mikhail Klassen[d], Ralph E. Pudritz[e]

[a] Hamburger Sternwarte, Universität Hamburg, Gojenbergsweg 112, Hamburg 21029, Germany
[b]Institut für Computergestützte Wissenschaften, Universität Zürich, Winterthurerstrasse 190, Zürich CH-8057, Switzerland
[c]Max-Planck-Institut für Astrophysik, Karl-Schwarzschild-Str. 1, Garching D-85748, Germany
[d]Department of Physics and Astronomy, McMaster University, 1280 Main Street W, Hamilton, ON L8S 4M1, Canada
[e]Origins Institute, McMaster University, 1280 Main Street W, Hamilton ON L8S 4M1, Canada

ABSTRACT

We present an algorithm for solving the radiative transfer problem on massively parallel computers using adaptive mesh refinement and domain decomposition. The solver is based on the method of characteristics which requires an adaptive raytracer that integrates the equation of radiative transfer. The radiation field is split into local and global components which are handled separately to overcome the non-locality problem. The solver is implemented in the framework of the magneto-hydrodynamics code FLASH and is coupled by an operator splitting step. The goal is the study of radiation in the context of star formation simulations with a focus on early disc formation and evolution. This requires a proper treatment of radiation physics that covers both the optically thin as well as the optically thick regimes and the transition region in particular. We successfully show the accuracy and feasibility of our method in a series of standard radiative transfer problems and two 3D collapse simulations resembling the early stages of protostar and disc formation.

Keywords: Radiative transfer, Hydrodynamics, Star formation

1. INTRODUCTION

Radiative feedback plays a crucial role in the process of star and disc formation, the evolution of circumstellar discs and the thermodynamics of the interstellar medium (ISM). Massive stars emit a large number of energetic UV photons and strongly determine the structure of giant molecular clouds (GMCs) by creating large bubbles of ionized gas (HII regions) (e.g. Peters, Banerjee, Klessen, Mac Low, Galván-Madrid, Keto, 2010, Walch, Whitworth, Bisbas, Wünsch, Hubber, 2012 and Dale, Ercolano, Bonnell, 2013). On smaller scales, low mass and intermediate mass stars also significantly influence their surroundings by radiative heating. By increasing the fragmentation scale, radiative heating can completely inhibit further fragmentation in a radius of several AU and prevent, e.g., the formation of a binary system (Price and Bate, 2010). Offner et al. (2009) investigate the initial mass function (IMF) and the star formation rate (SFR) by comparing 3D hydrodynamical simulations of low mass star formation with and without the effects of radiative transfer. They find that the thermal support of a protostar's accretion luminosity suppresses further fragmentation in the cloud core as well as in the protostellar disc. The SFR in their simulations is about half the value of the simulations without radiative transfer and the mass distribution of protostars of very low mass ($M_* < 0.1\,M_\odot$) is significantly reduced. Bate (2009) finds similar effects.

Regarding the formation and evolution of circumstellar discs, radiative feedback is indispensable to understand their fragmentation behavior, thermodynamics, and morphology (Chiang and Goldreich, 1997) and to model the infrared excess observed in their spectral energy distributions (SEDs) (e.g. Dullemond and Monnier, 2010). The initial formation of massive discs during the Class 0 phase has been investigated using hydrodynamical and magnetohydrodynamical (MHD) simulations (e.g. Yorke, Bodenheimer, Laughlin, 1993, Mellon, Li, 2008, Machida, Inutsuka, Matsumoto, 2010, Peters, Banerjee, Klessen, Mac Low, Galván-Madrid, Keto, 2010 and Seifried, Banerjee, Klessen, Duffin, Pudritz, 2011), and Seifried et al. (2013) emphasize the importance of turbulence to explain the formation of Keplerian discs even if strong magnetic fields are present.

Despite a large number of studies, the actual transition from the early self-gravitating protostellar disc (Class 0) to the Keplerian protostellar disc is still poorly understood. Recent observations (e.g. Tobin et al., 2012) indicate that Keplerian discs might form very early during the protostellar evolution and the analytic study by Forgan and Rice (2013) emphasizes the effects of radiative processes. However, the effects of radiative transfer have usually been neglected in MHD simulations so far or were substantially approximated (e.g. Yorke, Bodenheimer, Laughlin, 1993, Mellon, Li, 2008, Machida, Inutsuka, Matsumoto, 2010 and Seifried, Banerjee, Klessen, Duffin, Pudritz, 2011). The self-consistent modeling of the formation and early evolution of protostars and protostellar discs therefore creates the need for numerical methods to make 3D radiation MHD simulations feasible.

In this context, radiative transfer is a rather costly computation compared to solving Euler's equations. The reason for this is that the timescale of radiative transfer is usually much shorter than those of hydrodynamics and MHD because of the large speed of light compared to the sound speed of the gas in, e.g., a molecular cloud or the characteristic Alfvén wave speeds of the magnetic field. The short timescale on which radiation emerges throughout the complete computational domain makes radiative transfer a highly non-local problem compared to MHD which is determined completely by local thermodynamic properties of the gas. In this sense, hydrodynamics and radiative transfer are two very different numerical tasks and very challenging to solve consistently. Modern Eulerian MHD codes like FLASH (Fryxell et al., 2000) mostly solve the Euler equations on a grid with adaptive mesh refinement (AMR) to resolve fluid features on a wide range of length scales. These codes are parallelized by subdividing the computational domain into several subdomains each containing a fixed number of cells. Since the Euler equations describe local fluxes of mass, momentum and energy, all subdomains can be handled in parallel during a hydrodynamical time step. Between the time steps, boundary values of the subdomains are exchanged using the Message Passing Interface (MPI) for communication. In contrast, characteristics based radiative transfer codes are usually designed very differently. Instead of domain decomposition, these codes are parallelized exploiting the formal independence of the radiative transfer equation (RTE) on the solid angle. Resolving the anisotropy of the radiation field accurately requires a large set of characteristics each covering a discrete opening angle of the 4π unit sphere. If all radiative quantities are assumed to be fixed during one solution step, characteristics of different directions can be computed independently of each other which makes it ideal for parallelization. However, the spatial information of the computational domain with all radiative quantities has to be available for each processor computing a certain number of characteristics on the solid angle grid. This can be a severe drawback in terms of memory requirement if high spatial resolution is required or a large number of frequencies or both (e.g., synthetic stellar spectra). Solving both Euler's equations and the RTE consistently requires careful approximations to the radiative transfer problem to make the coupling of an MHD code with a radiative transfer code feasible. van Noort et al. (2002) present a radiation solver that is coupled to a hydrodynamical code using AMR and domain decomposition in 2D. The radiation solver uses short characteristics (SC) for integrating the RTE while boundary values are communicated between Lambda iteration steps. The focus of this approach lies on modeling the dynamics of scattering dominated stellar atmospheres. The SC approach allows for a fast converging Gauss–Seidel iteration scheme (e.g., Trujillo Bueno and Fabiani Bendicho, 1995), while non-local contributions have to be communicated by a successive exchange of boundary values between subdomains. This approach was also extended for 3D simulations (e.g., Hayek, Asplund, Carlsson, Trampedach, Collet, Gudiksen, Hansteen, Leenaarts, 2010 and Davis, Stone, Jiang, 2012). However, while the Gauss–Seidel short characteristics approach is well suited for highly scattering dominated regimes, it introduces a lot of numerical diffusion because a large number of upwind interpolations is necessary. Razoumov and Cardall (2005) implement a method that is as computationally cheap as the SC method but less diffusive. They create rays on each refinement level separately while their approach is fully threaded but not MPI parallelized. Recently, Tanaka et al. (2014) parallelized this approach using the *multiple wavefront method* by Nakamoto et al. (2001) based on a carefully chosen calculation sequence on a spatially decomposed domain. This method requires successive communication of boundary values. A similar approach is used with long characteristics (LC) in 3D by Heinemann et al. (2006) without AMR.

Another approach for including radiative transfer in hydrodynamical simulations is based on the moment equations (the angular integrated RTE) of the zeroth, first and second moment of the specific intensity. A moment-based scheme does not necessarily require to integrate along large sets of characteristics, however, the anisotropy of the radiation field has to approximated reasonably in order to close the set of moment equations for the mean intensity, radiative flux and pressure. A possible closure relation is the M1-closure used, e.g., in the HERACLES code (González et al., 2007). The closure relation can also be explicitly calculated using, e.g., a characteristics based approach which is known as the Variable Eddington Tensor (VET) method (e.g. Jiang et al., 2012). A common approach for star formation simulations is the diffusion approximation of the angular moment equations which assumes the radiation field to be completely isotropic. In regions of high opacities χ, the diffusion approximation is an expansion of the specific intensity in which all terms $\propto 1/\chi$ are neglected in the RTE. This leads to Eddington's approximation in which the isotropic radiation pressure is proportional to the radiation energy density. The moment equations of the radiative intensity themselves then form a set of hyperbolic equations, like Euler's equations. However, since those two hyperbolic systems would still have to be handled on their individual timescales, one can even make one further step and neglect the time dependence of the radiation flux by assuming it to be proportional to the gradient of the radiation energy (Fick's law). The moment equations can then be combined into a single diffusion equation for the energy of the radiation field. Because the flux in the diffusion approximation lost its finite propagation speed, one has to introduce a flux-limiter to avoid unphysical propagation speeds depending on the actual opacity. This *flux-limited diffusion approximation* (FLD) (Levermore and Pomraning, 1981) has been successfully used in radiation hydrodynamical star formation simulations coupled within Eulerian grid codes (e.g. Stone, Mihalas, Norman, 1992, Krumholz, Klein, McKee, 2007, Commerçon, Teyssier, Audit, Hennebelle, Chabrier, 2011, Flock, Fromang, González, Commerçon, 2013, Zhang, Tan, McKee, 2013 and Bryan, Norman, O'Shea, Abel, Wise, Turk, Reynolds, Collins, Wang, Skillman, Smith, Harkness, Bordner, Kim, Kuhlen, Xu, Goldbaum, Hummels, Kritsuk, Tasker, Skory, Simpson, Hahn, Oishi, So, Zhao, Cen, Li, The Enzo Collaboration, 2014) as well as smoothed particle hydrodynamics (SPH) codes (e.g. Bate et al., 2013). However, the diffusion approximation is only valid in optically thick regions where the radiation field becomes isotropic. Kuiper et al. (2010) have shown the significant drawbacks of exclusively using FLD in the transition regions from optically thick to optically thin regimes where the radiation field becomes highly anisotropic. Recent efforts have been made to combine raytracing methods with FLD solvers (Flock, Fromang, González, Commerçon, 2013, Klassen, Kuiper, Pudritz, Peters, Banerjee, Buntemeyer, 2014 and Kuiper, Klahr, Dullemond, Kley, Henning, 2010) to handle, at least, primary stellar or protostellar radiation separately from the FLD approximation and to avoid the stellar flux from diffusing into shadow regions.

Finally, Monte Carlo (MC) methods have become increasingly popular during the last decade, especially in post-processing MHD simulations. The MC method is a statistical approach and treats individual photons or photon packages by following its propagation path and computing absorption, emission and scattering probabilities. Several advances have been introduced, e.g., photon peel-off (Lucy, 1999), immediate reemission (Bjorkman and Wood, 2001) and diffusion approximations (Min et al., 2009) which make the MC method a powerful tool to calculate synthetic spectra, SEDs or polarization maps from the outcome of MHD simulations. The angular and frequency resolutions are, in principle, unlimited since the direction of propagation of a photon package and its frequency are chosen randomly from a continuous probability function. In that sense, the MC method always gives a quite reasonable result even in the limit of a small number of photon packages while a low resolution shows mainly up as statistical noise in the solution. But the statistical approach also has a severe drawback since we do not know in advance the exact path a photon package will travel, and how and when it is emitted or absorbed. Therefore, it is extremely difficult to implement on a decomposed domain. MC methods are extremely successful in post-processing the outcome of MHD simulations but are rarely used in combination with hydrodynamical simulations. Those approaches which does include MC methods (e.g. Acreman et al., 2010) are fairly restricted in their spatial resolution of the AMR grid, since each processor has to get a copy of the complete computational domain to be able to follow the path of an arbitrary photon package. For our approach, we therefore choose a discrete ordinate method using characteristics to integrate the RTE which requires a raytracer that works on an AMR grid with domain decomposition.

The paper is organized as follows: In Section 2, we give a brief introduction into the theory of radiative transfer as far as it concerns our method and we describe in detail the method of hybrid characteristics. We also describe the coupling between our radiation solver and the FLASH code. In Section 3, we show results from test calculations we performed to investigate the accuracy of our radiation code as well as the coupling to the FLASH code. In Section 4, we present results from 3D radiation hydrodynamical collapse simulations and the parallel scaling performance of our code is described in Section 5. In Section 6, we discuss our results and put it into context with other state-of-the-art radiation transfer methods.

2. THEORY AND NUMERICS

In this section, we describe the theory of radiative transfer (RT) that forms the basis of our solution method as well as the numerics. We describe the hydrodynamics only as it becomes important in the coupling with the radiation solver. For a more detailed description of the FLASH code and its capabilities we refer to Fryxell et al. (2000).

2.1. The Equation of Radiative Transfer

The theory of radiative transfer in this section is based on the work by Mihalas and Weibel Mihalas (1984) in the limit of geometrical optics. The energy of the radiation field is described by a scalar field of specific intensities $I(\mathbf{x}, \mathbf{n}(\vartheta, \phi), v)$, where $\mathbf{x}$ is the position in space, ϑ and ϕ define the direction of propagation $\mathbf{n}$, and v is the frequency. The radiative transfer equation (RTE) describes the change of the specific intensity during its propagation in a medium which is determined by an energy balance between emission and absorption processes. It reads

$$\frac{1}{c}\frac{\partial I_v}{\partial t} + \mathbf{n} \cdot \nabla I_v = \eta_v - \chi_v I_v.$$

$$(1)$$

where η_v denotes the emissivity (energy volume density per unit time and solig angle), χ_v is the extinction coefficient (1/length) and c is the speed of light. The specific intensity denotes the radiative energy flux *per solid angle* $d\Omega = d\Omega = \sin\vartheta\, d\vartheta\, d\phi$, thus in vacuum, it is constant along a line of sight.

Interaction processes between radiation and matter determine the extinction coefficient χ_v. However, for this work, we solve the time independent RTE since we assume the radiation field to emerge instantaneously throughout the entire computational domain during a hydrodynamical time step. Furthermore, we use the definition of the *source function* $S_v = \eta_v/\chi_v$ and rewrite the RTE in terms of the optical depth without explicit frequency dependence:

$$\frac{dI(\mathbf{n})}{d\tau(\mathbf{n})} = S - I(\mathbf{n}).$$

(2)

This form of the RTE describes the propgation of the specific intensity along a specific line element *ds* in the direction **n** and the optical depth element $d\tau = \chi ds$ respectively. This requires a proper paramerization depending on the coordinate system and, hence, the definition of $\mathbf{n} \cdot \nabla$. Note that the RTE in the form of Eq. (2) becomes a 1D ordinary differential equation. However, for the numerical solution in 3D, the optical depth element $d\tau$ is discretized and parameterized in Cartesian coordinates and the solution is obtained by integrating the RTE on a solid angle grid. The source function S is a more general form of Kirchoff's law. It describes the ratio of emission and extinction of radiative energy and allows arbitrary contributions from thermal emission as well as scattering contributions. In fact, the complexity of the model and the solution of the RTE depend strongly on the source function we choose to accurately describe the current radiation transfer problem (e.g. local thermodynamic equilibrium (LTE), non-LTE (NLTE), grey or non-grey, anisotropy, dust continuum radiation, line transfer, etc.). Describing the complete radiation field would require 6 dimensions, which makes it an extremely challenging task to compute and store a 3D solution. In order to handle the radiation field numerically, the intensity is only computed on the fly and accumulated in form of the solid angle averaged mean intensity

$$J = \frac{1}{4\pi} \oint_{4\pi} I \, d\Omega.$$

(3)

The mean intensity is the zeroth moment of the specific intensity and closely related to the radiative energy density $E_r = 4\pi J/c$. Depending on the model setup, the source function usually depends on the radiation field itself and the mean intensity becomes a part of the source function, which makes the RTE an integro-differential equation. In order to find a self-consistent solution, one has to invoke an iteration scheme of some form. Formally however, the RTE from Eq. (2) can be solved by the *formal solution*:

$$I(\tau_2) = I(\tau_1)\, e^{(\tau_2-\tau_1)} + \int_{\tau_1}^{\tau_2} S(\tau)\, e^{(\tau_2-\tau)} d\tau$$

$$(4)$$

The formal solution describes the intensity propagation along a line element with the optical depth $\Delta\tau = \tau_2 - \tau_1 \Delta\tau = \tau_2 - \tau_1$. It contains the incoming intensity $I(\tau_1)$, which is partially extinct and additional energy from emission processes. The RTE and the formal solution are linear in the intensity and allow us to split the radiation field in as many components as the solution method requires. This is a crucial part in our approach of solving the RTE on a decomposed computational domain such as the adaptive grid embedded in the FLASH code.

2.2. Numerical Radiative Transfer

Integrating the RTE along a set of rays of different directions **n** using Eq. (4) is based on the *method of characteristics*. It was first introduced into the radiation transfer community by Olson et al. (1986). The RTE is integrated for each cell in the computational domain and each direction by computing a stepwise formal solution along a ray, or *long characteristic* (LC), according to the discretized formal solution

$$I_i = I_{i-1}\, \exp\left(-\Delta\tau_{i-1}\right) + \Delta I_i,$$

$$(5)$$

where $\Delta\tau_i$ is the finite optical depth element given by a piecewise linear interpolation

$$\Delta\tau_i = \frac{1}{2}(\chi_{i-1} + \chi_i)\Delta s.$$

χ_i is the opacity at the discretization point s_i on the characteristic. ΔI_i is the discretized counterpart to the integral in the formal solution (Eq. (4)) and is solved by either a linear or parabolic interpolation according to

equation(6)

$$\Delta I_i = \alpha_i S_{i-1} + \beta_i S_i + \gamma_i S_{i+1}.$$

The coefficients α_i, β_i and γ_i depend on the optical depths between s_{i-1}, s_{i-1}, s_i and $s_{i+1} s_{i+1}$. They are given in Olson et al. (1986). Fig. 1 shows the geometrical situation of a characteristic passing through a homogenous grid at an arbitrary direction $n_i(\vartheta, \phi)$. Since the opacity and source function are stored in the cell centers of the finite volume FLASH grid (dashed lines), they are assumed to be constant inside the cell. However, since we use a parabolic interpolation [1] of the source function integral, we introduce a point-based RT grid which is based on the cell centers of the FLASH grid. These cell centers define the vertices from which we interpolate the values at the intersection points of the ray bilinearly. Consequently, the point-based RT grid is staggered by half a grid cell since the ray does not intersect with the grid faces of the finite volume grid but with the faces of the point based grid defined by the FLASH cell centers. The characteristic is traced on the RT-grid using the fast voxel traversal algorithm introduced by Amanatides and Woo (1987). The opacity χ_i and the source function S_i at the intersection points of the characteristic with the RT-grid are interpolated bilinearly from the adjacent vertices.

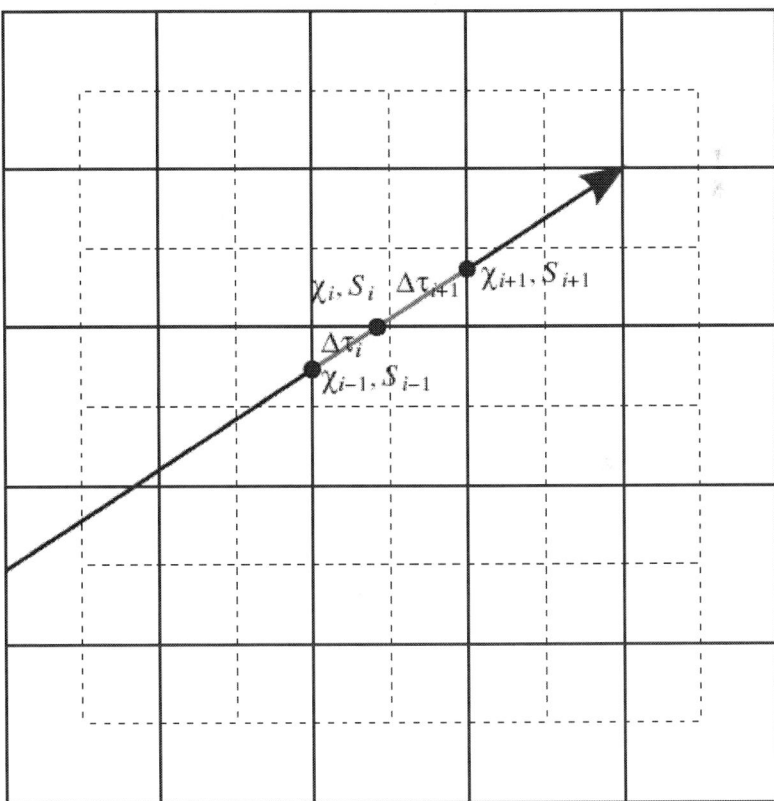

Figure 1. The staggered RT-grid (solid lines) defined by the cell centers of the underlying finite-volume hydro-grid (dashed lines); a long characteristic at an arbitrary direction is shown, which integrates the RTE for the hydro cell-center at the upper right corner of the domain.

While the RTE is integrated along each direction **n**, the mean intensity is computed by accumulating all intensities:

$$J = \frac{1}{4\pi} \sum_{\mathbf{n}} I(\mathbf{n}) \Delta\Omega.$$

$$(7)$$

which requires a discretization of the solid angle Ω. If no information about the anisotropy of the radiation field is available, one should choose a homogeneous discretization which is a non-trivial problem if one considers spherical coordinates on the unit sphere. For this purpose, we use the HEALPix (Hierarchical Equal Area isoLatitude Pixelization) scheme introduced by Górski et al. (2005). HEALPix ensures an optimal discretization of the unit sphere into a number of finite solid angles $\Delta\Omega$. The discretization is based on 12 base pixels which are subdivided depending on the required resolution level. Consequently, typical numbers of directions for the integration of the specific intensity are $N_{pix}=12,48,192$ or 768 (Appendix B)

Characteristics based radiative transfer is the attempt to approximate the radiative interaction of each cell with each other cell in the computational domain. Although the method of long characteristics is very accurate in doing this, it is rather inefficient as it requires to shoot a large number of rays for each cell to sample the radiation field accurately in 3D. An alternative is to use a short characteristics (SC) approach, in which only neighboing cells are used to interpolate incoming intensities from different directions. This requires to sweep the cells in an ordered fashion to make sure that all intensities which are required for interpolation are available. The SC approach introduces numerical diffusion because of the large number of interpolations involved but reduces the cost of the RT calculations by a factor of n_c (the total number of cells involved). Either way, the method of characteristics must invoke a raytracer, which samples radiative interactions between arbitrary regions in the computational domain.

2.3. Raytracing on The Decomposed AMR Grid

The parallel design of the FLASH framework, in which our solver is currently implemented, forbids to trace rays over the entire domain as it is necessary for the method of characteristics. FLASH invokes PARAMESH (Olson et al., 1999), and lately also the CHOMBO library,[2] for implementing an adaptive mesh refinement (AMR) grid. Paramesh uses a *block* structured AMR mesh, in which the fundamental data structure is a block containing cells which are logically indexed by a coordinate triple (i,j,k). The entire computational domain consists of a number of blocks of different physical sizes ordered hierarchically in an octree data structure. Blocks at the bottom of the tree structure, called *leaf blocks*, contain valid data and they cover the entire physical size of the computational domain. FLASH allows for massively parallel computation by invoking the Message Passing Interface (MPI) for the communication of ghost cell information between the blocks. Optimal load balancing is guaranteed by splitting the AMR tree equally between all available MPI tasks to ensure that each task receives more or less the same number of leaf blocks. E.g., the AMR tree of a star formation simulation typically requires more than 10 levels of spatial resolution with up to several 10^5 blocks each containing 8^3 cells, which is only made possible (in terms of cpu-time and memory requirements) by the parallelization described above.

The method of characteristics stays in direct contrast to the spatial parallelization of the AMR grid (Fig. 2). However, in order to account for non-local coupling of the radiation field, we adapt a raytracing technique originally developed by Rijkhorst et al. (2006) and improved by Peters et al. (2010), which uses a combination of local long characteristics and a global "short-characteristics-like" interpolation of outgoing intensities from the decomposed domains of the AMR grid. The basic idea is to split the radiation field in two components:

- •A local component uses long characteristics to compute only radiative contributions to both the cells inside a block as well as contributions that leave the block (*face values*). The computation is done in parallel and in accordance with the design of the block structured AMR grid.
- •A global component which is computed by communicating and accumulating face values (see Fig. 3). This step invokes raytracing over the block structure of the AMR tree and a linear interpolation of face values very similar to the SC method (but on the level of subdomains). After the communication of face values and the tree hierarchy, this step is also done in parallel.

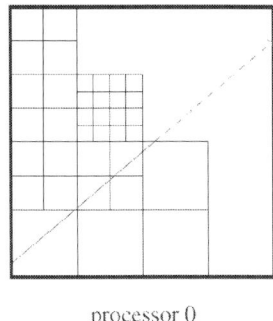 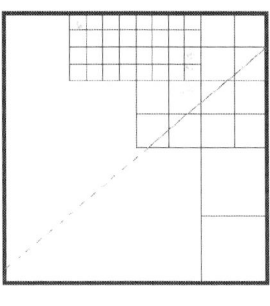

processor 0   processor 1

Figure 2. Example for a 2D AMR grid distributed over two processors without shared memory. The bold rectangle shows the boundary of the whole computational domain. Thin lines show the leaf blocks at different refinment levels that make up the whole subdomain a processor is assigned to. Raytracing through the domain is obviously restricted to the subdomain each processor is assigned to.

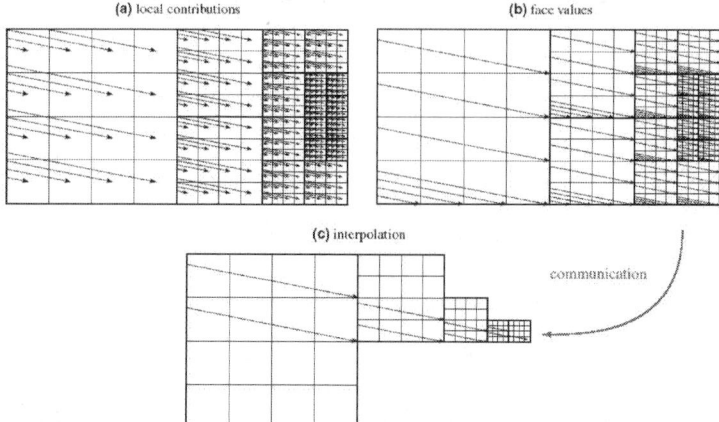

Figure 3. The basic steps of the hybrid characteristics method for parallel rays (compare to Rijkhorst et al. (2006)) in a 2D AMR domain that is refined from left to right. Bold lines show the boundaries of the patches at different AMR levels (in FLASH, these patches are called *blocks* and are distribted equally over the available number of MPI tasks). In this example, each block contains 4x4 cells (indicated by thin lines). (a) Local contributions as calculated with long characteristics. (b) The outgoing face values which are communicated. In fact, we communicate all face values even though they might be part of the same subdomain of a certain MPI process since we need them in the following interpolation step. (c) Example for the linear interpolation of face values for a particular target cell after the communication step. The linear interpolation requires to weight the face values from two rays. The weights depend on where a certain ray segment starts at the boundary of a block (which is a 4x4 cell patch at a certain refinement level) or subdomain respectively.

This approach, called *hybrid characteristics*, only needs to communicate the face values of the blocks and information about the AMR tree hierarchy but no 3D data. By this, the amount of communicated data is reduced significantly. Originally, this method was developed by Rijkhorst et al. (2006) to compute column densities only with respect to point sources for UV ionization. The original method requires to communicate the whole AMR tree structure at the highest level of spatial resolution during the raytracing step on the AMR block structure. This stands in contrast to the parallel design of the FLASH code and restricts the available range of refinement levels of the AMR tree substantially because of the large memory overhead. Peters et al. (2010) add some major improvements to the algorithm by introducing a walk through the AMR tree, which only requires the communication of basic AMR information and conserves the idea of shared memory parallelization.

However, the method was, originally, restricted to compute column densities along rays which originate at a certain point in the grid and used to represent, e.g., a stellar source. For this work, we removed this restriction and implemented a radiative transfer framework which is able to compute the radiation field independently from any point source by solving the RTE for large sets of characteristics along parallel rays. By combining our improvements with the original method, the solver cannot only account for the primary emission by point sources (as in Rijkhorst et al. (2006) and Peters et al. (2010)) but also for the reemitted, diffuse component of the radiation field. Fig. 4 shows a 2D example of a simple test setup with an irradiated central density clump using AMR. From the figure we can see the ability of the method to create sharp shadows and to transport incoming radiation over the entire domain.

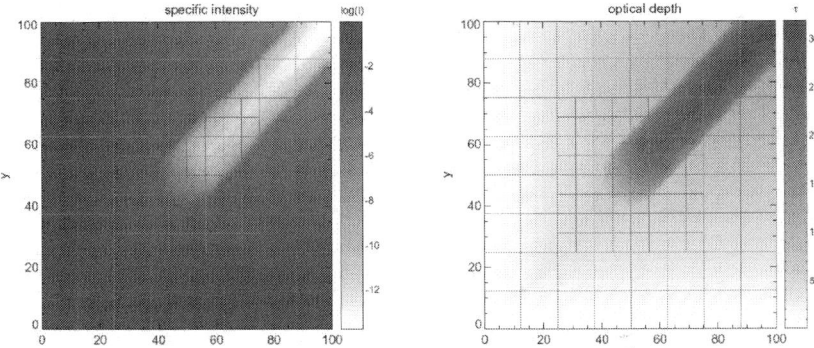

Figure 4. The specific intensity and the optical depth computed diagonally in the *xy*-plane with a central density clump. The source function is set to unity at the left and bottom outermost boundaries and zero everywhere else. The opacity of the central clump is one order of magnitude higher than the ambient opacity. The grid indicates the block structure of the AMR grid, units are arbitrary.

2.4. Coupling to the FLASH Code

Since our method is implemented in the FLASH framework, it is straightforward to couple the radiative transfer module to the hydrodynamical and MHD modules of the FLASH code. The coupling is done by accounting for radiative emission and absorption processes, which are determined by the thermal emission opacity $\chi_e = \kappa_e \rho$ and the thermal absorption opacity $\chi_a = \kappa_a \rho$. The opacities are calculated from mass specific cross sections κ_e and κ_a. Note that the total extinction coefficient χ, which is used for the solution of the RTE, may include an additional scattering opacity χ_s and therefore $\chi = \chi_a + \chi_s$. The coupling of both the radiation and the MHD solver is achieved by computing a source term according to Mihalas and Weibel

Mihalas (1984) which describes the total net gain or loss of energy due to radiative heating and cooling. It reads

$$Q_{rad} = 4\pi \chi (J - S) = 4\pi \chi_a (J - B).$$

(8)

This source term is computed from the time-independent solution of the radiation field as described in the previous sections and it is coupled to the MHD integrator in an operator splitting step. Hence, the set of compressible MHD equations in dimensionless form including gravitation and radiative energy exchange are those of

$$\frac{\partial \rho}{\partial t} + \nabla \cdot (\rho \mathbf{v}) = 0,$$

(9)

momentumconservation

$$\frac{\partial (\rho \mathbf{v})}{\partial t} + \nabla \cdot (\rho \mathbf{v} \otimes \mathbf{v} + p_* \mathbf{1} - \mathbf{B} \otimes \mathbf{B})$$

(10)

energyconservation

$$\frac{\partial E}{\partial t} + \nabla \cdot (\mathbf{v}(E + p_*) - \mathbf{B}(\mathbf{v} \cdot \mathbf{B})) = \rho \mathbf{v} \cdot \mathbf{g} + Q_{rad},$$

(11)

Andtheinductionequation

$$\frac{\partial \mathbf{B}}{\partial t} - \nabla \times (\mathbf{v} \times \mathbf{B}) = 0,$$

(12)

with the gas velocity field $\mathbf{v}$, the magnetic field vector $\mathbf{B}$ and the gravitational acceleration $\mathbf{g}$. p_* is the total pressure and E is the total energy density of a fluid element containing magnetic contributions according to

$$p_* = p + \frac{B^2}{2},$$

(13)

$$E = \frac{1}{2}\rho u^2 + e_{int}\rho + \frac{B^2}{2}.$$

(14)

with the gas density ρ, the thermal pressure p and the internal specific energy e_{int}. $\mathbf{1}$ denotes the unity matrix. The MHD equations are closed by an ideal gas equation of state (EOS) which relates the internal energy to the thermal gas pressure according to

$$p = (\gamma - 1)\rho e_{int}.$$

(15)

We assume $\gamma = 5/3\gamma = 5/3$ which corresponds to a mono atomic (hydrogen) gas. The temperature is also related to the internal energy by:

$$e_{int} = (\gamma - 1)^{-1} \frac{k_b}{\mu m_p} T,$$

(16)

where k_b is the Boltzmann constant, m_p the proton mass and μ the mean molecular weight of the gas.

Note that we solve the equations of MHD and RT successively by an operator splitting step and not simultaneously. Furthermore, for the following test cases, the thermal pressure dominates the hydrodynamics and it is several orders of magnitude larger than the radiation pressure, which we therefore neglect in the momentum Eq. (10). However, since our method explicitly computes the angular dependency of the radiation field, it is straightforward to couple it into the MHD equations.

2.4.1. Choosing the time step

The current coupling is done by an update of the internal gas energy e_{int} and temperature T respectively. Since we solve the time independent RTE, there is no update of the radiative energy or the source function during the solution of Euler's equations. Instead this is done in the following time step when the gas quantities have been updated. The update of the internal energy is done explicitly by

$$\Delta e_{int} = \Delta t \, Q_{rad}.$$

(17)

Due to the explicit update, we have to make some restrictions on the time step. The radiation field does not have an explicit influence on the CFL time step since the energy update is done after the solution of the MHD equations. Instead, we compute a cooling time step which is chosen if it is shorter than any other time step from a FLASH module. The cooling time step is chosen so that the energy contribution Δe_{int} does not exceed a fixed percentage of the internal energy. Otherwise, if the time step Δt is chosen too large, the total radiative energy could become negative (e.g., $\Delta e_r > e$). This leads to the following time step restriction:

$$\Delta t_c = \min \left(\frac{e_{int}}{|\Delta e_{int}|} \right)_i k_c \, \Delta t_{CFL},$$

(18)

wherek_c determines how much change in the internal energy is allowed, Δt_{CFL} is the CFL time step, and i denotes the indices of all grid cells in the computational domain. Because of the explicit energy update, the cooling time step is usually shorter than the CFL time step. So far, there is no subcycling involved and the FLASH code chooses a global minimum time step from all physics modules involved (including, e.g., self-gravity). The cooling time step highly depends on the absorption coefficient χ since it determines the optical depth of the medium and how much radiation is absorbed and emitted during a single time step. Typically the choice of $0.2 > k_c > 0.01$ is convenient as it produces accurate results (Section 3.3) and time steps about one oder of magnitude lower than the CFL time step.

2.5. The Lambda Formalism

Computing the radiation field in the form of the mean intensity in Eq. (7) requires a formal solution of the RTE in the way described above. Usually, this task is described in a rather compact form by using the *Lambda operator*:

$$J = \Lambda[S].$$

(19)

Formally, the Lambda operator for *one cell* in the computational domain contains the radiative contributions from each other cell. The construction of the operator would require to explicitly calculate the radiative coupling between a cell and each other cell. But this is far too costly concerning computation time and memory requirements. Instead, we do not construct the operator but we approximate the Lambda step from Eq. (19) by using the formal solution from Eq. (4) to compute the radiation field J from the source function S in the way described above. The accuracy of this approximation in a 2D or 3D computation depends crucially on the angular resolution, since it determines whether we actually "hit" each other cell during the angular integration of the mean intensity or not. We avoid this problem partly by calculating the radiation from point sources (e.g. a stellar source) explicitly for each cell by combining our method with the original hybrid characteristics method by Rijkhorst et al. (2006) and Peters et al. (2010). However, the Lambda step from Eq. 19 requires that we know the source function in advance. If we take the temperature from FLASH's hydro solver, we can compute the source function simply as being $S=B(T)$ then solve for the radiation field, couple it back to the hydro solver and we are done. This approach assumes the gas to be in a state of thermodynamical equilibrium but this is, of course, not always the case. If the radiation field is decoupled from the gas temperature, we do not know the source function in advance. The solution then requires some kind of iterative procedure to account for the non-local coupling of the radiation field with the gas. In the theory of radiative transfer, this iterative method is called *Lambda iteration*, which requires iterating over Eq. (19) until a self-consistent solution for $J(S$) is found. Strictly speaking, even in the LTE case with S=B(T),S=B(T), we have to iterate to find a temperature that is consistent with the internal energy of the gas since this determines the thermal emission. However, the Lambda iteration may need several hundreds of iteration steps, which is too costly and ineffective to be employed in a hydrodynamical simulation. One way of resolving this problem, is to partly solve Eq. (19) analytically by splitting the Lambda operator. These approaches, called *Accelerated Lambda iteration* (ALI), have been investigated and used extensively in the stellar atmosphere community (e.g. Trujillo Bueno and Fabiani Bendicho, 1995). We have implemented the most simple form of ALI, the local lambda operator, to solve radiative transfer problems even in regions of high optical depths and strong decoupling where the classical Lambda iteration usually fails (Appendix A).

3. TESTS

In this section, we show test results from the implementation of our radiation solver. The tests include time independent (Sections 3.1 and 3.2) as well as dynamical tests (Section 3.3) in 1D and 3D. We also show results from the combined FLASH/RT code in a series of 1D radiative shock calculations in Section 3.4.

3.1. 1D atmosphere

In the first test, we compute the radiation field in a grey, isothermal, scattering dominated 1D atmosphere. This test is typically used to verify a radiation solver's iterative performance and accuracy in a non-local thermodynamic equilibrium (NLTE) situation on a wide range of optical depths. It is also particularly useful to ensure that the solver accurately reproduces the diffusion limit in an optically thick regime, e.g., in the lower parts of the atmosphere. This test also requires the ALI method since the classical Lambda iteration fails to reproduce the solution in the case of strong scattering contributions. The amount of scattered radiation is quantified by the ratio of the thermal absorption coefficient to the total extinction coefficient according to

$$\epsilon = \frac{\chi_a}{\chi_a + \chi_s},$$

(20)

where we neglect the frequency dependence and $\epsilon\epsilon$ is the *photon destruction probability*. The grey source function in the atmosphere contains a thermal part and a scattering contribution, and it reads

$$S = \frac{\eta}{\chi} = \frac{\eta_s}{\chi_a + \chi_s} + \frac{\eta_e}{\chi_a + \chi_s},$$

(21)

$$= (1 - \epsilon)J + \epsilon B,$$

(22)

where we defined $J=\eta_s/\chi_s, J=\eta_s/\chi_s$, the thermal emission is $B=\eta_e/\chi_a, B=\eta_e/\chi_a$, and η_s and η_e denote the scattering and the thermal emissivity respectively. Since the atmosphere is isothermal, we assume that we know the temperature and normalize it so that $B=1B=1$. The crucial part in this test is to find the source function which has to be consistent with the mean intensity J which is

$$J = \frac{1}{2}(I_- + I_+),$$

(23)

where I_- and I_+ are the downward and upward (2 stream solution) integrated specific intensities respectively. Since we assume a uniform mass specific opacity κ and constant temperature T, the intensity is only a function of optical depth $d\tau=\chi dz, d\tau=\chi dz$, thermal emission B and the photon destruction probability $\epsilon\epsilon$. The mean intensity is then given by the analytic solution

$$J = B\left(1 - \frac{\exp(-\sqrt{\epsilon}\tau)}{1 + \sqrt{\epsilon}}\right).$$

(24)

The density ρ of the model increases exponentially with distance from the upper boundary and we assume that $\chi \propto \rho$ but with $\epsilon\epsilon$ being constant. There is no incoming radiation at the upper boundary of the atmosphere at $\tau=0\tau=0$ while at the lower boundary the incoming radiation is $I=BI=B$. The resulting model atmosphere provides an exponentially varying optical depth τ which resolves the transition region from the optically thick inner LTE-regions to the optically thin NLTE-regions at the outer boundary. We test the solver for a wide range of photon destruction probabilities from $\epsilon=10^{-1}\epsilon=10-1$ to $10^{-8}10-8$. The domain consists of 8 subdomains each containing 8 cells which results in a total spatial resolution of 64 cells. Fig. 5 shows the results. In the outer optically thin parts of the atmosphere, the scattering contribution in the source function becomes dominant since radiation leaves the atmosphere. The numerical solution is in excellent agreement with the analytic solution.

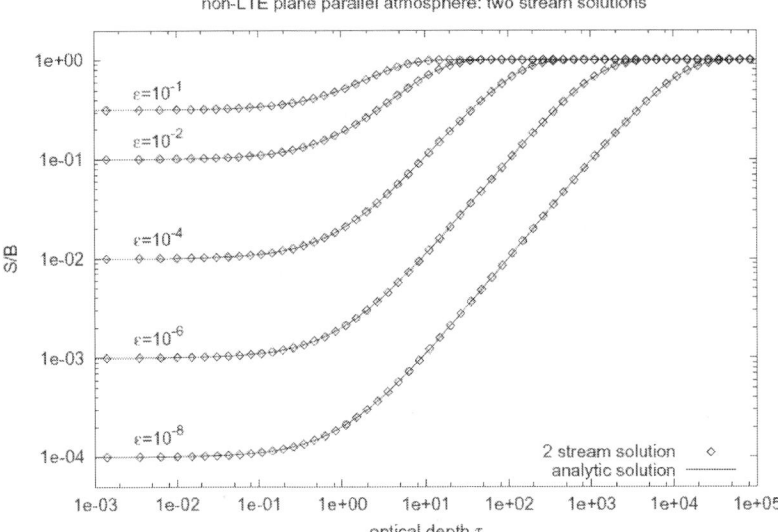

non-LTE plane parallel atmosphere: two stream solutions

Figure 5. Scattering dominated 1D atmosphere problem. The solutions from the radiation solver (symbols) are compared to the analytic solutions (lines) for five different photon destruction probabilities.

3.2. Hydrostatic Protostellar Disc

Cosmic dust is one of the most important constituents of the ISM. By mass, it makes up only a small fraction of typically about 1%, but dust has important radiative and chemical properties. Dust particles have strong continuum opacities which are highly frequency-dependent. Especially in the optical regime, dust absorbs light much more efficiently than in the infrared regime. That is why young protostars, which are surrounded by gaseous and dusty envelopes, are difficult to observe in the visible wavelengths but require infrared observations. Thermal absorption and reemission of radiation by dust (a process called *reprocession*) strongly determines the thermodynamical properties of a protostellar disc, especially in those regions where the disc is opaque to direct stellar radiation and dominated by thermal reemission of dust molecules. This is mainly the case near the equator of the disc because radial optical depths with respect to the central star are typically much larger than unity ($\tau_* \gg 1$). Therefore, modeling the temperature structure requires diffuse radiative transfer to be taken into account.

In this test setup, we combine emission from a point source with the solution of the RTE. The goal is to determine the self-consistent temperature structure of the gas in a protostellar disc. The setup is based on the benchmark by Pascucci et al. (2004), which is based on the theoretical work by Chiang and Goldreich (1997). We compare our solutions from a 3D calculation with the results from the Monte Carlo radiative transfer code RADMC-3D (Dullemond, 2012).

3.2.1. Thermal radiative transfer

A protostellar disc combines optically thick and thin regimes, which requires the computation of primary stellar radiation and the thermal reemission from dust molecules in the disc. Our approach follows the idea of splitting the radiation field in two components handling each separately. Following the work of Dullemond (2002), the first component we compute is the extinct stellar flux. This can be handled by using the original hybrid characteristics method, which computes the optical depth with respect to a central stellar source (τ_*). The extinct stellar flux F_* at a distance r from a star of luminosity L_* is given by

$$F_*(\mathbf{x}) = \frac{L_*}{4\pi r^2} \exp\left(-\tau_*(\mathbf{x})\right),$$

(25)

assuming that the star can be approximated as a point source. The amount of energy per unit time that is absorbed this way is determined by the absorption coefficient χ and given by

$$Q(\mathbf{x}) = \chi F_*(\mathbf{x}).$$

(26)

The reemitted radiation of the dust grains in the disc is treated as a secondary component of the radiation field. This component is computed with the general transfer algorithm using parallel rays. Assuming LTE, the dust grains will acquire an equilibrium temperature such that they emit exactly the same amount of energy which they absorb

$$\frac{\sigma}{\pi}\chi T^4 = \frac{Q}{4\pi} + \chi \frac{1}{4\pi}\oint_{4\pi} I\, d\Omega.$$

(27)

where I is the specific intensity of the reprocessed radiation field. The first term in Eq. (27) accounts for the direct stellar radiation while the second term describes the energy of the reprocessed radiation field. The transfer equation for reemitted radiation by dust grains is

$$\frac{\partial I}{\partial \tau} = \frac{\sigma}{\pi}T^4 - I.$$

(28)

Hence, the source function in this setup is the frequency integrated thermal emission from dust grains S=σSBπT4. The task at hand is to find a temperature that is consistent with the coupled set of Eqs. (27) and (28). This is done by iterating the equations until convergence is reached (Lambda-iteration).

3.2.2. The disc model

For the simulation setup we are following the benchmark test of Pascucci et al. (2004) which resembles a *flared disc*(Chiang and Goldreich, 1997). The idea is to define a radial gas surface density distribution and to assume that the vertical density structure is only determined by the hydrostatic equilibrium in the vertical direction. The gas density distribution is given by

$$\rho(r,z) = \rho_0 \, f_1(r) \, f_2(r), \quad f_2(r) = \exp\left(-\frac{\pi}{4}\left(\frac{z}{h(r)}\right)^2\right),$$

$$f_1(r) = \left(\frac{r}{r_d}\right)^{-1.0}, \qquad h(r) = z_d\left(\frac{r}{r_d}\right)^{1.125},$$

$$\tag{29}$$

where r is the radial distance in the disc midplane, z is the height above the disc, and ρ_0 is the gas density in the midplane at $r=r_d=500\text{AU}$ and $z=0z=0$. The outer disc radius is defined by $r_{out}=1000\text{AU}=2r_d$ and we crop the disc at an inner radius $r=r_{in}r=r_{in}$. z_d determines the height of the disc which we choose to be $0.25\,r_d$ consistent with Pascucci et al. (2004). We choose the central source to have solar properties with $M*=1M\odot$, $R*=1R\odot$ and $T*=5800\text{K}$. We use a grey opacity at the visible wavelength of $\lambda=550\lambda=550$ nm from the opacity tables used in Pascucci et al. (2004) ($\kappa=8736\text{cm2g}-1$).

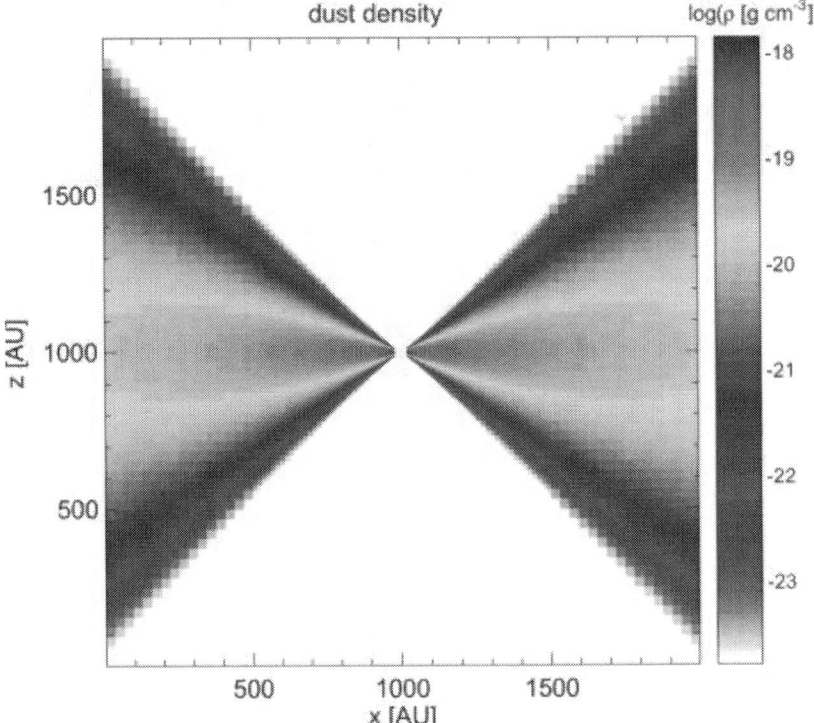

Figure **6.** The dust density in the *xz* -midplane for the Pascucci benchmark for a total optical depth of $\tau_{disc}=1$.τdisc=1.

In contrast to the Pascucci benchmark, we perform our calculations in 3D instead of 2D. Therefore, we cannot directly compare our results to the Pascucci results but instead use the results from RADMC-3D as a reference. We perform calculations for three cases of ρ_0 so that the total radial optical depth of the disc in the midplane varies from $\tau_{disc}=1$, $\tau_{disc}=10$ and $\tau_{disc}=100$. We do not explicitly distinguish between a gas and a dust temperature and assume both to be tightly coupled and the dust density is defined as a fixed fraction of the gas density (1%). The dust density distribution through the xz -midplane of the disc setup for the optically thin case ($\tau_{disc}=1$) is shown in Fig. 6.

The linear spatial resolution varies over 4 refinement levels from $\Delta x=31.25$AU in the outer regions to $\Delta x=1.953$AU in the center of the disc. The solid angle integration is performed using 768 directions (nSide=8).

3.2.3. Results

The resulting temperature structures and averaged midplane profiles are shown in Fig. 7. As it turns out, the accuracy of the solution is very sensible to the spatial resolution of the inner edge of the disc at $r=r_{in}$ which is a result of discretizing the inner circular rim on a Cartesian grid. Therefore, we increase the inner radius from $r_{in}=10$AU, 20AU to 40 AU for the three different setups to guarantee sufficient resolution at the point where the disc becomes optically thick.

In the optically thin case ($\tau_{disc}=1$), the midplane temperature is almost entirely dominated by the direct illumination of the central source. In the optically thick cases, the midplane temperature is dominated by the reprocessed radiation from dust in the photosphere of the disc, which is directly illuminated by the central source. At the point where the disc becomes optically thick, a bump emerges in the temperature profile since the dust distribution becomes dense enough to absorb a considerable amount of radiation from the central source. Our results are in excellent agreement with the reference computed by RADMC-3D and within the 10% range of the results from the different codes used for the Pascucci et al. (2004) benchmark.

However, the temperature structure in the left panel of Fig. 7 is sensitive to the angular resolution. Although the raytracer takes care of the known primary stellar radiation, the solution in the outer regions depend on whether the reprocessed radiation from the hot inner rim is accounted for correctly. Especially in the optically thick case ($\tau_{disc}=100$), single rays become visible in the temperature structure even for a large number of directions ($N_{pix}=768$) since not each cell is correctly connected to the hot inner rim in terms of radiative exchange. A larger number of directions is very costly, but an alternative is to also model the emission from such "hot spots" as a part of the primary emission. The problem is to identify these hot spots in the domain since they cannot be represented by point sources or sink particles. But once these regions are identified, their emission can be handled by an inverse raytracer similar to the approach for point sources (Peters et al., 2010). However, this approach requires an adaptive angular grid while our approach is only capable of using a homogenous angular resolution at the moment (Appendix B).

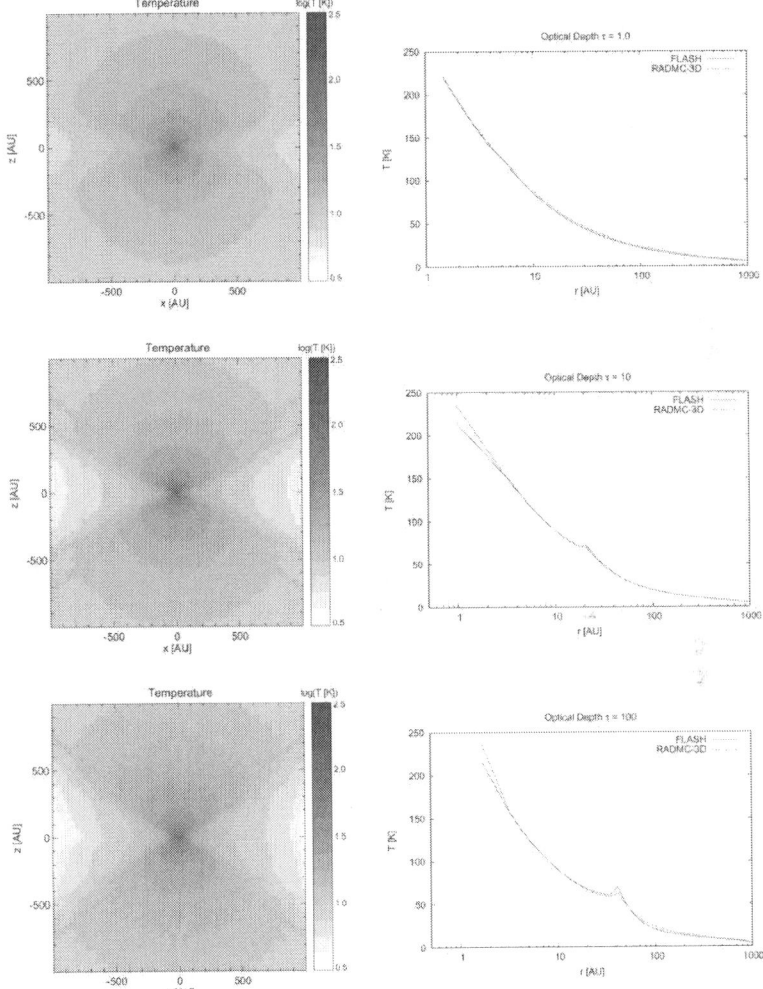

Figure 7. The solutions of the Pascucci et al. (2004) benchmark problem. Left column: the temperature structure through the *xz* -midplane of the disc for total radial optical depths of τ=1τ=1 (top), τ=10τ=10 (mid), and τ=100τ=100 (bottom). Right column: averaged temperature profiles in the *xy*-midplane in comparison with the solutions of RADMC-3D. Solutions obtained with FLASH/RT use 768 directions for the angular discretization. Monte Carlo computations with RADMC-3D were performed using 10^8 photon packages.

3.3. Diffusion Test

In this section, we show results from a time dependent radiative transfer calculation. Solving the time dependent RTE on the timescale of the speed of light would lead to time steps far too small for the use in a hydrodynamical simulation on astrophysical scales. However, since we are not interested in the dynamics of the propagation of the radiation itself but in its contribution to the energy budget of the gas, we assume the hydrodynamical timescale to be much larger than the timescale on which radiation is transported. This means that the radiation field emerges instantaneously everywhere, and we assume the solution of the time independent RTE as being convenient. Consequently, the time dependence of the radiation field originates exclusively from the coupling to the FLASH code using the energy source term from Eq. (8).

In this section, we show results from testing the evolution of the source term by following the propagation of the radiation field in a highly opaque medium. The source function is updated using a simple forward Euler time integration of the energy source term. Since the radiation field shows a diffusion like evolution in the limit of high optical depths, we compare our numerical solution to the analytic solution of the diffusion equation.

3.3.1. Setup

In this test, we investigate the ability of our solver to follow the flux of radiative energy into a highly opaque medium. In this case, the propagation of the radiation field can be described by the diffusion approximation, and we show that our approach reproduces the diffusion limit accurately. The diffusion approximation is derived from the moment equations of the RTE by invoking a closure relation between the radiative energy and the radiative pressure (e.g., the Eddington approximation). The radiation equations themselves then form a hyperbolic system. By neglecting the explicit time dependence of the radiative flux $\mathbf{F}$ and assuming that $\mathbf{F} \propto \nabla E_r$, the flux can be eliminated from the equations. The dynamics of the radiation field $J = cE_r/(4\pi)J=cE_r/(4\pi)$ can then be described in a single equation, the diffusion equation (Mihalas and Weibel Mihalas, 1984):

$$\frac{\partial J}{\partial t} - \nabla\left(\frac{c}{n\chi}\nabla J\right) = c\chi\,(S-J).$$

(30)

where n denotes the number of dimension. We do not allow any interaction of the radiation field with the hydrodynamics and only follow the propagation of the radiation field. Hence, the diffusion equation becomes homogeneous since $S = JS = J$. In this case, the solution to the diffusion equation is described by the Gaussian function

$$J_D(\mathbf{x}, t) = \frac{J_0}{(4\pi Dt)^{n/2}}\exp\left(-\frac{(\mathbf{x} - \mathbf{x}_0)^2}{4Dt}\right).$$

(31)

where J_0 denotes the initial mean intensity at $t=t_0, t=t_0, x_0$ is its initial position, and $D=c/(\eta\chi)D=c/(\eta\chi)$ is the diffusion coefficient. We use Eq. (31) to compute the initial conditions $J(x, t_0)$ for our test setup. We perform 1D and 3D computations with the initial conditions $J_0=J(x_0,t_0)=105 \text{ergs}-1\text{cm}-2\text{sr}-1$ with $t_0=10-11\text{s}$ in 3D and $t_0=10-10\text{s}$ in 1D respectively, The center of the Gaussian is at $x_0=0, x_0=0$, and we evolve the radiation field until $t=20\times t_0 t=20\times t_0$ is reached. The length of the computational domain is 1 cm with a homogeneous density distribution of $\rho=1\text{gcm}-1$ and a constant absorption coefficient $\kappa=1000\text{cm}2\text{g}-1$, which results in a highly optically thick medium. The temperature is constant and arbitrarily set to $T=1T=1K$. Since no heating or cooling is allowed, there is no hydrodynamical response from the medium and all hydrodynamical quantities are constant in space and time.

Since we solve the time-independent RTE, there is a problem in reproducing the time-dependent term in Eq. (30). Strictly speaking, the static source function vanishes since we do not couple the radiation field to the medium through which it propagates. Consequently, the mean intensity would also vanish in the time independent solution. However, the time dependence causes an effective contribution in the source function (e.g. Jack et al., 2012) if the time discretization is carried out implicitly in the RTE (1). This contribution depends on the specific intensities of the previous time step, is evolved through time, and describes the evolution of the radiation field. Since we do not account for this implicit contribution (which would require to store the complete scalar field of angle dependent specific intensities), we solve the problem by operator splitting using the right-hand side of Eq. (30) to calculate the new source function at the following time step. The evolution is done using a simple forward Euler time integration scheme of the form

$$S_n = S_{n-1} + \Delta t_n \, \chi \, c \, (J_{n-1} - S_{n-1}),$$

(32)

where Δt_n is the length of the current time step n. Therefore, the time step is restricted to be (Section 2.4.1)

$$\Delta t_n = \min \left(\frac{S_{n-1}}{(|S_{n-1} - S_{n-2}|)} \right)_i k_{\text{rad}} \Delta t_{n-1}.$$

(33)

where the min function denotes the minimum change in the source function with time from all cells i in the computational domain (FLASH does not support adaptive time stepping on a block level but rather uses a uniform global time step). k_{rad} limits the maximum change in the source function, and we found a value of $k_{\text{rad}} \approx 0.1$ to give stable and accurate results in 3D.

3.3.2. Results

The results of the 1D solutions are shown in Fig. 8. We compare the numerical results with the analytic solution given by Eq. (31) and found our results to be within 1% accuracy at a resolution larger than 32 cells. At the edge of the domain, the numerical solution deviates from the diffusion solution as radiative energy can leave the domain and we allow no irradiation from the outside. The results from the 3D computation are shown in Fig. 9 and compared to the diffusion solution along the three main axes of the domain. In the 3D case, the domain is subdivided by the AMR grid into 4 blocks in each dimension. Each block contains 8^3 cells giving a total linear resolution of 32 cells. In the 3D case, the setup consists of a Gaussian kernel around the origin which diffuses outwards. The solutions along each coordinate axis are obviously indistinguishable, emphasizing the accuracy and importance of the homogeneous angular HEALPix tessellation (Appendix B). The 3D computations were performed using 192 directions.

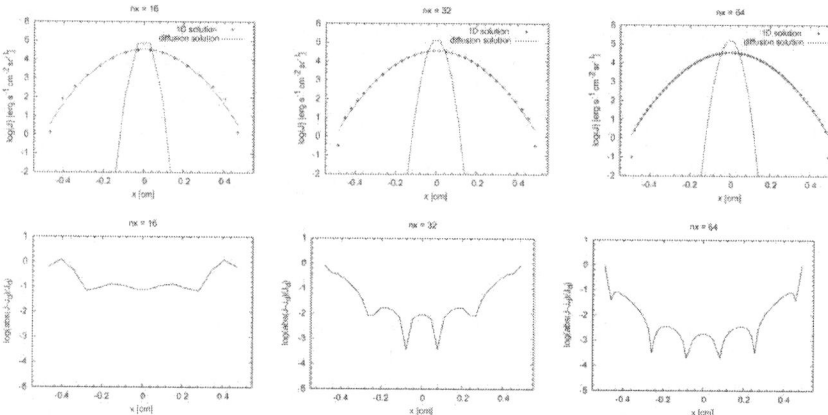

Figure 8. Results of the 1D diffusion test for different homogeneous spatial resolutions, *top:* nx = 16, *mid:* nx = 32, *right:* nx = 64. The dashed lines show the initial conditions at t=t$_0$=t0 determined by the Gaussian solution of the diffusion equation. The initial radiative energy (symbols) is evolved and diffuses outwards until t=20×t$_0$=20×t0 is reached and compared to the analytical solution (solid lines) of the homogeneous diffusion equation. For a sufficient spatial resolution, the numerical solution stays within 1% accuracy. At the edge of the domain, the radiation solver deviates from the diffusion solution as radiation leaves the domain while the diffusion solution is valid for an infinite domain.

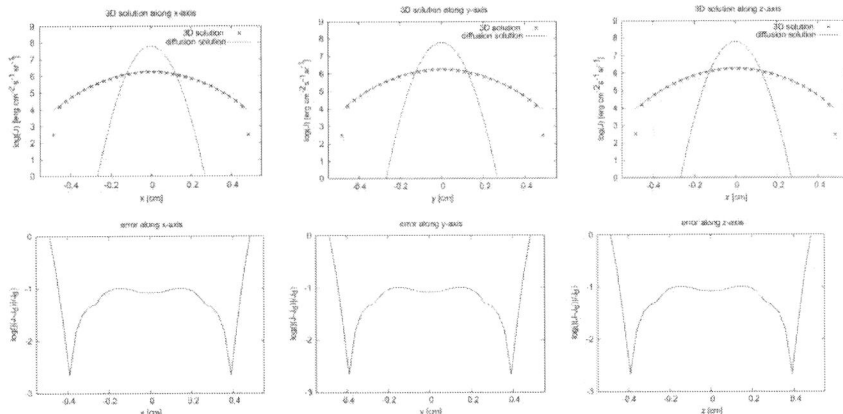

Figure 9. Results of the 3D diffusion test along the x-(*left*), y-(*mid*) and z-axis (*right*) of the simulation box with a homogeneous spatial resolution of nx=ny=nz=32 (symbols). *Top row:* The dashed lines show the initial conditions at t=t0=t0 determined by the Gaussian solution of the diffusion equation. Blue solid lines show the analytic solution according to Eq. 31. Symbols show the numerical results from our radiation solver. *Bottom row:* The relative error of the numerical solution. The 3D solution is not as accurate as the 1D results but still within 10% of the analytical solution. The obvious independence of the solution on the direction axis results from the homogeneous angular HEALPix tessellation. The calculations were performed using 192 directions. (For interpretation of the references to color in this figure legend, the reader is referred to the web version of this article).

3.4. 1D Non-Equilibrium Radiative Shocks

Testing the radiative transfer solver for radiative shock computations is the next crucial step and requires the combination of our radiation solver with the FLASH code. The source term is determined by the energy budget of absorption and emission processes. We recall the frequency integrated source term from Eq. (8) here:

$$Q_{rad} = 4\pi \chi_a (J - B),$$

(34)

which is coupled to the hydrodynamical solver by adding it to the right-hand side of the Euler equation for the internal gas energy. For this test case, the emission and absorption opacities are equal. Since the shock setup is used for test purposes, we neglect the magnetic field.

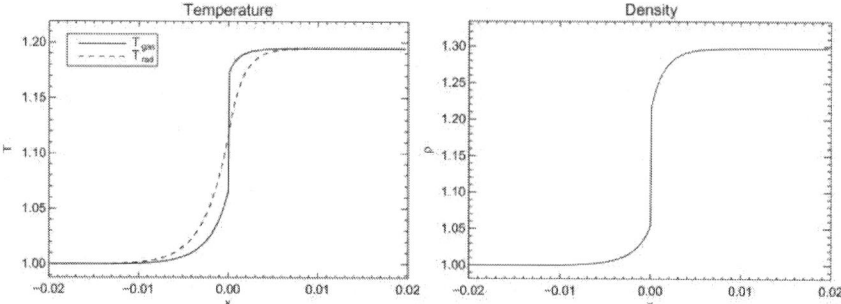

Figure 10. Normalized temperature and density profiles for the subcritical shock with $M_0=1.2M0=1.2$ in the equilibrium state after 10 ns. The gas is preheated on the upstream side and cools down on the downstream side of the hydro shock front.

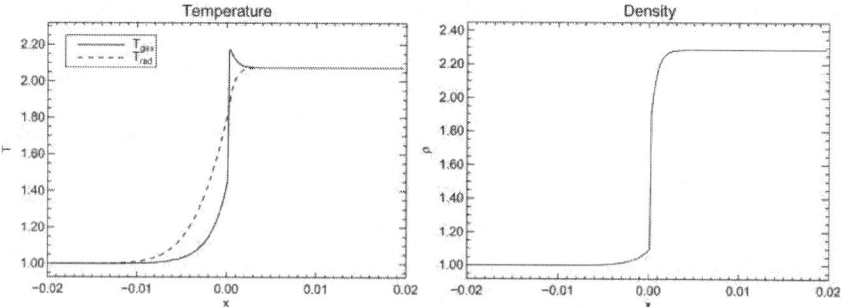

Figure 11. Same conditions as in Fig. 10 but with $M_0=2M0=2$. The maximum temperature at the shock begins to exceed the downstream equilibrium temperature which results in the Zel'dovich spike. Since the temperature at the upstream side of the shock is still well below the downstream temperature, the shock is subcritical.

3.4.1. Initial conditions

The initial conditions are consistent with the theoretical work of Lowrie and Edwards (2008). In their work, the jump conditions and the equations of radiation hydrodynamics are given in a dimensionless form. The equations are normalized using reference material quantities and a constant P_0 which arises from the normalization process and is given by

$$P_0 = \frac{\tilde{\alpha}_r \tilde{T}_0^4}{\tilde{\rho}_0 \tilde{a}_0^2}.$$

(35)

The quantities denoted with a tilde are the dimensional reference material attributes (temperature $\tilde{T}_0$, density $\tilde{\rho}_0$, sound speed $\tilde{a}_0$) and α_r is the radiation constant. The "0"-subscript indicates pre-shock state initial values. P_0 gives a measure for the relative importance of gas and radiation pressure or alternatively, the radiative energy to the material energy (Mihalas and Weibel Mihalas, 1984). For our test setups, we choose a grey non-equilibrium shock setup with Mach numbers of $M_0=1.2, M0=1.2, M_0=2 M0=2$ (subcritical), and $M_0=5 M0=5$ (supercritical), which we compute in the reference frame of the shock with $P_0=10^{-4} P0=10-4$ and $\gamma=5/3 \gamma=5/3$. Lowrie and Edwards (2008) give a dimensionless absorption and transmission cross section, which determine the radiative energy exchange and diffusivity of the radiating materials. Evaluating the dimensionless values gives an absorption coefficient of $\kappa_a \approx 423.0 \text{ cm}^2/\text{g}$ and a total extinction coefficient of $\chi \approx 788.0 \text{ cm}^2/\text{g}$, which results in an effective photon destruction probability of $\epsilon=\kappa_a/\chi \approx 0.5377 \epsilon=\kappa a/\chi \approx 0.5377$. The initial dimensionless pre-shock gas temperature T_0 and density ρ_0 are set to unity, the post-shock initial values (T_1, ρ_1) are computed using the Rankine–Hugoniot jump conditions. The actual dimensional initial conditions can then be calculated using their dimensional reference material values (for more details we refer to Lowrie and Edwards (2008)). Finally, the radiation temperature

$$T_r = \left(\frac{\pi}{\sigma_{SB}} J\right)^{1/4}$$

(36)

is initially in equilibrium with the gas temperature. For the radiation shock test problem, the source function is determined by a thermal emission and a diffusive part. This is equivalent to using the isotropic scattering source function

$$S = (1 - \epsilon)J + \epsilon B$$

(37)

with the appropriate photon destruction probability and a thermal energy contribution given by the frequency integrated Planck emission $B = \frac{\sigma_{SB}}{\pi} \tilde{T}^4$. Since the radiation field will not be not be in thermal equilibrium with the material throughout the simulation, we need to iterate until a consistent solution of the mean intensity J is found. However, since $\epsilon \approx 0.5377$ gives only a moderate scattering contribution and using the solution from the previous time step, the accelerated lambda iteration usually converges after 2 or 3 iteration steps.

3.4.2. Results
The shocks need a few nanoseconds to relax into a static equilibrium state. Figs 10, 11 and 12 show the resulting temperature and density profiles after 10 nanoseconds. Sufficiently far upstream (left) and downstream (right) of the hydrodynamical shock (at $x=0 x=0$), gas and radiation are in a thermodynamical equilibrium and the radiation temperature coincides with the gas temperature computed from the initial conditions. Since the total extinction coefficient χ is about twice the thermal absorption and emission coefficient, the temperature of the radiation field and the gas are out of equilibrium near the shock front.

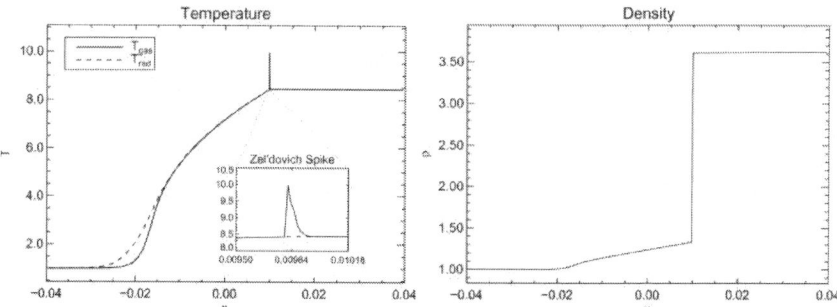

Figure 12. Same conditions as in Fig. 10 but with $M_0=5$. The temperature on the upstream side of the hydro shock front reaches the downstream equilibrium value. The Zel'dovich spike gets very narrow and the shock becomes supercritical.

The subcritical shock with $M_0=1.2$ (Fig. 10) shows a hydro shock but no spike in the radiation temperature. For $M_0=2$ the so called *Zel'Dovich spike* in the gas temperature appears for the first time as seen in Fig. 11. The spike appears since radiation is transported through the hydrodynamical shock and preheats the inflowing gas, which is initially in a thermal equilibrium with the radiation field in the upstream region. After the gas has passed the hydrodynamical shock, it cools down until the radiation field and the gas are again in thermal equilibrium on the downstream side of the shock. Since the upstream temperature at the shock front is still less than the downstream temperature the shock is subcritical. For $M_0=5$, the shock becomes supercritical, since the upstream gas is preheated until it reaches the downstream gas temperature even before passing the hydrodynamical shock front. The discontinuity in the gas temperature is then restricted to the narrow range of the Zel'dovich spike (Fig. 12). Our solutions resemble the semi-analytical results from Lowrie and Edwards (2008) and show the correct spike evolution. However, a closer look at the results show a slight deviation of the shock front from its initial position (at $x=0$). Especially in the supercritical case, the shock front drifts very slowly into the downstream direction. This drift is due to the absence of the radiation pressure in our approach, which becomes important for high Mach numbers (with a high downstream gas temperature). While the shock front drifts very slowly, the temperature and density profiles do not change since the radiation source term is still very well approximated in our approach.

4. 3D COLLAPSE SIMULATIONS

In this section, we show results from full 3D radiation hydrodynamical simulations performed with FLASH/RT. Since we aim to use our framework for the modeling of radiative feedback in star formation simulations, we show the capabilities of our method in two self-gravitating collapsing cloud simulations. We follow the collapse until the first hydrostatic core is formed and before the dissociation of hydrogen molecules start (the first collapse). In Section 4.2, we show results from a basic collapse simulation without rotation and compare the resulting profiles to other similar works. Afterwards, we show results from a more complex simulation including rotation and turbulence (Section 4.3) and compare the results to a simulation without modeling radiative transfer. The angular resolution of the radiative transfer calculations are the same for both collapse simulations, and we use 768 directions to compute the radiation field (nSide = 8 for the HEALPix tessellation).

4.1. Opacities

Since our solver does not yet support any frequency dependence, the source function S is only determined by the frequency-integrated thermal emission of the gas ($S=B=\sigma SBT4\pi$), and we neglect any scattering processes. Consequently, we have to use frequency-integrated mean dust opacities. For this purpose, we choose the Planck mean opacities by Semenov et al. (2003). In their work, the dust composition model takes into account the evaporation temperatures of ice, silicates, iron as well as their density dependencies. We coupled their subroutines[3] for computing temperature and density dependent dust opacities into FLASH, and we choose the input parameters for spherical homogeneous dust grains with a normal relative iron content in the silicates of $Fe/(Fe+Mg) = 0.3$.

4.2. Collapse without Rotation

In this section, we study the collapse of a spherical, homogeneous, and gravitationally unstable density distribution. The initial conditions do not contain any turbulence or density perturbations and hence, the results are spherically symmetric. This setup represents a common benchmark for the capabilities of a radiation hydrodynamical astrophysical computer code, and we compare our results to similar work done by Commerçon et al. (2011), Masunaga et al. (1998), and the pioneering simulations of Larson (1969).

4.2.1. Initial conditions
We start with highly gravitationally unstable initial conditions. The cloud core of one solar mass consists of a homogeneous sphere with radius $R0=7.07\times1016cm(\approx4725AU)$ and and density $\rho0=1.38\times10-18gcm-3$, which results in an initial free fall time of $t_{ff} \approx 56.67$ kyrs. The linear size of the 3D computational domain is four times the initial cloud radius R_0 in each dimension. The surrounding gas density is a hundred times less than the initial cloud density $\rho0$, and the cloud is initially in thermal equilibrium with the ambient gas at a temperature of $T0=10K$ resulting in an initial isothermal sound speed of $cs\approx0.195kms-1$. Since the cloud is initially not in pressure equilibrium with its surroundings, FLASH's hydrodynamical solver drives a weak shock wave

into the ambient gas which is soon dissipated. To prevent our radiation solver from resolving this shock in terms of radiative energy exchange (which would result in rather small time steps), we do not couple the radiation field to the hydrodynamics outside of R_0 but rather keep the ambient gas and radiation temperature fixed.

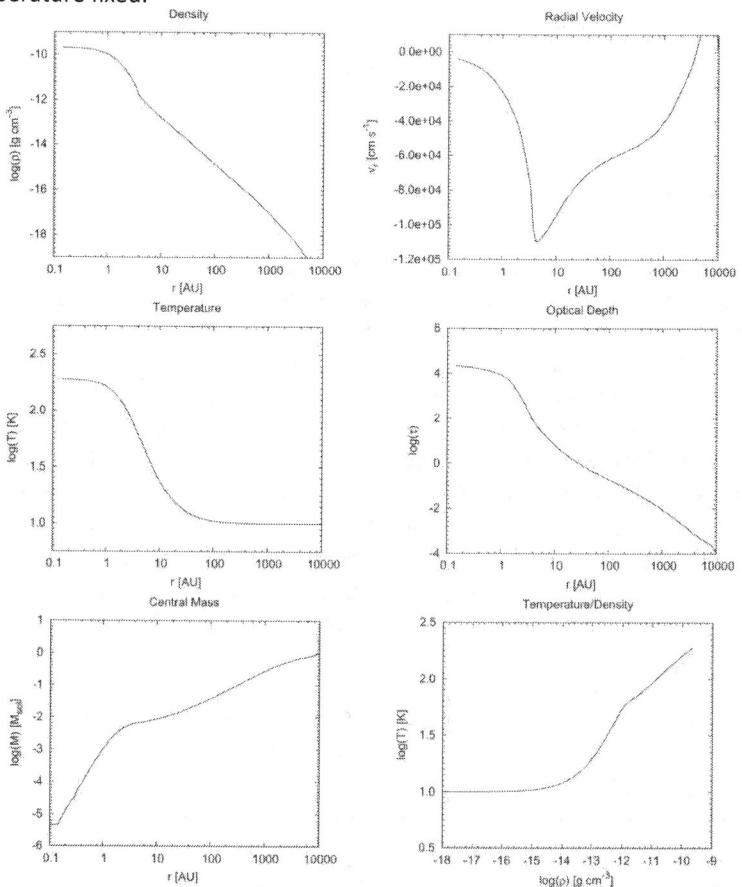

Figure 13. Profiles of the collapse simulation after t=1.036tff; the maximum density at the core center is ρc≈2.0×10−10gcm−3 with a temperature of Tc≈186K, a radius of Rfc≈4AU and a mass of Mc≈10−2M⊙.

The initial conditions result in a gravitationally unstable cloud core which contains nearly two Jeans masses. To ensure a proper resolution and avoid artificial fragmentation during the collapse, we use the Jeans condition by Truelove et al. (1997) as the refinement criterion of the AMR grid. In our case, we use at least Nj=9Nj=9 grid cells per Jeans length. To resolve the first hydrostatic core properly, we allow a maximum linear resolution of $\Delta x \approx$ 0.07 AU which requires the AMR grid to cover 11 levels of resolution.

The summarized initial conditions are:

$$Mass \quad M = 1.0\,M_\odot.$$
$$Density \quad \rho_0 = 1.38 \times 10^{-18}\,g\,cm^{-3}.$$
$$Temperature \quad T_0 = 10\,K.$$
$$Angular\ velocity \quad \Omega = 0.0\,rad\,s^{-1}.$$
$$Radius \quad R_0 = 7.07 \times 10^{16}\,cm.$$
$$Free\ fall\ time \quad t_{ff} = 56.67\,kyrs.$$

4.2.2. Results

The cloud core starts to collapse, and as soon as the maximum density in the cloud exceeds about 10–13gcm–3, the central regions of the cloud core become optically thick. At this point, the central temperature starts to rise rapidly and the following evolution proceeds almost adiabatically with more gas falling onto the central quasi-hydrostatic core. Since the simulation does not contain any rotation or turbulence, the 3D solution is spherically symmetric, and we present the results in the form of averaged radial profiles. The profiles for density, radial velocity, temperature, optical depth, and central mass after $1.036 \times t_{ff}$ are shown in Fig. 13. The resulting protostellar core has a mass of Mfc≈1×10–2M$_\odot$, a radius of $R_{fc} \approx 4$ AU, and a central temperature of T_c $\approx$ 186 K. The boundary of the core can be identified easily in the velocity profile, where there is a sudden decrease in the infall velocity (the accretion shock). Inside the core, the infall does not stop completely indicating that the core is only quasi-hydrostatic.

Table 1. Comparison of simulation results; R fcfc is the radius of the first core, M fcfc is the core mass, T fcfc the central temperature and T fcfc is the temperature at R fcfc.

Reference	R_{fc} [AU]	M_{fc}[M$_\odot$]	T_{fc}[K]	T_c [K]
This work	4	1×10^{-2}1×10–2	50	186
Commerçon et al. (2011)	8	2.1×10^{-2}2.1×10–2	81	396
Masunaga et al. (1998)	8	≈ $^{10-2}$10–2	60	200
Larson (1969)	4	1×10^{-2}1×10–2	–	170

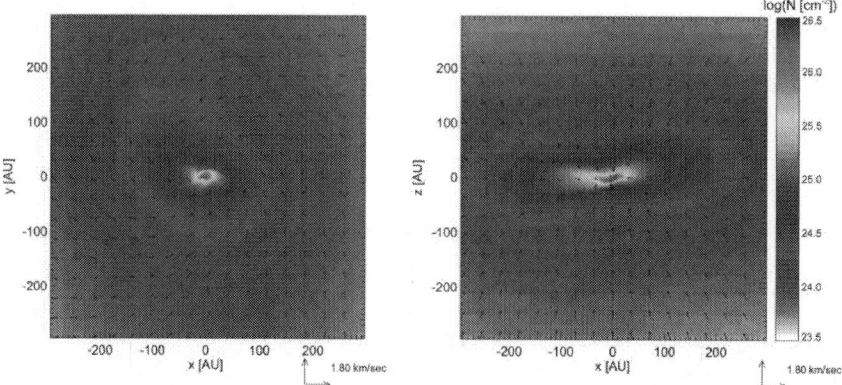

Figure 14. Column densities along the z- (left) and y-axis (right) of the simulation box after the formation of the first hydrostatic core at $t \approx 60$ kyrs $\approx$ 1.07 t_{ff} including rotation and turbulence. The rotational energy forces the gas to accumulate in a circumstellar disc (in the xy-plane) around the first core.

Our results are quantitatively very similar to those of Larson (1969) and qualitatively very similar to the more recent works by Masunaga et al. (1998) and Commerçon et al. (2011). Table 1 shows an overview of the characteristic temperature, mass and radius of the first core in comparison to these works (the common reference point is when the maximum central density of the first core reaches $\rho_{fc} \approx 2 \times 10{-}10 \mathrm{gcm}{-}3$). Apparently, our computations produce qualitatively similar results, although the methods invoked in the other works are quite different and use different initial conditions and opacity models.

4.3. Collapse With Rotation And Turbulence

This simulation run has very similar initial conditions as described in the previous section except that we add rotational and turbulent energy. The cloud is initially in a rigid body rotation around the z-axis at the center of the simulation box. The ratio of rotational and gravitational energy is given by

$$\beta = \frac{1}{3} \frac{R_0^3 \, \Omega_0^2}{G \, M_0}.$$

(38)

We choose $\beta=0.03\beta=0.03$ which gives an initial angular velocity of $\Omega 0=1.886 \times 10{-}13 \mathrm{rads}{-}1$ and agrees with typically observed values of molecular cloud cores (Goodman et al., 1993). In addition, we superimpose a turbulent velocity perturbation on the initial uniform angular velocity field. The construction of the velocity perturbation is based on the theory for incompressible turbulence by Kolmogorov, in which the kinetic energy E of the velocity fluctuation with wave number k is described by a power spectrum

$$E(k) \propto k^p. \tag{39}$$

The wave number k=2π/lk=2π/l is the inverse of the length scale *l* of a turbulent fluctuation (sometimes called *eddy*). In our case, the spectrum has a power law index of p=−2p=−2 resembling a Burgers type model of turbulent energy decay. The geometries and density distribution of the initial cloud core are the same as for the simulation without rotation and turbulence.

In addition to the simulation run with FLASH/RT, we also run the simulation without modeling radiative transfer. Instead, we use a barotropic EOS with a density-dependent effective adiabatic exponent γ that mimics radiative cooling. The internal energy/temperature is fixed at T0=10K as long as the gas density is less than ρ≈10−15gcm−3 (isothermal). Above this threshold density, the temperature rises slowly with γ=1.1γ=1.1 until the adiabatic exponent becomes γ=4/3γ=4/3 above ρ≈10−13gcm−3 (adiabatic). We ran the simulation including radiative transfer as well as the reference run with the barotropic EOS until the formation of the first hydrostatic core with a central density of ρfc≈10−11gcm−3. At this point, both simulations cover 9 different levels of resolution in the AMR grid with a maximum linear resolution of Δx ≈ 0.57 AU while the whole simulation box has a linear extent of 18, 903 AU.

The summarized initial conditions are:

$$
\begin{aligned}
\text{Mass} \quad & M = 1.0\,M_\odot, \\
\text{Density} \quad & \rho_0 = 1.38 \times 10^{-18}\,\mathrm{g\,cm}^{-3}, \\
\text{Temperature} \quad & T_0 = 10\,\mathrm{K}, \\
\text{Angular velocity} \quad & \Omega = 1.886 \times 10^{-13}\,\mathrm{rad\,s}^{-1}, \\
\frac{\text{Rotational energy}}{\text{Gravitational energy}} \quad & \beta = 0.03, \\
\text{Radius} \quad & R = 7.07 \times 10^{16}\,\mathrm{cm}, \\
\text{Free fall time} \quad & t_{\mathrm{ff}} = 56.67\,\mathrm{kyrs}.
\end{aligned}
$$

4.3.1. Results

The rotational energy and the superimposed turbulent velocity perturbations break the symmetry of the simulation. Fig. 14 shows the column densities along the main axes of the inner region where the dense first core has formed after about 60 kyrs (≈1.07 t_{ff}) with a maximum gas density of ρfc≈10−11gcm−3. Because of the additional rotational and turbulent energy, the formation of the first core is deferred and forms later than in the previous simulation (Section 4.2). The conservation of angular momentum causes the first core to be flattened roughly along the z-axis and the density distribution shows a flat disc-like structure revolving around the central compact hydrostatic core. The resulting density distribution is roughly the same as in the reference run without radiative transfer. The initial collapse which seeds the formation of the central core does mostly occur in the isothermal phase, hence, modeling radiative feedback does not influence the initial formation of the core significantly. However, Fig. 15 shows the resulting density weighted temperature averages along the main axes in the central regions around the first core (e.g. $\int \rho\, T\, dz / \int \rho\, dz$). The left column shows the results including radiative transfer (FLASH/RT) while the right column shows results from the reference run. The FLASH/RT model clearly shows how the central core heats the surrounding gas to a temperature roughly 30% higher than in the reference run (like in Price and Bate, 2010). The resulting temperature density distribution in comparison to the barotropic EOS is shown in Fig. 16.

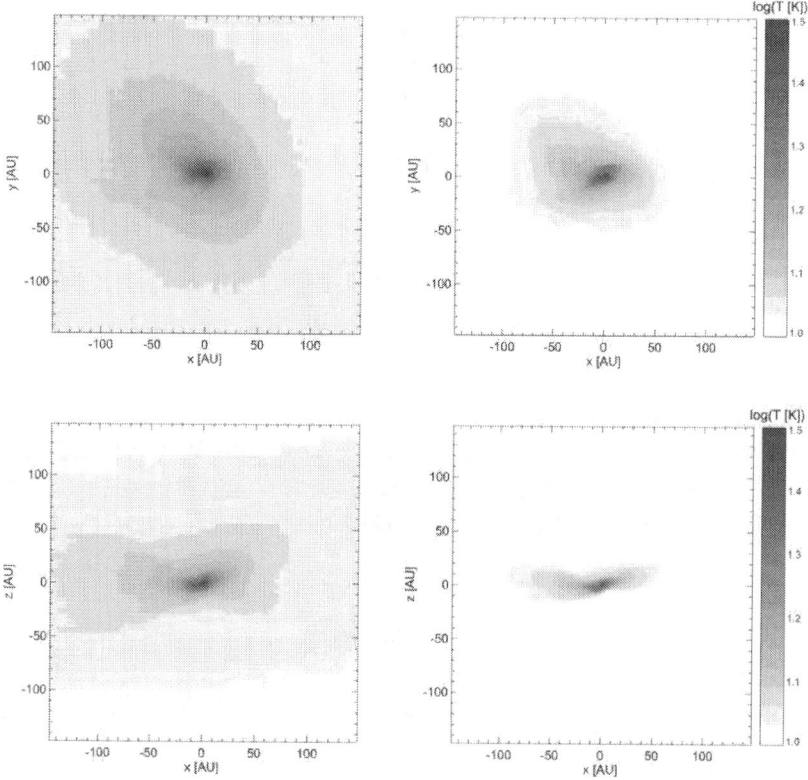

Figure 15. The plots show density weighted temperature averages (e.g., $\int \rho\, T\, dz / \int \rho\, dz$) from a collapse calculation including rotation and turbulence. *Left:* results from the FLASH/RT calculations including radiative transfer. *Right:* results from FLASH calculations using a barotropic EOS. The ambient gas temperature in the FLASH/RT models is about 30% higher.

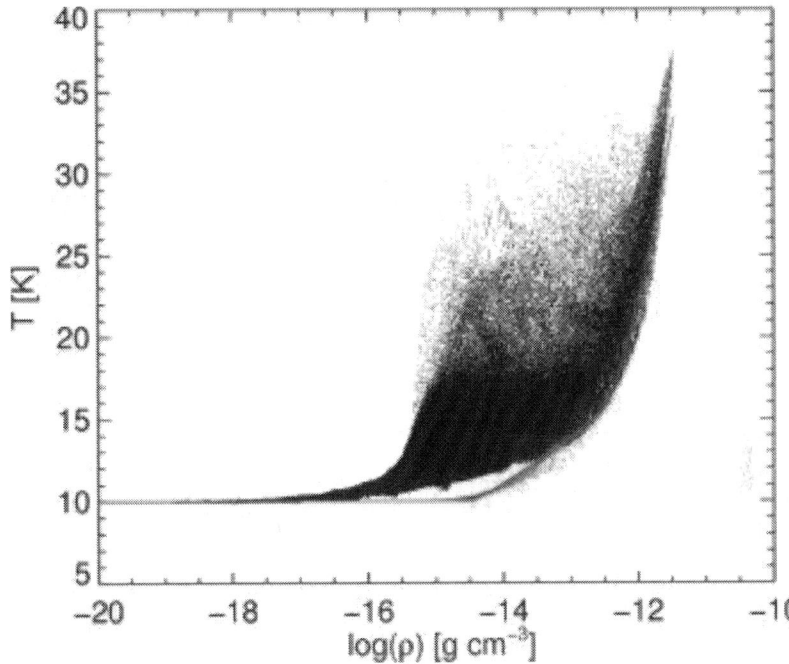

Figure 16. Temperature distribution with respect to the gas density in the simulation box at the end of the collapse simulation including rotation and turbulence. Black dots show the temperature distribution from the FLASH/RT run, red dots resemble the temperature density dependence of the barotropic EOS. (For interpretation of the references to color in this figure legend, the reader is referred to the web version of this article).

Table 2. Results from the scaling test normalized to a run with 96 cpus'; because of the increased communication overhead, each cpu should handle as many block as possible in terms of memory requirements.

No. of cores	Time [s]	Speedup	Blocks per cpu	Performace per Block [%]
96	86.06	1	37–39	100
144	60.6	1.42	25–26	95.2
192	48.01	1.79	18–20	89.6
240	41.11	2.09	14–16	82.6
288	34.93	2.46	12–13	81
336	32.98	2.6	10–12	75.5

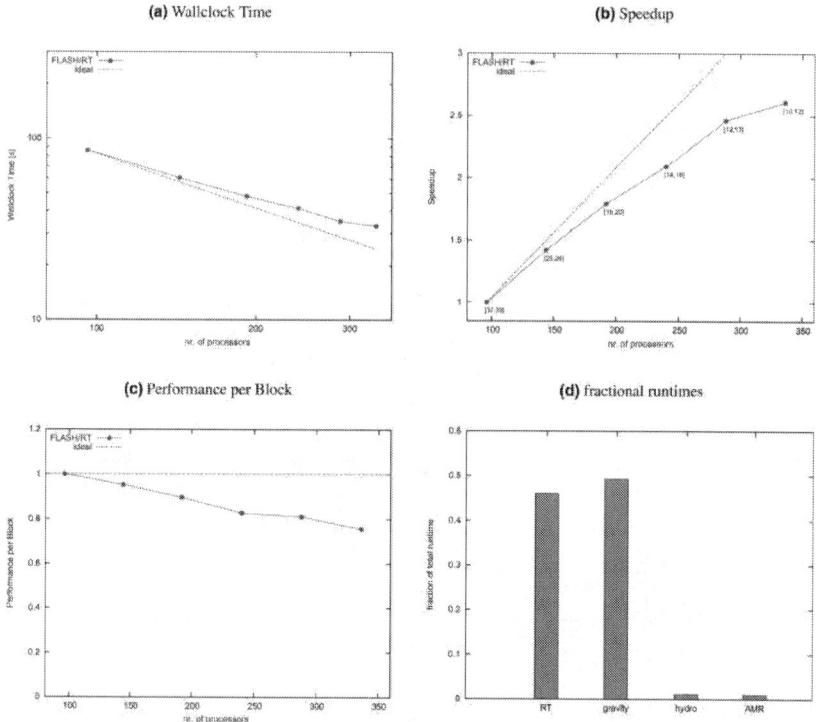

Figure 17. Results from the parallel scaling test. (a) The total wall-clock time for the formal solution averaged over 50 iteration cycles from the Pascucci disc benchmark is shown. (b) The speedup normalized to the wall-clock time using 96 cores is shown. The number in brackets denote the minimum and maximum blocks per core, which is the result from the Morton space-filling curve. (c) The performance per block is shown, which decreases by roughly 10% if the number of cores is doubled. (d) The fractional runtimes for the most costly steps of the collapse simulation (Section 4.2) is shown. All computations were performed using 192 directions.

Unfortunately, our FLASH/RT simulations are very costly (see Section 5 for more details) and currently, it is not feasible to continue these simulations without coupling the radiative transfer solver to a subgrid model for the formation of the central core, e.g., sink particles (Federrath et al., 2010). However, our current test simulations show the first stages of disc formation and the importance of modeling radiative transfer accurately. Since the thermodynamics of the gas significantly influence the fragmentation behavior, modeling radiative transfer is indispensable to study the further evolution of the protostar, the circumstellar disc, and the surrounding gas envelope.

5. PERFORMANCE

The FLASH code shows excellent scaling behavior on any computational infrastructure (e.g. Fryxell et al., 2000). For this work, the computations are clearly dominated by the solution of the RTE. Hence, the scaling behavior of the radiative transfer solver is crucial for the total performance of the FLASH/RT calculations. We investigate the scaling performance of our radiation code using the disc benchmark setup (see Section 3.2). We performed 50 formal solutions of the RTE using 192 directions on a spatial range covering 5 refinement levels. After the initial refinement, depending on the density structure and the radius, the computational domain consists of 3648 valid subdomains (leaf blocks) each containing 8^3 cells. The FLASH code distributes the blocks among all available MPI ranks using a Morton space-filling curve[4]. The scaling tests were run at The North-German Supercomputing Alliance in Berlin on the Cray XC30 "Gottfried" using 12-core Xeon IvyBridge processors. Fig. 17 and Table 2 show the scaling results for the computation of the formal solution of the RTE averaged over 50 cycles. The scaling is normalized to the wall-clock time using 96 cores (e.g., 8 Xeon IvyBridge processors). "Gottfried" provides 2 Xeon processors with 24 cores in total per computing node, hence, adding 24 cores to the computation will increase the communication overhead. Fig. 17b shows the speedup compared to a perfect scaling behavior. The radiation solver scales reasonably well considering the communication of non-local information, which is necessary for the solution of the RTE. Fig. 17c shows that doubling the number of cores decreases the performance per block by approximately 10%, which we consider also as reasonable.

The cost of the radiative transfer solver from a 3D collapse simulation (Fig. 17d) is comparable to the cost for the computation of the self-gravitational potential which is done by a Poisson tree-solver. However, the radiative transfer solver in this particular simulation uses a rather moderate angular resolution of 192 directions (using the HEALPix tessellation from Górski et al., 2005). For runs including rotation and turbulence, the angular resolution probably needs a much higher resolution of at least 768 directions or higher. Since the cost of the radiative transfer solver scales linearly with the number of directions, it dominates the entire simulation run compared to the calculation of self-gravity. So far, we have tested the FLASH/RT code on our own computing cluster in Hamburg (32 nodes with 2x Intel Xeon Hexa-Core CPUs, 2.40 GHz) and at the North-German Supercomputing Alliance in Berlin on the Cray XC30.

6. SUMMARY

We have implemented a new radiation transfer solver based on the method of hybrid characteristics. The solver successfully reproduces standard radiative transfer problems, including NLTE, thermal radiative transfer and the diffusion limit. We proved the feasibility of the method for 3D collapse simulations where radiative transfer is the dominant cooling process during the formation of the first protostellar core. In contrast to the FLD approximation, our method preserves the anisotropy of the radiation field which becomes crucial in the transition from optically thin to optically thick regions (e.g., a protostellar disc). The radiation solver is implemented in the framework of the MHD code FLASH which allows for a straight forward coupling of both codes (e.g., the collapse simulations.

However, the explicit energy coupling, as described in Section 2.4.1, puts rather strong limitations on the time step. A possible improvement can be achieved by combining the raytracer with the solution of the moment equation for the radiative energy. In contrast to the FLD approach, one can compute an angle dependent diffusion coefficient in the form of the variable Eddington tensor (VET) (Jiang et al., 2012) which can be achieved using our raytracer. The advantage is that the evolution of the radiative energy can be handled implicitly by solving the linearized moment equation for the radiation temperature, like in Commerçon et al. (2011), which resolves the problem of the time step restriction. The framework for this has already been implemented in FLASH by Klassen et al. (2014) and can be combined with our raytracer to implement the VET approach.

Our method is a generalized and enhanced implementation of the hybrid characteristics method by Rijkhorst et al. (2006) and Peters et al. (2010). The original implementation was restricted to direct irradiation from point sources and the integration of optical depths respectively. The FLASH code in combination with our radiative transfer framework allows for the solution of a much wider range of problems and can also very easily be extended to handle a more complex form of the radiative transfer equation. Our implementation fits very well into the parallel design of the FLASH code which is based on AMR with domain decomposition. Our method works within the AMR design of FLASH and is able to solve the 3D RTE on a wide range of scales which is indispensable for star formation simu-lations.

ACKNOWLEDGMENTS

L.B. acknowledges financial support by the Deutsche Forschungsgemeinschaft (DFG) mainly via the Emmy Noether Grant BA 3706/1-1*Theory of Massive Star Formation* and partly through the Graduiertenkolleg 1351 *Extrasolar Planets and their Host Stars* and the Priority Program 1573 *Physics of the Interstellar Medium*. T.P. acknowledges financial support through a Forschungskredit of the University of Zurich, Grant no. FK-13-112, and from the DFG Priority Program 1573 *Physics of the Interstellar Medium*. This work benefited from helpful discussions with Peter Hauschildt (Università Hamburg) and Stefan

Dreizler (Universität Göttingen). Most of the collapse simulations were carried out at The North-German Supercomputing Alliance in Berlin (*Gottfried*, HLRN).

REFERENCES

1. Acreman, Harries, Rundle, 2010 D.M. Acreman, T.J. Harries, D.A. Rundle MNRAS, 403 ,1143

2. Amanatides, Woo, 1987J. Amanatides, A. WooA fast voxel traversal algorithm for ray tracingG. Maréchal (Ed.), Proceedings of Eurographics 87, Elsevier Science Publishers pp. 3–10

3. Bate, 2009. M.R. BateMNRAS, 392),1363

4. Bate, Tricco, Price, 2013. M.R. Bate, T.S. Tricco, D.J. PriceMNRAS

5. Bjorkman, Wood, 2001. J.E. Bjorkman, K. WoodApJ, 554 ,615

6. Bryan, G.L. M.L. Norman, B.W. O'Shea, T. Abel, J.H. Wise, M.J. Turk, D.R. Reynolds, D.C. Collins, P. Wang, S.W. Skillman, B. Smith, R.P. Harkness, J. Bordner, J.-h. Kim, M. Kuhlen, H. Xu, N. Goldbaum, C. Hummels, A.G. Kritsuk, E. Tasker, S. Skory, C.M. Simpson, O. Hahn, J.S. Oishi, G.C. So, F. Zhao, R. Cen, Y. Li, The Enzo Collaboration ApJS, 211 (2014),19

7. Buntemeyer, Banerjee, Peters, Klassen, Pudritz, 2015. L. Buntemeyer, R. Banerjee, T. Peters, M. Klassen, R.E. PudritzRadiation Hydrodynamics using Characteristics on Adaptive Decomposed Domains for Massively Parallel Star Formation Simulations.

8. Chiang, Goldreich, 1997. E.I. Chiang, P. GoldreichApJ, 490 , 368 |

9. Commerçon, Teyssier, Audit, Hennebelle, Chabrier, 2011. B. Commerçon, R. Teyssier, E. Audit, P. Hennebelle, G. ChabrierA&A, 529 , A35+

10. Dale, Ercolano, Bonnell, 2013. J.E. Dale, B. Ercolano, I.A. BonnellMNRAS, 430 ,234

11. Davis, Stone, Jiang, 2012. S.W. Davis, J.M. Stone, Y.-F. JiangApJS, 199 , 9

12. Dullemond, 2002.C.P. DullemondA&A, 395 ,853

13. Dullemond.Dullemond, C. P., 2012. RADMC-3D: A Multi-Purpose Radiative Transfer Tool. Astrophysics Source Code Library.x1202.015.

14. Dullemond, Monnier, 2010. C.P. Dullemond, J.D. MonnierARA&A, 48 ,205

15. Federrath, Banerjee, Clark, Klessen, 2010. C. Federrath, R. Banerjee, P.C. Clark, R.S. KlessenApJ, 713 , 269

16. Flock, Fromang, González, Commerçon, 2013. M. Flock, S. Fromang, M. González, B. CommerçonA&A, 560 ,A43

17. D. Forgan, K. Rice. MNRAS, 433 , 1796

18. Fryxell, Olson, Ricker, Timmes, Zingale, Lamb, MacNeice, Rosner, Truran, Tufo, 2000. B. Fryxell, K. Olson, P. Ricker, F.X. Timmes, M. Zingale, D.Q. Lamb, P. MacNeice, R. Rosner, J.W. Truran, H. TufoApJS, 131 , 273

19. González, Audit, Huynh, 2007. M. González, E. Audit, P. HuynhA&A, 464 , 429

20. Goodman, Benson, Fuller, Myers, 1993. A.A. Goodman, P.J. Benson, G.A. Fuller, P.C. MyersApJ, 406 ,528

21. Górski, Hivon, Banday, Wandelt, Hansen, Reinecke, Bartelmann, 2005

22. K.M. Górski, E. Hivon, A.J. Banday, B.D. Wandelt, F.K. Hansen, M. Reinecke, M. BartelmannApJ, 622 ,759

23. Hayek, Asplund, Carlsson, Trampedach, Collet, Gudiksen, Hansteen, Leenaarts, 2010. W. Hayek, M. Asplund, M. Carlsson, R. Trampedach, R. Collet, B.V. Gudiksen, V.H. Hansteen, J. LeenaartsA&A, 517 , A49

24. Heinemann, Dobler, Nordlund, Brandenburg, 2006.T. Heinemann, W. Dobler, Å. Nordlund, A. Brandenburg. A&A, 448 , 731

25. Jack, Hauschildt, Baron, 2012. D. Jack, P.H. Hauschildt, E. BaronA&A, 546 , A39

26. Jiang, Stone, Davis, 2012. Y.-F. Jiang, J.M. Stone, S.W. DavisApJS, 199 , 14

27. Klassen, Kuiper, Pudritz, Peters, Banerjee, Buntemeyer, 2014. M. Klassen, R. Kuiper, R.E. Pudritz, T. Peters, R. Banerjee, L. Buntemeyer ApJ, 797 , 4

28. Kolmogorov.A. KolmogorovAkad. Nauk SSSR Dokl., 30 ,301

29. Krumholz, Klein, McKee, 2007. M.R. Krumholz, R.I. Klein, C.F. McKeeApJ, 656 , 959

30. Kuiper, Klahr, Dullemond, Kley, Henning, 2010. R. Kuiper, H. Klahr, C. Dullemond, W. Kley, T. HenningA&A, 511 , A81

31. Larson, 1969.R.B. LarsonMNRAS, 145 ,271+

32. Levermore, Pomraning, 1981. C.D. Levermore, G.C. PomraningApJ, 248 , 321

33. Lowrie, Edwards, 2008. R.B. Lowrie, J.D. EdwardsShock Waves, 18 , 129

34. Lucy, 1999.L.B. LucyA&A, 344 , 282

35. Machida, Inutsuka, Matsumoto, 2010. M.N. Machida, S.-i. Inutsuka, T. MatsumotoApJ, 724 , 1006

36. Masunaga, Miyama, Inutsuka, 1998. H. Masunaga, S.M. Miyama, S.-I. InutsukaApJ, 495 ,346

37. Mellon, Li, 2008. R.R. Mellon, Z. LiApJ, 681 ,1356

38. Mihalas, Weibel Mihalas, 1984. D. Mihalas, B. Weibel MihalasFoundations of radiation hydrodynamics, Oxford University Press, New York (1984)

39. Min, Dullemond, Dominik, de Koter, Hovenier, 2009.nM. Min, C.P. Dullemond, C. Dominik, A. de Koter, J.W. HovenierA&A, 497 , 155

40. Nakamoto, Umemura, Susa, 2001. T. Nakamoto, M. Umemura, H. SusaMNRAS, 321 , 593

41. Offner, Klein, McKee, Krumholz, 2009. S.S.R. Offner, R.I. Klein, C.F. McKee, M.R. KrumholzApJ, 703 ,

42. Olson, Auer, Buchler, 1986. G.L. Olson, L.H. Auer, J.R. BuchlerJ. Quant. Spec. Radiat. Transf., 35 , 431

43. Olson, MacNeice, Fryxell, Ricker, Timmes, Zingale, 1999. K.M. Olson, P. MacNeice, B. Fryxell, P. Ricker, F.X. Timmes, M. ZingaleBulletin of the American Astronomical Society, 31,1430

44. Pascucci, Wolf, Steinacker, Dullemond, Henning, Niccolini, Woitke, Lopez, 2004. I. Pascucci, S. Wolf, J. Steinacker, C.P. Dullemond, T. Henning, G. Niccolini, P. Woitke, B. Lopez A&A, 417, 793–805

45. Peters, Banerjee, Klessen, Mac Low, Galván-Madrid, Keto, 2010.T. Peters, R. Banerjee, R.S. Klessen, M. Mac Low, R. Galván-Madrid, E.R. KetoApJ, 711 , 1017

46. Price, Bate, 2010. D.J. Price, M.R. BateMagnetic fields and radiative feedback in the star formation process,in: G. Bertin, F. de Luca, G. Lodato, R. Pozzoli, M. Romé (Eds.), Proceedings of American Institute of Physics Conference Series, vol. 1242, American Institute of Physics Conference Series, 1002.0650 (2010), pp. 205–218

47. Razoumov, Cardall, 2005. A.O. Razoumov, C.Y. CardallMNRAS, 362 , 1413

48. Rijkhorst, Plewa, Dubey, Mellema, 2006. E.-J. Rijkhorst, T. Plewa, A. Dubey, G. MellemaA&A, 452 , 907

49. Seifried, Banerjee, Klessen, Duffin, Pudritz, 2011. D. Seifried, R. Banerjee, R.S. Klessen, D. Duffin, R.E. PudritzMNRAS, 417 , 1054

50. Seifried, Banerjee, Pudritz, Klessen, 2013. D. Seifried, R. Banerjee, R.E. Pudritz, R.S. KlessenMNRAS, 432 , 3320

51. Semenov, Henning, Helling, Ilgner, Sedlmayr, 2003. D. Semenov, T. Henning, C. Helling, M. Ilgner, E. SedlmayrA&A, 410 , 611

52. Stone, Mihalas, Norman, 1992. J.M. Stone, D. Mihalas, M.L. NormanApJS, 80 , 819

53. Tanaka, Yoshikawa, Okamoto, Hasegawa. Tanaka, S., Yoshikawa, K., Okamoto, T., Hasegawa, K., 2014.A new ray-tracing scheme for 3D diffuse radiation transfer on highly parallel architectures.ArXiv e-prints1410.0763.

54. Tobin, Hartmann, Chiang, Wilner, Looney, Loinard, Calvet, D'Alessio, 2012J.J. Tobin, L. Hartmann, H.-F. Chiang, D.J. Wilner, L.W. Looney, L. Loinard, N. Calvet, P. D'AlessioNature, 492 , 83

55. Truelove, Klein, McKee, Holliman, Howell, Greenough, 1997.J.K. Truelove, R.I. Klein, C.F. McKee, J.H. Holliman, L.H. Howell, J.A. GreenoughApJ, 489 , L179+

56. Trujillo Bueno, Fabiani Bendicho, 1995. J. Trujillo Bueno, P. Fabiani Bendicho, ApJ, 455 , 646

57. van Noort, Hubeny, Lanz, 2002. M. van Noort, I. Hubeny, T. Lanz, ApJ, 568 , 1066

58. Walch, Whitworth, Bisbas, Wünsch, Hubber, 2012, S.K. Walch, A.P. Whitworth, T. Bisbas, R. Wünsch, D. Hubber, MNRAS, 427 , 625

59. Yorke, Bodenheimer, Laughlin, 1993. H.W. Yorke, P. Bodenheimer, G. Laughlin, ApJ, 411 , 274

60. Zhang, Tan, McKee, 2013. Y. Zhang, J.C. Tan, C.F. McKee, ApJ, 766 , 86

Index

A

Aberration, 208, 214, 216

Accommodation, 11, 23, 24, 247, 262, 273

adaptive optical systems, 32, 41, 42, 43, 81, 84

adaptive optics, 1, 2, 4, 8, 10, 12, 13, 16, 17, 18, 19, 21, 22, 23, 24, 26, 27, 28, 29, 30, 32, 39, 68, 69, 74, 75, 86, 88, 92, 93, 108, 109, 111, 112, 113, 114, 115, 117, 118, 119, 120, 121, 122, 123, 125, 141, 142, 144, 146, 147, 148, 149, 150, 151, 152, 161, 166, 169, 170, 171, 174, 175, 176, 177, 184, 185, 186, 190, 194, 207, 245, 255, 273, 274, 275, 276, 277, 278, 279, 280, 281, 282, 283, 284

Adaptive Optics, 1, 4, 7, 11, 31, 83, 86, 95, 99, 114, 125, 147, 148, 149, 151, 152, 161, 169, 170, 171, 175, 176, 177, 180, 184, 186, 189, 207, 208, 216, 220, 245, 260, 261, 271, 277, 279, 280, 282

algorithm, 10, 40, 43, 44, 56, 58, 59, 62, 63, 66, 79, 112, 126, 127, 128, 129, 130, 132, 133, 134, 140, 143, 144, 160, 223, 229, 230, 234, 235, 236, 237, 238, 239, 241, 285, 293, 296, 304, 325

astigmatism., 6, 8, 11, 135, 139, 140, 263

astronomical, 5, 6, 20, 38, 69, 85, 90, 93, 125, 126, 176, 177, 180, 272

atmosphere, 39

atmospheric turbulence, 6, 74, 85, 151, 160, 162, 167, 176, 207, 208, 209, 212, 217, 222, 226, 227, 280

B

biomedical imaging techniques, 84

C

cell phones, 171, 189, 190, 194, 197, 202, 203, 204

D

Diabetic retinopathy (DR), 100

F

forks, 32, 33, 35, 41

Free space optics, 221

H

Hydrodynamics, 169, 285, 286, 325

K

Kolmogorov, 40, 161, 162, 216, 218, 318, 326

L

lab-on-a-chip readout, 190

laboratory, 32, 42, 43, 152, 163, 168, 170, 184, 270

liquid crystals, 52, 92, 171, 172, 173, 174, 175, 179, 180, 181, 182, 183, 184, 185

Liquid Crystals, 174

O

Ophthalmoscopes, 260

optical chemical sensing, 190

optical sensors, 84

P

Prototype, 208

R

Radiative transfer, 286

Radio over free space optics, 221

Retina, 247, 273, 277, 283

root-mean-square (RMS), 95

S

San Fernando Observatory (SFO), 152

Solar Atmosphere, 208

spatial light modulators, 43, 92, 171, 176, 180, 184, 185, 186, 256

Spectral Resolution, 208

Star formation, 286

T

tomography, 7, 21, 22, 86, 117, 119, 120, 122, 123, 141, 142, 143, 146, 147, 150, 250, 256, 261, 275, 276, 277, 278, 279, 280, 282, 283, 284

U

ubiquitous sensing, 190

W

Wavefront, 92

wavefront correction, 92, 114, 126, 141, 149, 171, 216, 217, 253, 264

Wavelength-division multiplexing, 221